AF738158

Greening of Industry Networks Studies

Volume 11

Series Editors
Diego A. Vázquez-Brust, Faculty of Business and Law, University of Portsmouth, Portsmouth, Hampshire, UK
Joseph Sarkis, Worcester Polytechnic Institute, School of Business, Worcester, MA, USA

This series aims to improve our understanding of how shifts in industrial regimes, trade, and technology are not only creating significant environmental and social impacts and inequities around the world but also opportunities for sustainable economic growth. It focuses on issues of industrial development, environment and society, and sustainable economy, dealing with the UN Sustainable Development Goals.

The series publishes high-quality peer-reviewed monographs, edited volumes and graduate textbooks related to the theme of sustainable economic development. The topics covered include industry networks, sustainable economic growth, supply chains, supply chain management, sustainable economy, green logistics, green growth, sustainable firms, collaboration for sustainability, corporate sustainability, corporate social responsibility, green banking, operations management, engineering management, sustainable value chains, sustainable business, circular economy, sustainable production, sustainable consumption, waste, resource management, productivity strategies, sustainable management, resource conservation, sustainability strategies, innovation processes, sustainable innovation, renewable resources, among others. Contributors to the series are researchers, business strategists, civil society thought leaders, and government policy makers.

The series is associated with ***The Greening of Industry Network***, an international network of professionals from research, education, business, government and civil society organizations located in more than 50 countries, with the goal of aligning industrial development strategies with sustainable development principles. The network stimulates, coordinates and connects high quality research to policies, strategies and actions in ways that contribute to a more sustainable society. Through a variety of arenas, it provides an open forum for creative debate to engage researchers, business, workers, government, consumers and other actors in developing a shared understanding of the changes required for creating a more sustainable future.

Now indexed by Scopus!

Niamat Ullah Ibne Hossain
Editor

Data Analytics for Supply Chain Networks

Springer

Editor
Niamat Ullah Ibne Hossain
Engineering Management, College of
Engineering and Computer Science
Arkansas State University
Jonesboro, AR, USA

ISSN 2543-0246 ISSN 2543-0254 (electronic)
Greening of Industry Networks Studies
ISBN 978-3-031-29822-6 ISBN 978-3-031-29823-3 (eBook)
https://doi.org/10.1007/978-3-031-29823-3

© The Editor(s) (if applicable) and The Author(s), under exclusive license to Springer Nature Switzerland AG 2023
This work is subject to copyright. All rights are solely and exclusively licensed by the Publisher, whether the whole or part of the material is concerned, specifically the rights of translation, reprinting, reuse of illustrations, recitation, broadcasting, reproduction on microfilms or in any other physical way, and transmission or information storage and retrieval, electronic adaptation, computer software, or by similar or dissimilar methodology now known or hereafter developed.
The use of general descriptive names, registered names, trademarks, service marks, etc. in this publication does not imply, even in the absence of a specific statement, that such names are exempt from the relevant protective laws and regulations and therefore free for general use.
The publisher, the authors, and the editors are safe to assume that the advice and information in this book are believed to be true and accurate at the date of publication. Neither the publisher nor the authors or the editors give a warranty, expressed or implied, with respect to the material contained herein or for any errors or omissions that may have been made. The publisher remains neutral with regard to jurisdictional claims in published maps and institutional affiliations.

This Springer imprint is published by the registered company Springer Nature Switzerland AG
The registered company address is: Gewerbestrasse 11, 6330 Cham, Switzerland

Contents

1 **Data Analytics Applications in Supply Chain Resilience and Sustainability Management: The State of the Art and a Way Forward** 1
Ziaul Haque Munim, Ornela Vladi, and Niamat Ullah Ibne Hossain

2 **Enhancing the Viability of Green Supply Chain Management Initiatives Leveraging Data Fusion Technique** 15
Sazid Rahman, Niamat Ullah Ibne Hossain, Adhir Roy, and Md. Saiful Islam

3 **Supply Chain Sustainability and Supply Chain Resilience: A Performance Measurement Framework with Empirical Validation** 49
Marta Negri, Enrico Cagno, Claudia Colicchia, Leonardo Coppellotti, and Caterina Baldassarri

4 **An Assessment of Decision-Making in Resilient and Sustainable Projects Between Literature and Practice** 107
Shahriar Tanvir Alam, Md Muzahid Khan, and Moddassir Khan Nayeem

5 **Barriers for Sustainable Supply Chain Management and Their Overcoming Strategies in Context of the Indian Automobile Industry** 129
Sanjoy Kumar Paul, Koppiahraj Karuppiah, Bathrinath Sankaranarayanan, and Syed Mithun Ali

6 **Prioritizing Sustainability Criteria of Green Supply Chains Using the Best-Worst Method** 167
Ertugrul Ayyildiz and Aslihan Yildiz

7 Economic Performance Analysis of a Resilient and Sustainable Supply Chain: Adoption of Electric Vehicles as a Sustainable Logistics Option 197
Md Muzahid Khan, Md Doulotuzzaman Xames, Priom Mahmud, Shahriar Tanvir Alam, and Niamat Ullah Ibne Hossain

8 Integrating Circular Economy and Reverse Logistics for Achieving Sustainable Dairy Operations 211
Mohammad Shamsuddoha, Tasnuba Nasir, and Niamat Ullah Ibne Hossain

9 The Impact of Big Data Analytics Capabilities on the Sustainability of Maritime Firms 227
Ziaul Haque Munim, Mariia Dushenko, Hyeran Raaness, and Tine Viveka Westerberg

10 Smart Transportation Logistics: Achieving Supply Chain Efficiency with Green Initiatives 243
Mohammad Shamsuddoha, Mohammad Abul Kashem, and Tasnuba Nasir

Index 259

Chapter 1
Data Analytics Applications in Supply Chain Resilience and Sustainability Management: The State of the Art and a Way Forward

Ziaul Haque Munim, Ornela Vladi, and Niamat Ullah Ibne Hossain

Abstract Digital technologies have become a cause célèbre among academics and practitioners as great tools capable of building resilient and sustainable supply chains. This chapter provides an overview of the recent developments in data analytics applications in supply chain resilience (SCR) and supply chain sustainability (SCS). There has been an exponential growth of literature on data analytics applications for SCR and SCS, with a particularly notable increase observed since 2015. In this systematic literature review, we find that both SCR and SCS research are concentrated around five main themes: (1) data analytics capabilities, (2) role of Industry 4.0, (3) blockchain adoption, (4) big data analytics, and (5) machine learning. Closed-loop supply chain design and circular economy are of unique focus in SCS, while digital twin only emerged as a research theme in SCR research. The underlying themes in SCR contexts are more dispersed than in SCS, mostly due to the comparative maturity of SCS research. In light of the promising developments in data analytics applications for SCR and SCS, promising avenues for future inquiry are the design of effective data-sharing incentive mechanisms, and the utilization of big data from social media platforms, yielding valuable insights for both research and practitioners.

Z. H. Munim (✉)
Faculty of Technology, Natural and Maritime Sciences, University of South-Eastern Norway, Horten, Norway
e-mail: Ziaul.h.munim@usn.no

O. Vladi
Department of International Economics, Government and Business, Copenhagen Business School, Copenhagen, Denmark
e-mail: ov.egb@cbs.dk

N. U. Ibne Hossain
Engineering Management, College of Engineering and Computer Science, Arkansas State University, Jonesboro, AR, USA
e-mail: nibnehossain@astate.edu

© The Author(s), under exclusive license to Springer Nature Switzerland AG 2023
N. U. Ibne Hossain (ed.), *Data Analytics for Supply Chain Networks*, Greening of Industry Networks Studies 11, https://doi.org/10.1007/978-3-031-29823-3_1

Keywords Data analytics · Artificial intelligence · Supply chain management · Supply chain resilience · Sustainable supply chain

1.1 Introduction

The rapid deployment of digital technologies and sensors in business and production processes has exposed organizations to the large volumes of data generated daily. In addition to the large volumes, data are also generated at an extremely high speed (in fraction of seconds and minutes) and a variety of formats. This is commonly referred to as the 3Vs of big data—volume, velocity, and variety. Other characteristics such as validity, variability, value, vagueness, veracity, venue, and vocabulary have since been introduced to provide a more comprehensive understanding of big data (Tsai et al. 2015). With the advancement in data analytics tools and techniques, multiple applications of big data in supply chain management (SCM) have been leveraged by organizations across industries. Some of the widely adopted applications are in manufacturing, warehousing, logistics management and transportation, procurement, and demand planning (Nguyen et al. 2018). SCM activities such as supplier selection and management, sourcing risk and cost optimization, production planning and control, product R & D, order picking, inventory control, and demand forecasting have improved significantly through big data analytics (Nguyen et al. 2018). Although the SCM literature devoted a lot of attention to big data analytics (BDA), less effort has been dedicated to synthesizing these contributions. This chapter addresses this gap by revisiting the concept of data analytics in SCM literature, identifying key conversational landmarks, and discussing future research avenues.

The core idea of a data analytics system is to collect data from various predetermined sources, processes them to extract valuable information, and finally use them to help organizations and management make informed decisions. In essence, data analytics involves three core activities: (1) input generation, (2) data analysis, and (3) output generation (Tsai et al. 2015). Input generation encompasses various functions including data collection, cleaning, selection, preprocessing, and transformation. Data analysis refers to the use of domain- or context-specific methods and tools, including artificial intelligence algorithms and statistical methods for creating valuable information. Finally, outputs are generated and evaluated according to performance metrics such as error rates or accuracy. This chapter provides an overview of the use of data analytics in the context of supply chain resilience and sustainability literature.

Supply chain resilience (SCR) and supply chain sustainability (SCS) have been defined by many scholars. Tukamuhabwa et al. (2015) have presented 23 definitions of SCR, while Ahi and Searcy (2013) have presented 12 definitions of SCS and 22 definitions of green supply chain management (GSCM). In this chapter, SCS and GSCM are considered synonyms. SCR can be broadly defined as the ability of a supply chain to return to its original or improved state within an optimal time and cost frame following a disruptive event. To achieve SCR, supply chains must develop specific capabilities that enable them to proactively prepare for disruptions and promptly react when they unfold. On the other hand, SCS can be defined as a supply chain function that considers the three dimensions of sustainability, namely, economic, environmental, and social performance dimensions, across the entire supply chain.

In the following sections of this chapter, we provide an overview of the published research on data analytics in SCR and SCS using the science mapping tool in the R software (Aria and Cuccurullo 2017). First, we examine publication trends in these topics including the number of publications by corresponding author countries. Second, we present the most impactful studies in both contexts, measured by normalized total citations. Third, we revisit the definition of resilient and sustainable supply chains based on the most impactful studies. Fourth, we map the underlying themes and concepts using keyword co-occurrence metrics. Finally, we identify the trending topics and suggest avenues for future research.

1.2 Publication Trends

Research on data analytics in SCS and SCR have only received attention in recent years, more specifically since 2015 and 2019, respectively (see Fig. 1.1). To provide a comprehensive overview of the literature, we extracted the bibliometric data of published journal articles using the Scopus database. Two separate literature searches were performed, one focusing on data analytics in SCR and the other on data analytics in SCS.[1] Overall, we found 216 and 1104 relevant publications on SCR and SCS, respectively. One of the reasons for having a higher number of journal articles related to the SCS is that it is already a relatively mature field of study, although data analytics applications have gained momentum only since 2014. On the other hand, SCR-related publications that focus on the use of data analytics for resilience were largely driven by the COVID-19 pandemic outbreak.

The first studies that utilized data analytics in the context of resilience and sustainability in SC appeared in 2000 and 1999, respectively. Based on our sample data, Lin and Pai (2000) and Azapagic and Clift (1999) were the first studies to utilize data analytics in SC resilience and sustainability, respectively. While the former used multiagent simulation to improve process adaptability of order fulfillment process by integrating reinforcement learning, the latter used multi-objective optimization of a number of environmental parameters that were derived from life cycle analysis (LCA). Since then, applications of data analytics have been developed in both contexts as evident in recent published studies, for example, Belhadi et al. (2022), Saurabh and Dey (2021), Verma et al. (2021), and Wilson et al. (2021).

[1] *Boolean term for resilience in SC*: ("data analytics" OR "big data" OR "artificial intelligence" OR "machine learning" OR "neural network*" OR "data mining" OR "text mining" OR "algorithm") AND ("resilien*") AND "supply chain*".

Boolean term for sustainability in SC: ("data analytics" OR "big data" OR "artificial intelligence" OR "machine learning" OR "neural network*" OR "data mining" OR "text mining" OR "algorithm") AND ("sustainab*" OR "green") AND "supply chain*".

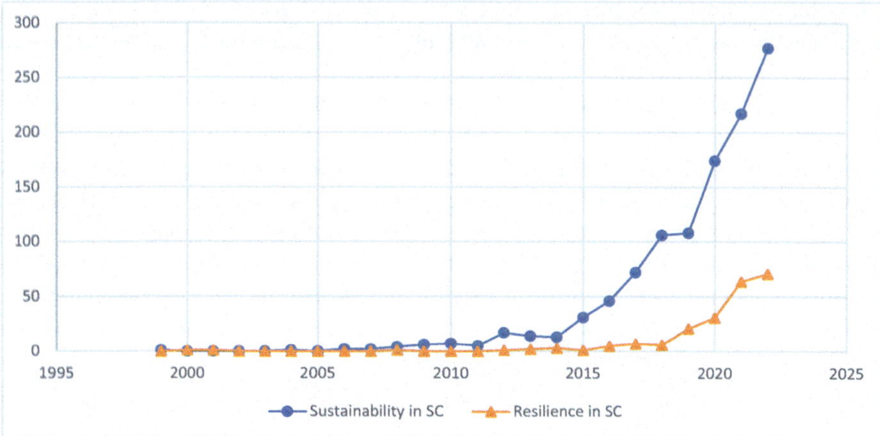

Fig. 1.1 Annual publication growth

The most productive countries based on the corresponding author's country are presented in Fig. 1.2a, b for data analytics in SCR and SCS, respectively. The leading four countries are the same in both contexts, which are China, Iran, the United States, and India. Germany and France emerged as highly productive countries in SCR, while Korea made it to the list in SCS, following the United Kingdom and France.

Figure 1.2 shows the distribution of studies authored by scholars from a single country and multiple countries. The analysis reveals that in SCR research, scholars from Austria, Poland, and Indonesia have not collaborated with foreign institutions. In contrast, the majority of the published articles by scholars based in the United Kingdom, France, Australia, and Denmark have co-authored with at least one scholar based in another foreign institution. Regarding the use of data analytics in SCS research, except for the United Kingdom, Denmark, and France, in the rest of the publications, there is a higher share of single-country publications (SCPs) in comparison to multiple-country publications (MCPs).

1.3 Most Impactful Studies

The top 15 most impactful publications, measured through the normalized total citation (NTC), are presented in Tables 1.1 and 1.2 for data analytics in SCR and SCS, respectively. The NTC is calculated by dividing the actual total citations of an article by the expected citation rate for articles. According to Table 1.1, the most impactful studies in data analytics in SCR are Ivanov and Dolgui (2021), Dubey et al. (2021), and Belhadi et al. (2021). Similarly, from Table 1.2, the most impactful studies in data analytics in SCS are Tirkolaee et al. (2022), Kusiak (2018), and Bag et al. (2021).

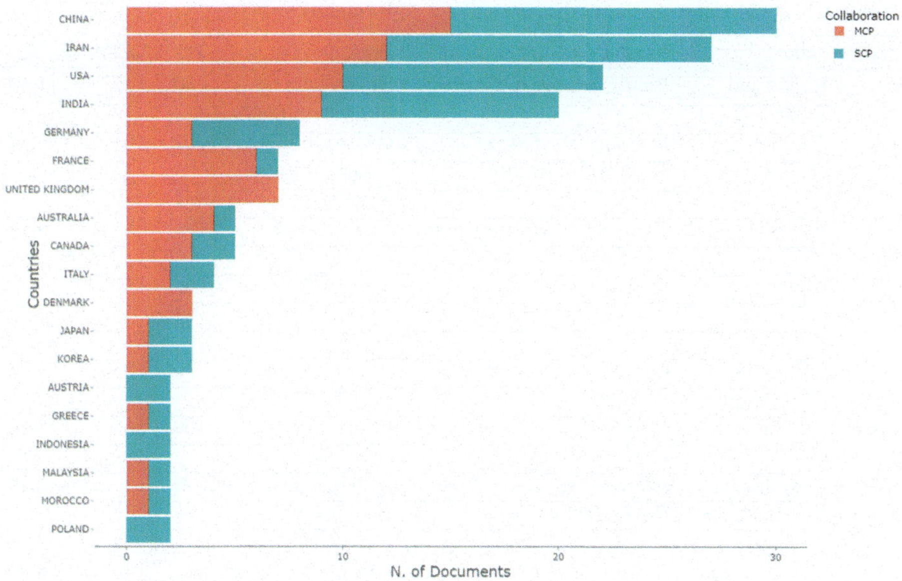

(a) Country production for SCR

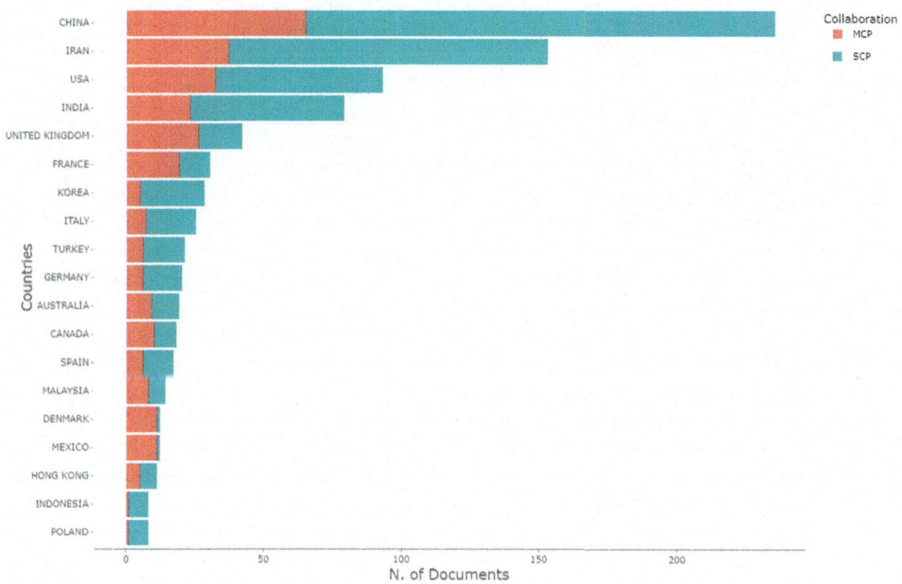

(b) Country production for SCS

Fig. 1.2 Corresponding author's country publications. (*MCP* multiple-country publications, *SCP* single-country publications)

Table 1.1 Highly cited studies in data analytics in SCR (ranked by NTC)

No.	Article	Journal	Topic	TC	TC/Y	NTC
1	Ivanov and Dolgui (2021)	*Prod Planning & Control*	Digital twin for risk and resilience	230	115	11.49
2	Dubey et al. (2021)	*International Journal of Production Research*	DA capabilities as component of resilience	208	104	10.39
3	Belhadi et al. (2021)	*Technological Forecasting and Social Change*	SC resilience during COVID	196	98	9.79
4	Ivanov et al. (2019)	*International Journal of Production Research*	Industry 4.0 impact on ripple effect	565	141.25	9.4
5	Mahmoudi et al. (2021)	*Operations Management Research*	Gresilient supplier selection	20	20	8.66
6	Wamba et al. (2020)	*International Journal of Production Economics*	Effects of big data analytics	169	56.33	8.21
7	Belhadi et al. (2022)	*International Journal of Production Research*	AI-based decision-making framework	15	15	6.49
8	Modgil et al. (2021)	*International Journal of Physical Distribution & Logistics Management*	AI applications in SC resilience during COVID	14	14.	6.06
9	Ghazal and Alzoubi (2022)	*Intelligent Automation and Soft Computing*	ML application in SC collaboration	14	14	6.06
10	Verma et al. (2021)	*IEEE Journal of Biomedical and Health Informatics*	Blockchain-based COVID vaccine distribution	10	10	4.33
11	Hosseini and Ivanov (2019)	*Expert Systems with Applications*	Resilience measure incorporating ripple effects	80	26.67	3.89
12	Papadopoulos et al. (2017)	*Journal of Cleaner Production*	Big data in disaster resilience for SC	323	53.83	3.65
13	Naz et al. (2021)	*Operations Management Research*	AI as resilience enabler	8	8	3.46
14	Vali-Siar and Roghanian (2022)	*Sustainable Production and Consumption*	SC design under COVID disruption	8	8	3.46
15	Bechtsis et al. (2022)	*International Journal of Production Research*	Incentive framework based on data sharing	8	8	3.46

Out of the top 15 most influential publications presented in Tables 1.1 and 1.2, only two articles appeared to have influenced both streams of research: Mahmoudi et al. (2021) and Papadopoulos et al. (2017). These studies investigated both resilience and sustainability in supply chain using data analytics. Despite the substantial conceptual and

Table 1.2 Highly cited studies in data analytics in SCS (ranked by NTC)

No.	Article	Journal	Topic	TC	TC/Y	NTC
1	Tirkolaee et al. (2022)	*Journal of Cleaner Production*	Closed-loop SC design	33	33	14.17
2	Kusiak (2018)	*International Journal of Production Research*	Smart manufacturing	563	112.6	14.05
3	Bag et al. (2021)	*Technological Forecasting and Social Change*	Big data and AI manufacturing	134	67.0	13.71
4	Garai and Sarkar (2022)	*Journal of Cleaner Production*	Reverse logistics on closed-loop SC	26	26	11.17
5	Saurabh and Dey (2021)	*Journal of Cleaner Production*	Blockchain in agri-food SC	95	47.5	9.72
6	El-Kassar and Singh (2019)	*Technological Forecasting and Social Change*	Big data in green innovation	325	81.25	8.90
7	Sazvar et al. (2021)	*Annals of Operations Research*	Closed-loop SC design	20	20	8.59
8	Mahmoudi et al. (2021)	*Operations Management Research*	Gresilient supplier selection	20	20	8.59
9	Wong et al. (2020)	*International Journal of Information Management*	Blockchain in SMEs' SC	185	61.67	8.43
10	Bag et al. (2020)	*Resources, Conservation & Recycling*	Big data analytics in SC	169	56.33	7.7
11	Sharma et al. (2020)	*Computers & Operations Research*	ML applications in agriculture SC	155	51.67	7.06
12	Wilson et al. (2021)	*Management of Environmental Quality*	AI in circular economy	16	16	6.87
13	Goodarzian et al. (2021)	*Engineering Applications of Artificial Intelligence*	Sustainable medical supply chain during COVID	67	33.5	6.85
14	Papadopoulos et al. (2017)	*Journal of Cleaner Production*	Big data in disaster resilience for SC	323	53.83	6.81
15	Li et al. (2020)	*International Journal of Production Economics*	Digital technologies in the context of Industry 4.0	147	49	6.7

empirical literature suggesting that these two concepts are intertwined, the number of studies that cover both concepts are relatively low. Hence, more research is needed. One of the concepts that bring the two together is gresilient supply chains, that is, green and resilient supply chains. Behnam Fahimnia was one of the first scholars to bring the two concepts together (Fahimnia and Jabbarzadeh 2016; Fahimnia et al. 2018). Papadopoulos et al. (2017) investigated disaster resilience for SCS and the role of big data using the Nepal earthquake in 2015 as a case study. This publication analyzed disaster relief activities using big data extracted from news, tweets, Facebook, Instagram,

Google+, YouTube, etc. To analyze such data, the authors employed advanced data analytics. Mahmoudi et al. (2021) proposed a decision-making framework for gresilient supplier selection in the post-COVID era.

In Table 1.1, we observe that concepts such as digital twins, Industry 4.0, ripple effects, artificial intelligence, machine learning, and big data dominate the data analytics applications in SCR. Similarly, in Table 1.2, we observe that closed-loop supply chain designs, smart manufacturing, applications of big data, artificial intelligence, machine learning, blockchain technologies, digital technologies, and Industry 4.0 are the core topics in data analytics applications in SCS. While majority of the concepts are common across data analytics applications in SCR and SCS, SCR has a growing attention toward data-sharing incentives (Bechtsis et al. 2022) and SCS towards circular economy (Wilson et al. 2021).

1.4 Underlying Themes in Data Analytics in Supply Chain in Resilience and Sustainability

We mapped the underlying themes and concepts in data analytics in SCR and SCS in Fig. 1.3. There are three core themes in data analytics in SCR (see Fig. 1.3a). The most dominant theme is the *"green" cluster* that uses data analytics tools such as integer programming, multi-objective optimization, computer simulations, decision support systems, and artificial intelligence for resilient supply network design, product design, and supplier selection. The *purple theme* indicates the use of big data, data analytics, machine learning, and blockchain technologies in information management and risk assessment, particularly in the context of resilience of Industry 4.0 and disaster relief systems. The *red* theme refers to the use of stochastic systems, forecasting, genetic algorithms, heuristic algorithms, and particle swarm optimization.

There are two core themes in data analytics in SCS (see Fig. 1.3b). The *blue* theme indicates the use of particle swarm optimization, heuristic algorithms, multi-objective programming, integer programming, nonlinear programming, scholastic systems, and life cycle analysis in closed-loop supply chain designs and environmental impact assessments such as reduction of carbon dioxide emissions and greenhouse gases. The *red* theme focuses more on sustainable development-related decision-making problems using data mining, big data, artificial intelligence, and algorithms. Overall, the application of data analytics tools and techniques is similar across SCR and SCS.

1.5 Trending Topics and Future Research

We identify the trending topics in data analytics in SCR and SCS by mapping the keyword frequencies in a timeline basis. Referring to Fig. 1.4a, artificial intelligence, data analytics, and security issues have emerged as trending topics in SCR during the

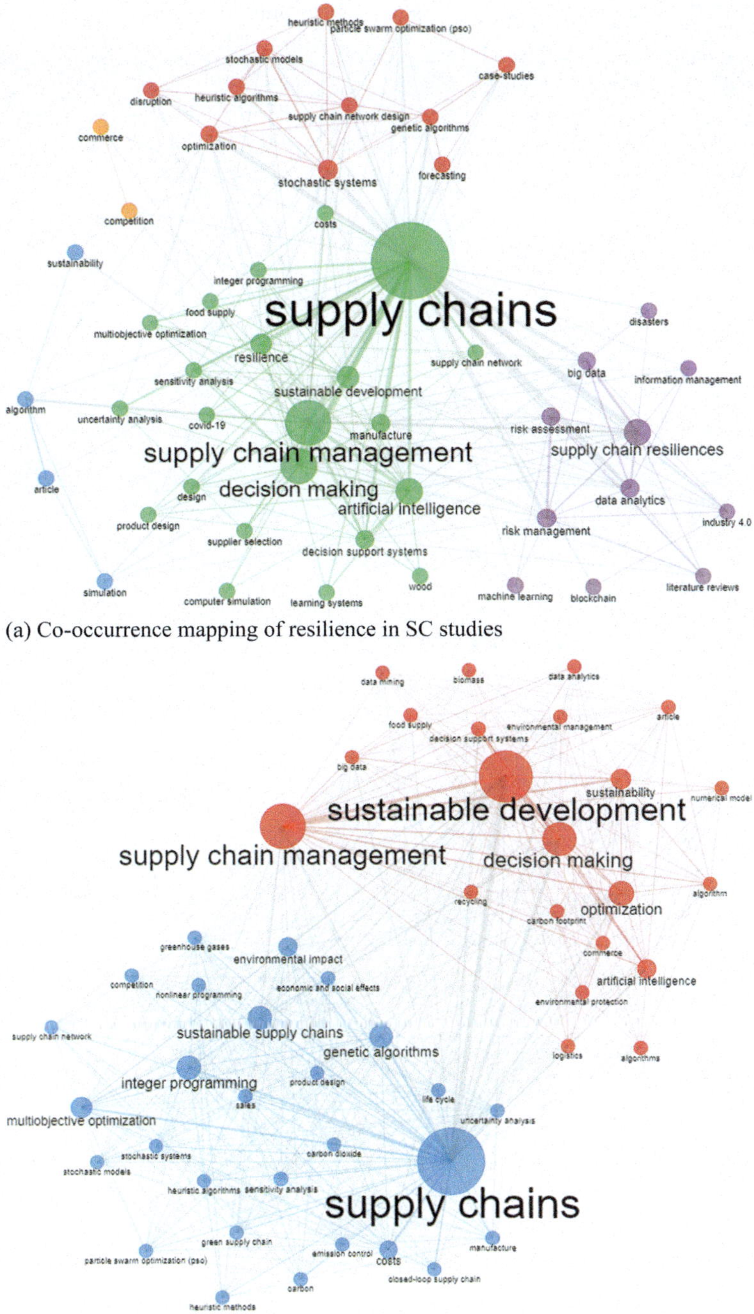

(a) Co-occurrence mapping of resilience in SC studies

(b) Co-occurrence mapping of sustainability in SC studies

Fig. 1.3 Thematic mapping using keyword co-occurrence

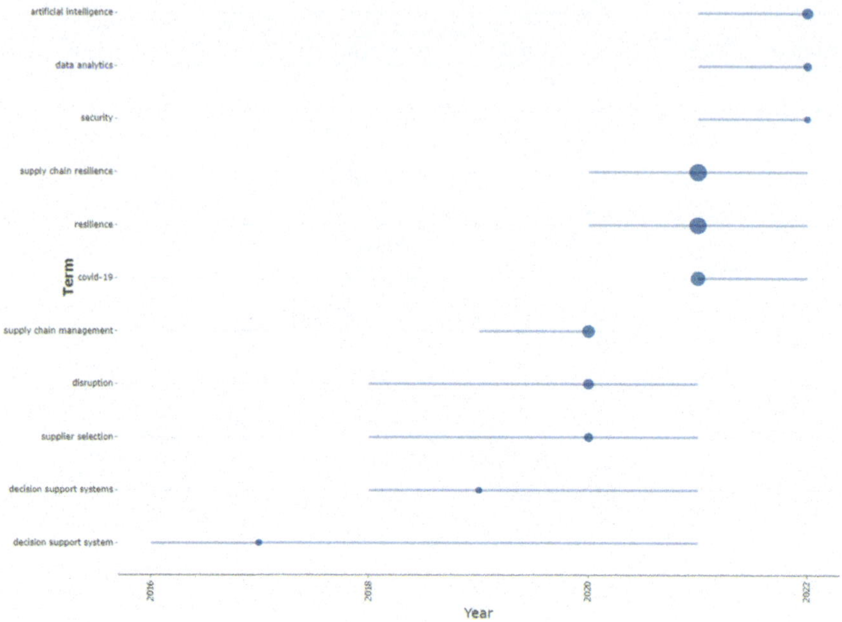

(a) Trending topics in DA in resilience in SC

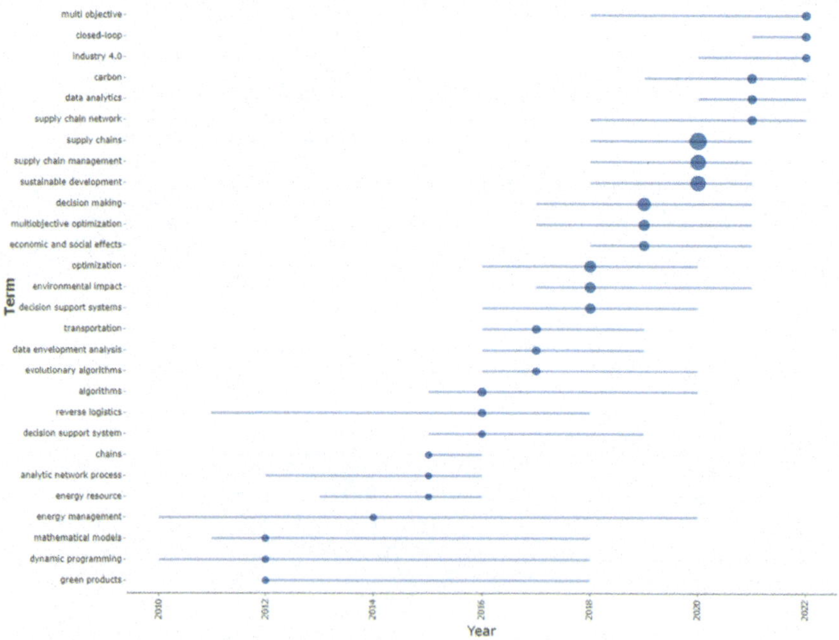

(b) Trending topics in DA in sustainability in SC

Fig. 1.4 Trending topic analysis

last couple of years. Resilient supplier selection, disruption risk assessment, and decision support systems seem to be trending topics in the SCR research since 2018.

We have observed different patterns in the SCS literature, with closed-loop supply chain design, multi-objective programming, and Industry 4.0 emerging as trending topics in most recent years (Fig. 1.4b). Furthermore, we observe a declining trend in SCS in the use of analytic network process (ANP), data envelopment analysis (DEA), and dynamic programming.

Along with the trending topics outlined in Fig. 1.4, we have identified two themes that hold potential for future research. The first theme is focused on social media data analytics (Papadopoulos et al. 2017), while the second theme explores data-sharing incentive mechanism (Bechtsis et al. 2022). Data scraping, harvesting, or extraction using online sources such as websites or social media and their processing using different tools such as machine learning or natural language processing for informed decision-making processes remains largely underexplored. With the advancements in data analytics tools and techniques, the analysis of such big data is likely to become easier and facilitate their implementation in supply chain management.

Modern manufacturing processes and overall supply chain management practices have become smarter. Yet, there exists a great potential to further explore and make use of the large volumes of data produced by the deployment of sensors and digital technology manufacturing equipment, distribution centers, warehouses, transportation systems, and retail point-of-sale systems. This, in turn, can make SCR and SCS decision-making process more informed and create a shared value. Nevertheless, enormous data security and data privacy-related issues might emerge as a result of data processing and sharing, mainly due to the lack of structured incentive mechanism in place that can regulate the proportional distribution of benefits among different stakeholders. Exploring the trade-offs and optimal mechanisms for data sharing at supply chain level remains a promising yet underexplored research avenue.

1.6 Conclusion

Data analytics in SCR and SCS has only recently started receiving significant attention from academics and practitioners. Notably, the applications of data analytics in SCR have increased following the COVID-19 pandemic outbreak. Although SCS is a more mature research domain compared to SCR, the applications of data analytics in SCS have increased only since 2014.

This chapter provides an overview of the data analytics in SCR and SCS through the lens of science mapping tools and techniques. Our analysis reveals that scholars from China, Iran, the United States, and India have been the most productive in terms of total research output for both SCR and SCS data analytics applications. Common themes across both domains include Industry 4.0, big data, blockchain, and artificial intelligence. However, we also identify unique areas of focus for each domain. Specifically, digital twins and data-sharing incentives have emerged as unique areas for SCR, while closed-loop supply chain design and circular economy have emerged as trending topics in SCS research. Future research should focus further on the role of autonomous systems and technologies including the Internet of Things (IoT), drones, and servitization in supply chain resilience and sustainability management.

References

Ahi P, Searcy C (2013) A comparative literature analysis of definitions for green and sustainable supply chain management. J Clean Prod 52:329–341

Aria M, Cuccurullo C (2017) bibliometrix: an R-tool for comprehensive science mapping analysis. J Informet 11(4):959–975

Azapagic A, Clift R (1999) The application of life cycle assessment to process optimisation. Comput Chem Eng 23(10):1509–1526

Bag S, Wood LC, Xu L, Dhamija P, Kayikci Y (2020) Big data analytics as an operational excellence approach to enhance sustainable supply chain performance. Resour Conserv Recycl 153:104559

Bag S, Pretorius JHC, Gupta S, Dwivedi YK (2021) Role of institutional pressures and resources in the adoption of big data analytics powered artificial intelligence, sustainable manufacturing practices and circular economy capabilities. Technol Forecast Soc Chang 163:120420

Bechtsis D, Tsolakis N, Iakovou E, Vlachos D (2022) Data-driven secure, resilient and sustainable supply chains: gaps, opportunities, and a new generalised data sharing and data monetisation framework. Int J Prod Res 60(14):4397–4417

Belhadi A, Kamble S, Jabbour CJC, Gunasekaran A, Ndubisi NO, Venkatesh M (2021) Manufacturing and service supply chain resilience to the COVID-19 outbreak: lessons learned from the automobile and airline industries. Technol Forecast Soc Chang 163:120447. Retrieved from https://www.ncbi.nlm.nih.gov/pmc/articles/PMC7836947/pdf/main.pdf

Belhadi A, Kamble S, Fosso Wamba S, Queiroz MM (2022) Building supply-chain resilience: an artificial intelligence-based technique and decision-making framework. Int J Prod Res 60(14): 4487–4507

Dubey R, Gunasekaran A, Childe SJ, Fosso Wamba S, Roubaud D, Foropon C (2021) Empirical investigation of data analytics capability and organizational flexibility as complements to supply chain resilience. Int J Prod Res 59(1):110–128

El-Kassar A-N, Singh SK (2019) Green innovation and organizational performance: the influence of big data and the moderating role of management commitment and HR practices. Technol Forecast Soc Chang 144:483–498

Fahimnia B, Jabbarzadeh A (2016) Marrying supply chain sustainability and resilience: a match made in heaven. Transp Res E: Logist Transp Rev 91:306–324

Fahimnia B, Jabbarzadeh A, Sarkis J (2018) Greening versus resilience: a supply chain design perspective. Transp Res E: Logist Transp Rev 119:129–148

Garai A, Sarkar B (2022) Economically independent reverse logistics of customer-centric closed-loop supply chain for herbal medicines and biofuel. J Clean Prod 334:129977

Ghazal T, Alzoubi H (2022) Fusion-based supply chain collaboration using machine learning techniques. Intell Autom Soft Comput 31(3):1671–1687

Goodarzian F, Taleizadeh AA, Ghasemi P, Abraham A (2021) An integrated sustainable medical supply chain network during COVID-19. Eng Appl Artif Intell 100:104188. Retrieved from https://www.ncbi.nlm.nih.gov/pmc/articles/PMC7890303/pdf/main.pdf

Hosseini S, Ivanov D (2019) A new resilience measure for supply networks with the ripple effect considerations: a Bayesian network approach. Ann Oper Res 319(14):1–27

Ivanov D, Dolgui A (2021) A digital supply chain twin for managing the disruption risks and resilience in the era of industry 4.0. Prod Plan Control 32(9):775–788

Ivanov D, Dolgui A, Sokolov B (2019) The impact of digital technology and Industry 4.0 on the ripple effect and supply chain risk analytics. Int J Prod Res 57(3):829–846

Kusiak A (2018) Smart manufacturing. Int J Prod Res 56(1–2):508–517

Li Y, Dai J, Cui L (2020) The impact of digital technologies on economic and environmental performance in the context of industry 4.0: a moderated mediation model. Int J Prod Econ 229: 107777

Lin F-R, Pai Y-H (2000) Using multi-agent simulation and learning to design new business processes. IEEE Trans Syst Man Cybern A Syst Hum 30(3):380–384

Mahmoudi A, Javed SA, Mardani A (2021) Gresilient supplier selection through fuzzy ordinal priority approach: decision-making in post-COVID era. Oper Manag Res 15(3):1–25

Modgil S, Gupta S, Stekelorum R, Laguir I (2021) AI technologies and their impact on supply chain resilience during COVID-19. Int J Phys Distrib Logist Manag. https://doi.org/10.1108/IJPDLM-12-2020-0434

Naz F, Kumar A, Majumdar A, Agrawal R (2021) Is artificial intelligence an enabler of supply chain resiliency post COVID-19? An exploratory state-of-the-art review for future research. Oper Manag Res 15:1–21

Nguyen T, Li Z, Spiegler V, Ieromonachou P, Lin Y (2018) Big data analytics in supply chain management: a state-of-the-art literature review. Comput Oper Res 98:254–264

Papadopoulos T, Gunasekaran A, Dubey R, Altay N, Childe SJ, Fosso-Wamba S (2017) The role of big data in explaining disaster resilience in supply chains for sustainability. J Clean Prod 142: 1108–1118

Saurabh S, Dey K (2021) Blockchain technology adoption, architecture, and sustainable agri-food supply chains. J Clean Prod 284:124731

Sazvar Z, Zokaee M, Tavakkoli-Moghaddam R, Salari SA-S, Nayeri S (2021) Designing a sustainable closed-loop pharmaceutical supply chain in a competitive market considering demand uncertainty, manufacturer's brand and waste management. Ann Oper Res 315:1–32

Sharma R, Kamble SS, Gunasekaran A, Kumar V, Kumar A (2020) A systematic literature review on machine learning applications for sustainable agriculture supply chain performance. Comput Oper Res 119:104926

Tirkolaee EB, Goli A, Ghasemi P, Goodarzian F (2022) Designing a sustainable closed-loop supply chain network of face masks during the COVID-19 pandemic: Pareto-based algorithms. J Clean Prod 333:130056. Retrieved from https://www.ncbi.nlm.nih.gov/pmc/articles/PMC8671674/pdf/main.pdf

Tsai C-W, Lai C-F, Chao H-C, Vasilakos AV (2015) Big data analytics: a survey. J Big data 2(1): 1–32

Tukamuhabwa BR, Stevenson M, Busby J, Zorzini M (2015) Supply chain resilience: definition, review and theoretical foundations for further study. Int J Prod Res 53(18):5592–5623

Vali-Siar MM, Roghanian E (2022) Sustainable, resilient and responsive mixed supply chain network design under hybrid uncertainty with considering COVID-19 pandemic disruption. Sustain Prod Consum 30:278–300. Retrieved from https://www.ncbi.nlm.nih.gov/pmc/articles/PMC8651491/pdf/main.pdf

Verma A, Bhattacharya P, Zuhair M, Tanwar S, Kumar N (2021) Vacochain: Blockchain-based 5g-assisted uav vaccine distribution scheme for future pandemics. IEEE J Biomed Health Inform 26(5):1997–2007

Wamba SF, Dubey R, Gunasekaran A, Akter S (2020) The performance effects of big data analytics and supply chain ambidexterity: the moderating effect of environmental dynamism. Int J Prod Econ 222:107498

Wilson M, Paschen J, Pitt L (2021) The circular economy meets artificial intelligence (AI): understanding the opportunities of AI for reverse logistics. Manag Environ Qual Int J 33(1): 9–25

Wong L-W, Leong L-Y, Hew J-J, Tan GW-H, Ooi K-B (2020) Time to seize the digital evolution: adoption of blockchain in operations and supply chain management among Malaysian SMEs. Int J Inf Manag 52:101997

Chapter 2
Enhancing the Viability of Green Supply Chain Management Initiatives Leveraging Data Fusion Technique

Sazid Rahman, Niamat Ullah Ibne Hossain, Adhir Roy, and Md. Saiful Islam

Abstract The assessment of viability for such GSCM initiatives is a crucial task to ensure strategic picking of the right initiatives and eliminate the harmful ones. A holistic assessment framework for viability assessment requires managers to rely on information sources that generate qualitative and subjective data, even often incomplete. To deal with the unique characteristics of subjective data, a data fusion technique, comprised of an integrated approach of *analytical hierarchy process (AHP) and the Dempster-Shafer theory*, is applied in a *hierarchical evidential reasoning (HER)* framework. This *HER* framework is developed using the factors affecting the viability of GSCM initiatives. The result of such a proposed methodology is obtained in the form of an index of viability of GSCM initiatives throughout the supply chain. In addition to the *Dempster-Shafer theory*, the *Yagers' recursive rule* of combination is applied to check the credibility of the calculation from DST. The proposed methodology can also be adopted and materialized by the practitioners to assess the condition state of the green management of any supply chain network.

Keywords Green supply chain management · Viability index · Hierarchical evidential reasoning (HER) · Analytical hierarchy process (AHP) · Dempster-Shafer theory · Yager's rule

---

S. Rahman
Industrial Engineering, University of Arkansas, Fayetteville, AR, USA

N. U. Ibne Hossain (✉)
Engineering Management, College of Engineering and Computer Science, Arkansas State University, Jonesboro, AR, USA
e-mail: niamat.hossain@gmail.com

A. Roy · M. Saiful Islam
Department of Industrial Engineering and Management, Khulna University of Engineering and Technology, Khulna, Bangladesh

© The Author(s), under exclusive license to Springer Nature Switzerland AG 2023
N. U. Ibne Hossain (ed.), *Data Analytics for Supply Chain Networks*, Greening of Industry Networks Studies 11, https://doi.org/10.1007/978-3-031-29823-3_2

2.1 Introduction

Green supply chain management (GSCM) has recently garnered a great amount of interest from practitioners and researchers alike. The increased importance to GSCM that is currently being given is the result of the deleterious impacts of different businesses on the environment. These deleterious impacts (e.g., waste overflow, depletion of raw material resources, environment pollution, etc.) come from the environmentally insecure conditions pertaining to the businesses. However, bringing about environmental sustainability or establishing green practices is not the only incentive that is pushing the concept of GSCM. Good business sense, in terms of the goal of getting higher profitability, is another reason (Wilkerson 2005). In fact, compliance with environmental regulations and government legislation is a major driver for customers while choosing manufacturers and suppliers. Government policies are not the only factor that works to impose environmental green practices. According to several researchers, green supply chain practices have been adopted with the view to have positive impacts on an organization's financial as well as environmental performance (Zhang and Yang 2016; Qi et al. 2009). Due to customer demand, different companies are adopting green practices in their supply chain (Kassinis and Vafea 2006). Even the business stakeholders can act as a motivating factor by pressuring the organizations or firms to adopt and maintain environmentally friendly green supply chain practices (Zhang and Yang 2016). Moreover, suppliers might even feel the need to discontinue doing business with the manufacturers in order to protect their own public image or to maintain their own compliance with environmental regulations if the manufacturing firm is known for causing pollution and following nongreen supply chain practices (Rivera-Camino 2007).

Bangladesh is among the leading manufacturers for the global fashion industry. The textile industry is the major force for Bangladesh's economic growth and stability (Anisul Huq et al. 2014; Chowdhury et al. 2019; Islam et al. 2019). The garments or textile industry has achieved this growth through relying on exporting to European or American customers. The exports in this industry approximately amount to 80% of the total national output for Bangladesh (Chowdhury et al. 2019). The key factor for the success of Bangladesh's textile industry is the low wages, which makes the manufacturing cost much lower than if the clothes were to be manufactured in most other countries. While there is considerable profit to be gained from the country's textile industry, ostensibly making it a competitive market, the factor in which Bangladesh lags behind others is compliance. For example, the collapse of Rana Plaza in April 2013, with the death of around 1000 workers, brought the issue of compliance to the forefront of textile industries. Not just the Bangladesh government but also the customers of these export-based textile companies started to value the state of compliance while doing business with the manufacturers (Asgari and Hoque 2013; Barua and Ansary 2017; Naciti 2019). This also brings the issue of being compliant with environmental regulations. Textile companies, at present, are required to maintain strict adherence to environmental regulations in order to be competitive in the market.

In order to maintain adherence to international and local environmental legislations, a good assessment of an organization's current scenario is pivotal. As the literature reveals, the most popular method for determining a business' or supply chain's environmental responsibility is measuring the organization's carbon emission rate. But a textile company might be the culprit of pollution in other ways than just releasing carbon emission. In the context of Bangladesh's textile industry, a holistic consideration of the supply chain in terms of being environmentally responsible or *green* needs to be done. The issue with adopting green supply chain management (GSCM) practices or initiatives, however, is that implementing such initiatives is not always favorable for all organizations. The introduction of such green practices or initiatives sometimes has to face some severe barriers that might cripple the financial structure of the organization. Hence, there is a need for assessing the condition of the organization so that it can be clearly ascertained if the business can really successfully launch a GSCM initiative and sustain it.

As the prime objective is to develop a holistic assessment framework, just identifying the internal or functional factors of an organization will not be sufficient. Consideration from both the internal and the external point of view is necessary while identifying the factors of GSCM initiatives. Even though existing literature offers a plethora of discussion over this matter, the proper identification of factors regarding the viability of GSCM initiatives with both external and internal points of view is still paramount for a holistic assessment framework.

To achieve the primary goal of developing an assessment framework, we identified the factors of green supply chain management and formulated a holistic assessment model. This study makes two very important contributions to the field of green supply chain management. Our paper tries to answer two important research questions to bridge the gap in the literature pertaining to green supply chain management:

Q1. What are the internal and external factors for green supply chain management initiatives?

Q2. How can a supply chain be assessed for being viable of adopting GSCM initiatives in the context of Bangladesh's textile industry?

While trying to answer the first question, we identified some factors for GSCM initiatives in two categories: internal factors and external factors. Then after collecting data in the form of experts' opinions, we calculated an overall index as the measure of being viable of introducing GSCM initiatives. Due to the subjective nature of the data, we implemented the Dempster-Shafer theory (DST) as it can deal with the effects of uncertain, incomplete, or subjective data.

2.2 Review of Literature

2.2.1 *Barriers and Factors for GSCM Initiative Implementation*

These days, the importance of green supply chain management is apparent to everyone. Green et al. (2012) explored the impact of GSCM practices. The 2012 study also developed a theoretical model demonstrating the relationship among suppliers, manufacturers, and customers when such GSCM practices are operational. The study further went on to prove that the economic and environmental performance improves with the tuning of GSCM practices or initiatives. The topic of green supply chain management has garnered so much interest from practitioners and researchers alike that there is enough analysis on the pertaining literature. Fahiminia et al. (2015) used rigorous bibliometric analysis to find out the research scope regarding the GSCM initiatives. Sarkis et al. (2011) also ventured on this research area to identify the opportunities for future research and organized the literature review in nine categories. Chin et al. (2015) provided a conceptual framework capable of determining the relation among environmental collaboration, sustainability, and GSCM. One key area that has gained the focus of researchers is the barriers and factors impacting GSCM initiatives. Mathiyazhagan et al. (2013) used an interpretive structural modeling approach to analyze the barriers to GSCM implementation. Govindan et al. (2014) studied and identified some obstacles to GSM practices with a view to determine the effectiveness of procurement processes. Testa and Iraldo (2010) prepared a comparative study between GSCM practices and other managerial practices and found that it is beneficial for an organization to give preference to the improvement of environmental performance. Shang et al. (2010) focused specifically on electronic manufacturing organizations and identified six factors that act as barriers to GSCM implementation. These six factors are green manufacturing and packaging, green suppliers, green marketing, green eco-design, green stock, and environmental participation. Zhu et al. (2012) proposed a model that measures the performance of internal practices and external factors of GSCM. Ninlawan (2010) further improved this by performing an assessment of the GSCM practices and chose an electronic manufacturing company to test the model. Even though they performed an assessment model for GSM practices, the model assesses the GSCM based on some practice that has already been adopted and operated. The gap in the literature on this topic is that, at least to our best of knowledge, there is no assessment framework established to act as a decision-making tool for managers before taking any GSCM initiative. The potential financial ramifications of adopting GSCM initiatives without having a proper assessment of the viability of such an initiative make the formulation of such an assessment methodology an absolute necessity. For instance, failing to successfully implement GSCM initiatives in the supply chain may result in potential sunk cost that could be catastrophic for small and medium enterprises (Carter and Rogers 2008). All these studies emphasize that a holistic approach for the assessment for viability of GSCM initiatives needs to be

Table 2.1 A summary of the literature review

Authors	Contribution to the literature	Approach
Green et al. (2012)	Empirically assess performance of adopted GSCM practices	A structural equation modeling methodology
Fahiminia et al. (2015)	Identification of key research topics as well as the interrelations and collaboration patterns	Rigorous bibliometric tools
Sarkis et al. (2011)	Categorizing GSCM literature into nine broad organizational theories	Critical evaluation of research scopes
Chin et al. (2015)	Explore the relation among sustainability performance, GSCM, and environmental collaboration	Empirical study using advanced structural equation modeling approach
Mathiyazhagan et al. (2013)	Analysis on barriers for implementing GSCM initiatives	Interpretive structural modeling approach
Govindan et al. (2014)	Analysis on barriers for implementing GSCM	Analytic hierarchy process approach
Testa and Iraldo (2010)	Assessment of determinants and motivations for GSCM implementation	Qualitative analysis
Shang et al. (2010)	Investigation of GSCM capability dimensions and performance	Comparative assessment between four groups of organizations
Zhu et al. (2012)	Exploration of internal and external GSCM practices on the basis of environmental, operational, and economic performance	Survey-based empirical analysis
Ninlawan (2010)	Evaluation of green supply chain management performance	Survey-based qualitative assessment
Carter and Rogers (2008)	Exploring the relation between social, environmental, and economical performance of SC	Developing a conceptual framework

taken. The challenges with such a consideration are that the data required for such an approach would mostly be qualitative. To cope with this challenge, DST is used in this study for the computation process. Table 2.1 summarizes the common themes associated with the GSCM.

2.2.2 *Computational Approach for Assessment Methodology*

As the qualitative nature of the data requires a computational approach that can deal with uncertainties and incompleteness in the data, the Dempster-Shafer theory is an ideal candidate for this approach. It is a general probabilistic theory that follows the principle of the evidential reasoning approach. The advantages associated with DST include a high degree of flexibility, the capability to account for the uncertainty or ignorance in the data, and the capability of combining the qualitative data with expert knowledge (Silva et al. 2019). Wherever there is a level of uncertainty involved in the data, DST can act as a powerful tool to effectively dissipate such uncertainty measures and get a meaningful quantitative value as a result of computation (Wu et al. 2005). DST has also an advantage over other probabilistic theories as it allows the consideration of data that are not even complete (Kang et al. 2019). In terms of decision-making tools, the combination of AHP and DST is an ideal computational approach if there is qualitative data involved (Beynon et al. 2000). Green et al. (2012) signified the qualitative subjective data as the information required to better assess the GSCM initiatives from a holistic point of view. Thus, DST is singled out as the primary computational tool for this study.

Although there is very little shortage of literature on GSCM practices, the discussions in the previous sections clearly identify an important existing gap in the extant literature. There has been no computational framework that has been modeled to assess the viability of GSCM initiatives before the initiatives are taken. To fill this gap in the literature, this study makes a contribution to the literature by exploring the idea of establishing an index value for viability assessment in terms of organization's condition:

- Identification of a set of factors that affects the successful implementation of GSCM initiatives.
- Developing a computational framework that assesses the viability of GSCM initiatives' adoption while being able to utilize qualitative data for the computation.

2.3 Evaluation Approach

The evaluation approach used in this paper includes three methods, namely, the analytical hierarchical process, the Dempster-Shafer Theory, and the Yager's rule. This integrated approach combining these methods is formulated to obtain the index of viability condition for introducing GSCM initiatives.

2.3.1 Analytical Hierarchical Process (AHP)

One of the most widely used techniques in the studies of multi-criteria decision-making problems is the analytical hierarchical process (AHP), which was proposed by Saaty (1990). The usefulness of this technique is that it generates a quantitative comparison or a reliable ranking of multiple alternatives. This is exactly the same purpose for which AHP is used in this study. The criteria that were found through literature review and experts' opinions to have impacts on GSCM are rated using this AHP technique. In order to complete the process of this MCDM technique, a pairwise comparison is conducted for this set of criteria. The steps that were followed in the AHP technique are described as follows:

1. At the start of the AHP technique, the objective and goals of the problem are identified.
2. Once the goal-setting step is complete, the hierarchical structuring needs to be done. This tripartite, hierarchical structure includes *objectives* at the top, *criteria* at the intermediate levels, and, finally, *alternatives* at the lowest level.
3. At this stage of AHP, a pairwise comparison is done to generate a *(n x n)* matrix for the lower tiers. Relative priorities between the criteria are used to make the pairwise comparisons. This process is also carried on for the intermediate levels.
4. As the comparisons are done pairwise, there is a reciprocal nature in the comparison. This makes each matrix in step 3 needing to be an $n(n - 1)/2$ number of comparison judgments.
5. The calculation of the *consistency index (CI)* is the next stage after pairwise comparisons. The calculation of *consistency index (CI)* uses eigenvalue λ_{max}, where $CI = (\lambda_{max} - n)/(n - 1)$, where n is the size of the matrix. *Consistency index (CI)* is divided by *random consistency index (RI)* to get the *consistency ratio*. The value of *consistency ratio* determines whether the judgment that was made is consistent or not. The judgment matrix can be considered acceptable if the value of the *consistency ratio* ≤ 0.1. Improved judgment needs to be taken if the original judgment is found to be inconsistent.
6. The final step in AHP is the normalization of eigenvectors using the weights of respective criteria.

2.3.2 Dempster-Shafer Theory

2.3.2.1 Frame of Discernment and Basic Probability Assignment

The Dempster-Shafer theory is a data fusion technique that was first introduced by Arthur Dempster (1967) and later improved by Glenn Shafer (1976). This method accounts for uncertainty and ambiguity in subjective data. DST addresses the uncertainty of data using belief functions. Evidences from all sources are combined in a data fusion technique to get an overall degree of belief. The fusion of data allows DST to tackle the uncertainty and ambiguity of the data. Compared to other techniques available in the extant literature, DST can deal with these uncertainties efficiently. For instance, the provision of allocating evidence to not only disjointed but also non-disjointed data sets makes DST capable of tracing and demonstrating the uncertainties in the data more effectively compared to the Bayesian theory (Bappy et al. 2019). The partial or incomplete knowledge is clearly stated as such in case of DST. This is not something that can be found in Bayesian probabilistic reasoning. The provision of setting a probability measure to the frame of discernment is another characteristic of DST that should be taken into account (Deng and Chan 2011).

The term frame of discernment (θ)in DST can be defined as a set of mutually exclusive hypotheses that can be written as follows: $\theta = \{H_1, H_2, H_3\}$, where H_1, H_2, and H_3 are three hypotheses. A set, termed as the power set, contains all the subsets of (θ,) and this power set is denoted by 2^θ. For the example used prior, the power set is constituted with the following elements:

$$2^\theta = \{\varnothing, \{H_1\}, \{H_2\}, \{H_3\}, \{H_1, H_2\}, \{H_1, H_3\}, \{H_2, H_3\}, \{H_1, H_2, H_3\}\} \quad (2.1)$$

Every element of this power set is known as focal elements, while the $\{H_1, H_2, H_3\}$ element denotes ignorance as it doesn't provide any specific information. A degree of belief ranging from 0 to 1 can be assigned to this focal element where 0 is attributed to no belief and 1 is attributed to complete belief. The degree of belief for each hypothesis is called a *basic probability assignment (BPA)* or *mass function (m)* such that $0 \leq m(A) \leq 1$, $\forall A \epsilon \theta$; $m(\varnothing) = 0$ (Bappy et al. 2019). The proposition or hypothesis *m*(*A*) satisfies the following properties, as presented in Eq. (2.1):

$$\sum_{A \epsilon \theta} m(A) = 1 \quad (2.2)$$

The incompleteness of information can be overcome by assigning a nonzero probability mass to the union of two or more classes. For all $A \in 2^\theta$, two parameters, namely, support *Su*(*A*) and plausibility *Pl*(*A*), can be defined as:

$$Su(A) = \sum_{B_1 \subseteq A} m(B_1) \quad (2.3)$$

$$Pl(A) = \sum_{A \cap B_2 \neq \varnothing} m(B_2) = 1 - Su(\overline{A}) \quad (2.4)$$

where B_1, $B_2 \in 2^\theta$, and $\overline{A}$ denote the complement of hypothesis A. The above equations define the support of a class as the sum of probability masses assigned to that class by a source of data and information. Support *Su*(*A*) is a measure of how much the evidence contradicts a proposition. On the other hand, the plausibility of a class can be defined as the sum of all probability masses that are not assigned to the complement of the class (Yang et al. 2020).

2.3.2.2 DS Rule of Combination

Information or evidence from different sources can be combined in the Dempster-Shafer theory or evidence theory. For example, if two different sources are a subset of θ, i.e., $A \subseteq \theta$, then the two BPAs obtained from two sources, namely, $m_1(A)$ and $m_2(A)$, can be combined through the following equation:

$$m_{12}(A) = m_1(A) \oplus m_2(A) \quad (2.5)$$

$$= \begin{cases} 0, & \text{and when } A = \varnothing \\ \sum_{X \cap Y = A, \quad \forall X, Y \epsilon \theta} m_1(X) m_2(Y), & \text{and when } A \neq \varnothing \end{cases} \quad (2.6)$$

where $= \sum_{X \cap Y = \varnothing, \quad \forall X, Y \in \theta} m_1(X)m_2(Y)$. The factor K represents the conflict between two subsets, X and Y. Because of the cumulative property of the DST, the order in which the combination takes place does not matter. More precisely, the result of the combination of the evidence would remain the same irrespective of orders. This yields the generalized form of combination as follows:

$$m_{12\ldots n}(A) = m_1(A) \oplus m_2(A) \oplus m_3(A) \oplus \ldots\ldots\ldots \oplus m_n(A) \tag{2.7}$$

In this study, the basic criteria (factors) are aggregated to assess the cyber resilience of a supply chain to obtain a cyber-resilience (CR) index, which is streamlined according to the DS rule of combination. This can be written as in Eq. (2.7):

$$\mathrm{CR} = X_1 \oplus X_2 \oplus X_3 \oplus \ldots\ldots\ldots \oplus X_n \tag{2.8}$$

where X_1, X_2, X_3, ………, X_n are the contributing factors in the assessment and n represents the total number of factors that are being used. $S(X_i)$ represents the evaluation of parameter X_i, while $m(X_i)$ represents the basic probability assignment for each factor X_i. The weight attributed to these factors is denoted by w_i for each factor X_i. To avoid any computational complexity, this combination rule is used recursively (Bappy et al. 2019).

2.3.3 *Yager's Recursive Rule*

It is a similar combination technique to the *DS combination rule*. The *difference* between them is the absence of the normalization by nonconflicting evidence (Bappy et al. 2019). The factor K represents conflicting evidence as this information is considered as the level of ignorance throughout the combination procedure (Bappy et al. 2019).

2.4 Methodology

This section discusses the assessment model that is developed to generate the viability index of GSCM initiatives. The assessment methodology constitutes four major phases, namely, a preliminary phase, a data collection phase, a GSCM condition assessment phase, and a result comparison phase. This methodology of all four phases is depicted in Fig. 2.1.

The first stage of this methodology is the preliminary phase which involves identifying the factors of GSCM initiatives. On the platform of "Web of Science," pertaining literature are studied and selected for the purpose of the identification of

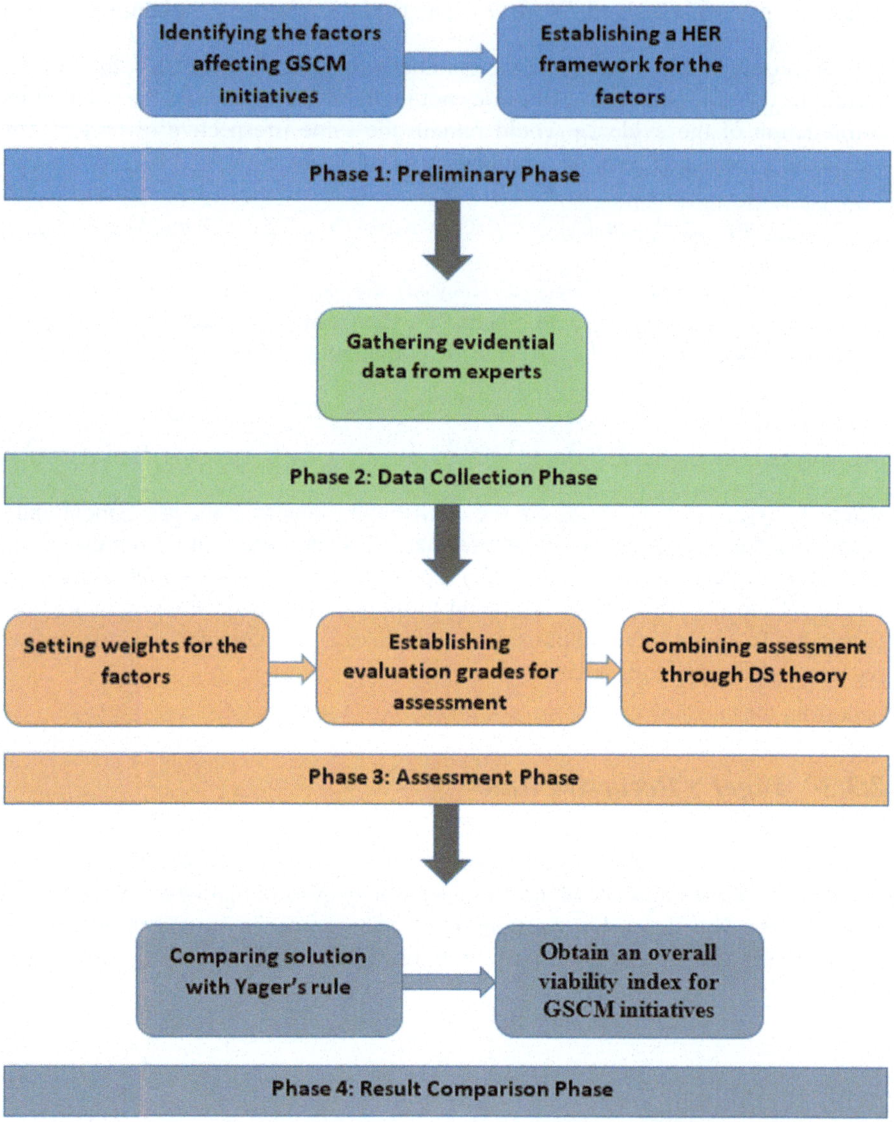

Fig. 2.1 Different phases of the methodology of viability assessment for GSCM initiatives

factors. Peer-reviewed papers, chapters of pertaining books, and conference proceedings are the contents that were searched and studied on the database to achieve a wide range of perspectives on topics related to the assessment of GSCM of the textile industry supply chain. The initial process of searching for papers in the database and then screening those documents resulted in more than 100 papers. After the initial screening, a more thorough and detailed screening based on some keywords made the list of papers much shorter. Then based on the abstract of the papers on hand, another screening process was conducted to cut the list of papers to only the most relevant of them. Finally, papers that are related to GSCM were selected after a thorough reading to identify the factors of GSCM initiatives. Then, experts from the textile industry were contacted to get their opinions on the factors selected through this identification process. A hierarchical decision tree was structured comprising these factors, as illustrated in Fig. 2.2, which includes three distinct levels. The upper level consists of only the goal or objective of the assessment, which generates a viability index for GSCM initiatives, whereas the lower two levels comprise the factors of GSCM initiatives. The lowest level is the level that describes the basic factors, while the level above consists of two general factors, namely, internal factors and external factors.

The data collection phase describes the gathering of evidential data from the relevant sources. The historical data is often difficult to obtain in this aspect. So, subjective data in the form of opinions from experts were collected to complete the calculation of the assessment. That is, in order to avoid bias, experts of different levels of experience were approached to get opinions. Due to the nature of this data, in order to get a meaningful result, a special approach is required that can cope with the uncertainty or ambiguity of the subjective data. DST satisfies this requirement.

The assessment phase includes the actual calculation for the assessment of GSCM condition and generation of the index. At first, from the experience of the experts, they assigned their beliefs as the subjective data to the factors of GSCM initiatives. Then, from the subjective data, an AHP model was used to assign the weights to the factors. A set of evaluation grades in the form of condition state were taken as measures in the assessment process. The most prevalent evaluation grade set in the literature was chosen for this study, and these were poor (P), fairly poor (F), average (A), good (G), and excellent (E) (Kong et al. 2015). Based on the condition for each

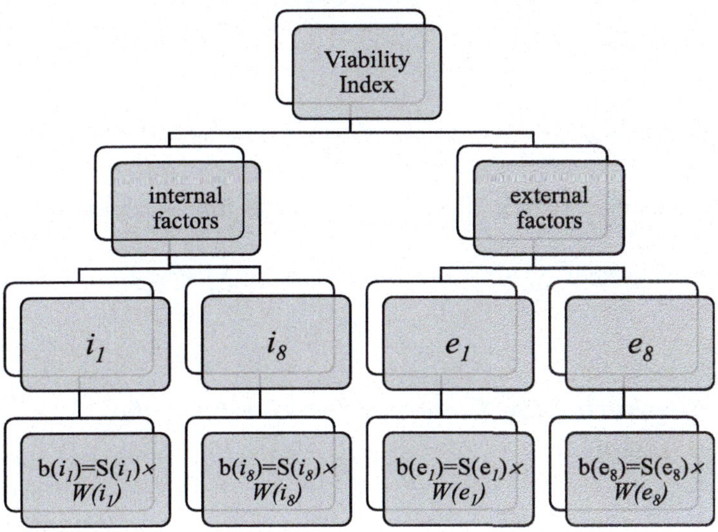

Fig. 2.2 GSCM assessment framework for HER model

factor, experts assigned their beliefs for the evaluation grades for each factor selected. The combination of these subjective belief values is done through the DST. The order of combination does not affect the end result of the assessment process according to the basic concept of DST. This assessment produces the probability of the condition states of supply chain in terms of viability to introduce the GSCM initiatives. In addition to this, further calculation for utility perspective assessment generates a numerical index value for viability. For the utility perspective assessment, the evaluation grades for GSCM factors were quantified as $u(P) = 0$, $u(F) = 0.25$, $u(A) = 0.50$, $u(G) = 0.75$, and $u(E) = 1$.

Through the combination of evaluation grades for each GSCM factor, a GSCM index is then calculated. The result obtained through this process is then validated using the Yager's rule by comparing results from both calculations.

2.5 GSCM Initiatives' Model Formulation

The hierarchical evidential reasoning (HER) framework is used for the assessment. The factors of GSCM initiatives that were identified are structured in an HER framework, and then the data are combined using an evidence theory to compute an overall index for GSCM initiatives. H_n is used to denote the evaluation grades that are attributed to a BPA by the expert, while $S(X_i)$ represents the assessment of criteria X_i as follows:

$$S(X_i) = \{(H_1, P_1^i), (H_2, P_2^i), (H_3, P_3^i), \ldots\ldots\ldots, (H_n, P_n^i)\} \tag{2.9}$$

where P_1^i represents the degree of belief that the *attribute* X_i is assessed to the grade of H_n and H_n denotes a set of evaluation grades, namely, poor, fairly poor, average, good, and excellent. Therefore, the first criterion for its assessment would be as follows:

$$S(X_1) = \{(H_1, P_1^1), (H_2, P_2^1), (H_3, P_3^1), (H_4, P_4^1), (H_5, P_5^1)\} \tag{2.10}$$

Considering that $0 \leq P_n^1 \leq 1$ and $\sum_{n=1}^{5} P_n^i \leq 1$ holds true. An assessment is considered to be complete if $\sum_{n=1}^{5} P_n^i = 1$ and incomplete if $\sum_{n=1}^{5} P_n^i < 1$.

2.5.1 *Structuring the Hierarchy for Viability Index Assessment*

After conducting an extensive review of the pertaining literature, two types of factors affecting adoption of GSCM initiatives were identified. The first is "internal factors," and the other one is "external factors" on the "Web of factors Science" database, which was our preferred medium to conduct the probe. The factors that were identified as "external factors" are the ones that the organization had minimal or no amount of control over at all. These "external factors" are *green purchasing*, *government and policy regulations*, *market/consumer*, *competitors*, *social*, *inbound logistics*, *outbound logistics*, and *customer awareness*. The description of these factors is provided below:

I. Green Purchasing: GSCM initiatives depend greatly on the suppliers. Suppliers can have a positive impact on GSCM initiatives if they practice good environmental management and they are certified (Diabat and Govindan 2011). Conversely, if the suppliers are found to be not practicing good environmental practices, they can have a negative impact on GSCM initiatives. The buying company's environmental ability greatly depends on the environmental compliance of the supplier (Zhu et al. 2005). Purchasing from environmentally compliant sources is termed as green purchasing (Walker and Brammer 2009). Businesses could even collaborate with suppliers to manufacture environmentally sustainable products (Zhu et al. 2005).

II. Government and Policy Regulation: Government regulations could impact GSCM initiatives through imposing a legal binding on the businesses and subjecting them to penalties in the event of failure to comply with such regulations (Lee and Klassen 2008). It is not economically preferable for businesses to spend on environmental measures (Diabat and Govindan 2011). Without government regulations, businesses would lack the impetus to take the necessary steps to ensure environmental sustainability (Khiewnavawongsa and Schmidt 2013).

III. Market/Consumer: Consumers can have an indirect effect on GSCM initiatives as consumer demand often drives businesses to follow certain organizational cultures (Khiewnavawongsa and Schmidt 2013). That is how consumers can determine what an organization needs to do to survive in the market. Consumers who belong to affluent groups may opt for products that are advertised as more environmentally sustainable (Zhu et al. 2005). Consumers who earn less might not concern themselves with anything other than the cost of the product (Zhu et al. 2005). So, the product's target consumer base impacts GSCM initiatives greatly (Lin et al. 2013).

IV. Competitors: Consumers are now becoming more conscious of the fact that there is a genuine need for being environmentally sustainable (Kumar and Chandrakar 2012). Having consumers who are more aware of environmental issues may drive the competitors to adopt GSCM initiatives, making competitors an important factor (Diabat and Govindan 2011). Competitors may even force organizations to discard GSCM practices as they might focus more on economic goals and making survival in the competitive market difficult for any organization which is actually following the environmental regulations (Dhull and Narwal 2016). In this case, pressure created by the competitor on pricing can act as the deterrent (Dhull and Narwal 2016).

V. Social: Good organizations now give a lot of importance to the image they present to the public as well as profitability (Darnall et al. 2008). In order to achieve this goal, organizations often get involved in some socially beneficial projects. An example of this is Hewlett-Packard creating their corporate social responsibility programs (Darnall et al. 2008). Having GSCM initiatives may also be considered an organization's social responsibility (Van Rensburg 2015). That is why social responsibility of the organization can be considered a factor affecting GSCM initiatives.

VI. Inbound Logistics: Wu and Dunn (1995) described inbound logistics as the functions of the business that are performed in order to receive and store materials. Inbound logistics include functions such as supplier management, inventory management, green transportation, and green purchasing (Sellitto et al. 2015).

VII. Outbound Logistics: Huynh (2013) explains outbound logistics as the collection of functions that are performed for the delivery of the finished product that the business produces to the point of its consumption. Outbound logistics include designing distribution networks, outbound transportation, and components of marketing (Kumar and Chandrakar 2012). Among these downstream functions, green outbound distribution or transportation could be considered as one of the most important factors that drives the effect of GSCM initiatives as it is directly responsible for carbon dioxide emission (Huynh 2013).

VIII. Environmental Organizations: This is the one external factor that was identified through the experts' opinions. Different environmental organizations function to create pressure on government and businesses to adopt green practices in supply chains. The impact of such pressure may lead businesses to adopting GSCM initiatives.

On the other hand, factors that were identified as "internal factors" are the ones on which the organization had at least some amount of control. These "internal factors" are *management support and commitment*, *organizational structure and strategy*, *cost*, *reverse logistics*, *eco-design*, *investment recovery*, *organizational learning*, and *recycling and reuse efforts*. The description of these factors is provided below:

I. Management Support and Commitment: A critical internal factor for adoption of GSCM initiative is the level of support offered by the management (Lee and Klassen 2008). Support from management can act as a strong motivating incentive for employees. This makes it easier to align business goals with GSCM initiatives improving the probability of success of those initiatives (Dhull and Narwal 2016). The opposite can also be true as the lack of management support might demotivate employees resulting in the failure of sustainable initiatives (Kumar and Chandrakar 2012).

II. Organizational Structure and Strategy: Organizational structure and strategies can act as both a driver and a barrier for GSCM initiatives (Lee and Klassen 2008). Proactive strategies aiming toward sustainability will facilitate the adoption of any GSCM initiative (Lee and Klassen 2008). Conversely, organizational structure may hinder such an adoption. For instance, small businesses producing very low carbon footprint may not find it feasible to introduce GSCM initiative in the supply chain (Lee and Klassen 2008).

III. Cost: Khiewnavawongsa and Schmidt (2013) identified financial implications as one of the main deterrents for businesses in adopting GSCM initiatives. The cost incurred for introducing GSCM initiatives can increase or decrease an organization's cost structure (Diabat and Govindan 2011). For instance, environmentally friendly materials, manufacturing, and packaging mostly cost more than normal process structure and materials (Zhu et al. 2005).

IV. Reverse Logistics: Reverse logistics defines the disposal and then the recovery of raw materials, finished goods, work in process, and scrap after consumption of the product (Jumadi and Zailani 2010). Due to the high cost (operation) of reverse logistics, businesses often choose to outsource such operations of the supply chain to minimize cost (Jumadi and Zailani 2010). This reduces organization's control over that specific function of the supply chain and impacts the effective adoption of new GSCM initiatives (Narayana et al. 2014).

V. Eco-design: Eco-design directly focuses on the core functional designs that can be altered or modified to improve the sustainability condition (Green et al. 2012). The success of eco-design depends on cross-functional cooperation and collaboration between different functions of the supply chain (Kumar and Chandrakar 2012).

VI. Investment Recovery: Investment recovery can be termed as the concept of converting excess assets into revenue for the business (Zhu et al. 2005). In terms of supply chain, the stagnant assets remaining in the organization should be strategically utilized to extract maximum value out of them (Zsidisin and Siferd 2001).

VII. Organizational Learning: Organizational learning is the historical data or analysis that teaches organizations to better strategize for future (Wood and Reynolds 2013). The effective deployment of the GSCM initiatives requires a better organizational learning structure (Wood and Reynolds 2013). The success of GSCM initiatives may require continuous assessment of historical data and changing the strategy over time (Dalkir 2005).

VIII. Recycling and Reuse Efforts. This is the internal factor of GSCM initiatives that was identified through experts' opinions. A management's efforts toward ensuring recycling and reusing scrap materials are a critical factor for GSCM initiatives. A proper management structure that is capable of ensuring the best recycling effort eases the obstacles for GSCM initiative adoption a great deal.

These factors for introducing GSCM initiatives in the supply chain were then structured according to the HER framework consisting of three levels, as illustrated in Fig. 2.2. The upper level of the hierarchy represents the goal of the assessment, which is the viability index (VI) for GSCM initiatives. The middle level consists of the assessment for two general factors: internal factors and external factors. Finally, the lowest level or echelon of the hierarchy structure consists of the eight basic factors for each of the two general factors. By combining the assessment of these factors, the overall viability index for GSCM initiatives was determined, and this combination was performed through the computational approaches, DST and Yager's rule.

2.5.2 *Factor's Weights and Assessment Grades with Utility*

Each identified factor of GSCM initiatives certainly does not have the exact same level of impact over the GSCM initiatives for the organization selected. In order to get a precise and accurate assessment, the relative importance for each factor needs to be considered. This is done by attributing weights to these factors according to their relative level of impact. AHP is a prevalent technique applied to obtain weights of multiple criteria from the subjective judgments of experts. In this study, the AHP method is used to assign weights to the factors of GSCM initiatives. These weights are then used to get a proper assessment of these factors through the evaluation grades and the belief function assigned to them. This allows the assessment to yield a unified utility value for the viability of GSCM initiatives. Weights for external and internal factors are illustrated in Tables 2.2 and 2.3, respectively.

Table 2.2 Weights attributed to the external factors

	Factors	Weights
External factors (0.67)	Green purchasing	0.06
	Government and policy regulation	0.32
	Market/consumer	0.03
	Competitors	0.03
	Social	0.1
	Inbound logistics	0.17
	Outbound logistics	0.09
	Customer awareness	0.21

Table 2.3 Weights attributed to the internal factors

	Factors	Weights
Internal factors (0.33)	Management support and commitment	0.05
	Organizational structure and strategy	0.13
	Cost	0.21
	Reverse logistics	0.08
	Eco-design	0.11
	Investment recovery	0.36
	Organizational learning	0.03
	Recycling and reuse efforts	0.04

2.6 Analytical Illustration Through a Case Study

This section is structured so as to demonstrate the analytical illustration of our proposed model. In order to make a sustainable decision, an organization needs to have a clear idea of the viability of the GSCM initiatives they would take. Our proposed model provides an analytical approach to quantify the viability in the form of an index measurement. This paper's analytical illustration is presented in three steps. First, a case is selected to illustrate the evaluation process, followed by the explanation of the data collection procedure, and finally an interpretive evaluation is provided to illustrate the implications of the results obtained through the proposed model.

A textile company from Bangladesh is identified to test the applicability of our model. The organization selected produces a mixture of textile products. While manufacturing their products, they rely on different functions, such as purchasing, manufacturing process, control process, distribution, etc., and on some technologies that produce harmful wastes in the environment. Chemical wastes released from the manufacturing process created environmental issues that require good GSCM initiatives as well as transportation that produces carbon dioxide in the air. This makes the organization an ideal candidate to test our proposed model. Subjective data that are necessary for the assessment were collected from the experts within the organization.

2.6.1 Data Collection

Data collection for this evaluation was done in two steps, the first one being the identification of the weights and the second one being the evaluation of the condition of the factors. As previously mentioned, initially, the GSCM initiative factors were identified by conducting a thorough examination of the relevant literature and taking opinions on the factors from experts. To be more precise with the identification of the factors, a thorough search was conducted on the "Web of Science" database. Based on the literature review conducted and experts' opinions taken, 16 different factors of two types were selected, each type consisting of 8 factors. In the second phase, data in the form of degree of belief measures were collected considering the evaluation grades of the selected factors for viability assessment.

Through extensive communication, the proper data source is identified for the assessment calculation. A textile manufacturing company is identified as that data source to collect all the relevant information regarding the assessment of viability of GSCM initiatives. The selected organization's middle- to top-level employees were questioned on the factors to get the required data. As the DST takes belief data or the basic probability assignment, this belief data was collected from the employees based on the selected evaluation grades. The application of DST in this study can be justified as this data is collected in the form of employees' opinions which are subjective in nature. This data is characterized as subjective since the experts'

Table 2.4 Subjective data from an expert regarding the condition of factors for GSCM initiatives

		Evaluation grades				
Factors	Basic factors	Poor	Fairly Poor	Average	Good	Excellent
Internal factors	*Eco-design*	0	0.2	0.6	0.2	0
	Cost	0	0.1	0.4	0.5	0
	Management support and commitment	0	0	0.1	0.9	0
	Organizational structure and strategy	0	0	0.3	0.5	0.1
	Organizational learning	0	0.2	0.6	0.1	0.1
	Green manufacturing	0	0	0.05	0.6	0.35
	Reverse logistics	0.2	0	0.3	0.5	0
	Recycling and reuse efforts	0	0.3	0.7	0	0
External factors	*Government policy and legislation*	0	0	0.2	0.8	0
	Green purchasing	0	0	0.15	0.85	0
	Inbound logistics	0	0.2	0.8	0	0
	Outbound logistics	0	0	0.2	0.5	0.3
	Competitors	0	0	0.2	0.4	0.4
	Market/consumptions	0	0.5	0.5	0	0
	Social	0	0	0.2	0.7	0
	Customer awareness	0	0	0.2	0.5	0.3

opinions, conversely, stem from their experience, knowledge, and education. The collected data according to the evaluation grades for each of the GSCM initiative factors are illustrated in Table 2.4.

For instance, when asked, the experts evaluated the factor "green purchasing" and attributed a belief measure to the evaluation grades based on the condition of the factor for the supply chain in question. This yielded evaluation grades for "green purchasing" to be *(poor, 0)*, *(fairly poor, 0)*, *(average, 0.15)*, *(good, 0.85)*, and *(excellent, 0)*. These values assigned to the evaluation grades by the experts represent the evidence supporting that the factor "green purchasing" is, at the time of the assessment, *0% poor*, *30% fairly poor*, *50% average*, *15% good*, and *0% excellent.* The unaccounted 5% can be attributed to the ignorance or uncertainty in the assessment. Experts subjectively judged these factors and assigned the degree of belief measures to the evaluation grades based on organizations' conditions regarding each factor.

2.6.2 Combining the Assessments Using DST

The subjective judgments depicted in Table 2.4 in the form of degree of belief are combined to assess the viability of GSCM initiatives. From the initial observation of

Table 2.4, it can be assumed that the value of the viability index should lie above the range of average. The precise calculation for assessment requires the consideration of relative importance of all 16 factors affecting GSCM initiatives. The weights illustrated in Tables 2.2 and 2.3 for each factor are used for this purpose. The Dempster-Shafer's recursive rule of combination is used here as a data fusion tool in a hierarchical evidential reasoning structure. First, data combination is performed to get the combined degree of belief for all internal factors by combining eight basic internal factors, namely, eco-design, cost, management support and commitment, organizational structure and strategy, organizational learning, green manufacturing, reverse logistics, and recycle and reuse efforts as illustrated in Table 2.3 and denoted by i_1, i_2, i_3, i_4, i_5, i_6, i_7, and i_8, respectively. Second, data combination for external factors is done in the exact same way by combining eight basic external factors, namely, government policy and legislation, green purchasing, inbound logistics, outbound logistics, competitors, market/consumption, social, and customer awareness as illustrated in Table 2.3 and denoted by e_1, e_2, e_3, e_4, e_5, e_6, e_7, and e_8, respectively. As defined in Eq. 2.11, let $\text{IF} = i_{1\oplus}\, i_{2\oplus}\, i_{3\oplus}\, i_{4\oplus}\, i_{5\oplus}\, i_{6\oplus}\, i_{7\oplus}\, i_8$, where $\oplus$ denotes the aggregation of two different factors. Similarly, let $\text{EF} = e_{1\oplus}\, e_{2\oplus}\, e_{3\oplus}\, e_{4\oplus}\, e_{5\oplus}\, e_{6\oplus}\, e_{7\oplus}\, e_{8.}$ The assessment of viability of GSCM initiative introduction considering all the factors can be illustrated through Eqs. 2.11, 2.12, and 2.13 as seen below:

$$
\begin{aligned}
&S(\text{Internal Factors}) = \\
&\quad S(\text{Eco} - \text{design} \oplus \text{Cost} \oplus \text{Management Support and Commitment} \\
&\qquad \oplus \text{Organizational Structure and Strategy} \\
&\qquad \oplus \text{Organizational Learning} \oplus \text{Green Manufacturing} \\
&\qquad \oplus \text{Reverse Logistics} \oplus \text{Recycle and Reuse Efforts}).
\end{aligned} \tag{2.11}
$$

$$
\begin{aligned}
&S(\text{External Factors}\ ) = \\
&\quad S(\text{Government Policy and Legislation} \oplus \text{Green Purchasing} \\
&\qquad \oplus \text{Inbound Logistics} \oplus \text{Outbound Logistics} \oplus \text{Competitors} \\
&\qquad \oplus \text{Market/Consumption} \oplus \text{Social Customer Awareness}\ )
\end{aligned} \tag{2.12}
$$

$$
S(\text{Viability Index}\ ) = S(\text{Internal Factors} \oplus \text{External Factors}) \tag{2.13}
$$

2.6.2.1 Interpreting the Evaluation of Viability Index for GSCM Initiatives Using DST

Each external factor for GSCM initiatives of textile supply chain is subjectively judged and attributed a percentage degree of belief on the evaluation grades (H_n, P^i_n) as presented below:

$$S(e_1)=\{(P,0.00),(FP,0.00),(A,0.20),(G,0.80),(E,0.00)\}$$
$$S(e_2)=\{(P,0.00),(FP,0.00),(A,0.15),(G,0.85),(E,0.00)\}$$
$$S(e_3)=\{(P,0.00),(FP,0.20),(A,0.80),(G,0.00),(E,0.00)\}$$
$$S(e_4)=\{(P,0.00),(FP,0.00),(A,0.20),(G,0.50),(E,0.30)\}$$
$$S(e_5)=\{(P,0.00),(FP,0.00),(A,0.20),(G,0.40),(E,0.40)\}$$
$$S(e_6)=\{(P,0.00),(FP,0.50),(A,0.50),(G,0.00),(E,0.00)\}$$
$$S(e_7)=\{(P,0.00),(FP,0.00),(A,0.20),(G,0.70),(E,0.00)\}$$
$$S(e_8)=\{(P,0.00),(FP,0.00),(A,0.20),(G,0.50),(E,0.30)\}$$

Each internal factor for GSCM initiatives of textile supply chain is subjectively judged and attributed a percentage degree of belief on the evaluation grades $\left(H_n, P_n^i\right)$ as presented below:

$$S(i_1)=\{(P,0.00),(FP,0.20),(A,0.60),(G,0.20),(E,0.00)\}$$
$$S(i_2)=\{(P,0.00),(FP,0.10),(A,0.40),(G,0.50),(E,0.00)\}$$
$$S(i_3)=\{(P,0.00),(FP,0.00),(A,0.10),(G,0.90),(E,0.00)\}$$
$$S(i_4)=\{(P,0.00),(FP,0.00),(A,0.30),(G,0.50),(E,0.10)\}$$
$$S(i_5)=\{(P,0.00),(FP,0.20),(A,0.60),(G,0.10),(E,0.10)\}$$
$$S(i_6)=\{(P,0.00),(FP,0.00),(A,0.05),(G,0.60),(E,0.35)\}$$
$$S(i_7)=\{(P,0.20),(FP,0.00),(A,0.30),(G,0.50),(E,0.00)\}$$
$$S(i_8)=\{(P,0.00),(FP,0.30),(A,0.70),(G,0.00),(E,0.00)\}$$

The relative weights for internal and external factors are 0.67 and 0.37.

The relative weights (W_i) calculated for these internal attributes are $W_i = \{W_{i1}, W_{i2}, W_{i3}, W_{i4}, W_{i5}, W_{i6}, W_{i7}, W_{i8}\} = \{0.05, 0.13, 0.21, 0.08, 0.11, 0.36, 0.03, 0.04\}$.

The relative weights (W_e) calculated for these external attributes are $W_e = \{W_{e1}, W_{e2}, W_{e3}, W_{e4}, W_{e5}, W_{e6}, W_{e7}, W_{e8}\} = \{0.06, 0.32, 0.03, 0.03, 0.10, 0.17, 0.09, 0.21\}$.

The multiplication of belief data with the respective weights assigned yields the *BPA or mass function* ($m(X_i)$) for each factor. The summation of *mass function* evaluation grades for the factors may not be equal to 1. If there is such a lag, then the difference between 1 and summation of *mass function* is attributed as ignorance or uncertainty in the data.

Assessing the Evaluation Grades for Internal Factors Using DST

$$\text{Internal factor index} = S(i_1)\times W(i_1) \oplus S(i_2)\times W(i_2) \oplus S(i_3)\times W(i_3) \oplus \ldots\ldots\ldots \oplus S(i_8)\times W(i_8) \tag{2.14}$$

The BPAs determined through the multiplication of the conditional ratings and respective weights are:

$$m(i_1) = \{0.000, 0.010, 0.030, 0.010, 0.000, 0.950\}$$
$$m(i_2) = \{0.000, 0.013, 0.052, 0.065, 0.000, 0.870\}$$
$$m(i_3) = \{0.000, 0.000, 0.021, 0.189, 0.000, 0.790\}$$
$$m(i_4) = \{0.000, 0.000, 0.024, 0.040, 0.008, 0.928\}$$
$$m(i_5) = \{0.000, 0.022, 0.066, 0.011, 0.011, 0.890\}$$
$$m(i_6) = \{0.000, 0.000, 0.018, 0.216, 0.126, 0.640\}$$
$$m(i_7) = \{0.006, 0.000, 0.009, 0.015, 0.000, 0.970\}$$
$$m(i_8) = \{0.000, 0.012, 0.028, 0.000, 0.000, 0.960\}$$

The BPA obtained from the aggregation of BPAs of two different factors is denoted as M. The initial aggregation is taken to be $M_1 = m_1$. The following aggregation of the remaining factors is then continued as the following approach:

$$\begin{aligned} K_2 &= \left(1 - \sum_{n=1}^{4} \sum_{p=1, p\neq n}^{4} M_1^n m_2^p\right)^{-1} \\ &= \left[1 - \left(0 + \ldots + m_1^{H_2} m_2^{H_1} + m_1^{H_2} m_2^{H_3} + m_1^{H_2} m_2^{H_4} + \right.\right. \\ &\qquad \left.\left. m_1^{H_2} m_2^{H_5} + m_1^{H_3} m_2^{H_2} + m_1^{H_3} m_2^{H_4} + \ldots + 0 + \ldots + 0\right)\right]^{-1} \\ &= [1 - (0.010 \times 0.013 + 0.010 \times 0.052 + 0.010 \times 0.065 + 0.030 \times 0.013 + 0.030 \\ &\qquad \times 0.052 + 0.030 \times 0.065 + 0.010 \times 0.013 + 0.010 \times 0.052 + 0.010 \times 0.065)]^{-1} \\ &= 1.004177 \end{aligned} \tag{2.15}$$

After calculating the factor K, the combined BPA for the first two attributes can be computed as follows:

$$\begin{aligned} M_2^{H_1} &= K_2 \times \left(m_1^{H_1} m_2^{H_1} + m_1^{H_1} m_2^{H} + m_1^{H} m_2^{H_1}\right) \\ &= 1.004177 \times (0.000 \times 0.000 + 0.000 \times 0.870 + 0.950 \times 0.000) = 0.000 \\ M_2^{H_2} &= K_2 \times \left(m_1^{H_2} m_2^{H_2} + m_1^{H_2} m_2^{H} + m_1^{H} m_2^{H_2}\right) \\ &= 1.021024 \times (0.108 \times 0.020 + 0.108 \times 0.870 + 0.950 \times 0.658) = 0.021268 \\ M_2^{H_3} &= K_2 \times \left(m_1^{H_3} m_2^{H_3} + m_1^{H_3} m_2^{H} + m_1^{H} m_2^{H_3}\right) \\ &= 1.012734 \times (0.018 \times 0.030 + 0.018 \times 0.870 + 0.950 \times 0.658) = 0.077382 \\ M_2^{H_4} &= K_2 \times \left(m_1^{H_4} m_2^{H_4} + m_1^{H_4} m_2^{H} + m_1^{H} m_2^{H_4}\right) \\ &= 1.032311 \times (0.054 \times 0.130 + 0.054 \times 0.870 + 0.950 \times 0.658) = 0.071397 \\ M_2^{H_5} &= K_2 \times \left(m_1^{H_5} m_2^{H_5} + m_1^{H_5} m_2^{H} + m_1^{H} m_2^{H_5}\right) \\ &= 1.104883 \times (0.000 \times 0.020 + 0.000 \times 0.870 + 0.950 \times 0.658) = 0.000 \\ M_2^{H} &= K_2 \times \left(m_1^{H} m_2^{H}\right) \\ &= 1.011224 \times (0.870 \times 0.950) = 0.829953 \end{aligned}$$

The remaining internal factors are aggregated following the same process as the aggregation of the first two factors. The third internal factor is aggregated with the combination of the first two factors, and subsequently, the process is followed for all other factors. After the completion of aggregation for all internal factors, the combined BPA for all internal factors of GSCM initiatives is obtained as follows:

$$M_8 = \{0.002, 0.025, 0.125, 0.363, 0.086, 0.386116\}$$

This illustrates that $M_8^{H_1} = 0.002$, $M_8^{H_2} = 0.025$, $M_8^{H_3} = 0.125$, $M_8^{H_4} = 0.363$, $M_8^{H_5} = 0.086$, and $M_8^{H} = 0.386116$.

M_8^H can be termed as the level of ignorance that is inherent in the data, and it remains through the combination of belief functions as DST allows these ignorance or uncertainties to be taken into the calculation. Performing a normalization will dissipate the ignorance to every evaluation grade proportionally. This normalization is done according to the following equation:

$$\{H_n\} : P_n = \frac{M_8^{H_n}}{1 - M_8^H}; n = 1, 2, .., N \tag{2.16}$$

For the first evaluation grade, this results in $\{H_1\} : P_1 = \frac{0.002}{1 - 0.386116} = 0.004$, similar for other evaluation grades, namely, $P_2 = 0.041$, $P_3 = 0.204$, $P_4 = 0.591$, and $P_5 = 0.140$.

Hence, the final condition ratings found through DST for the internal factors are *{(poor, 0.004), (fairly poor, 0.041), (average, 0.204), (good, 0.591),* and *(excellent, 0.140)}*.

Overall Assessment for GSCM Initiative Factors Using DST

The previous section served the purpose of illustrating the calculation for determining the condition state based on the belief data pertaining to the internal factors of GSCM initiatives. Similarly, the belief data for external factors are used to determine the condition state in terms of viability of GSCM initiatives. Combined condition ratings for internal and external factors are then finally combined to determine the final condition rating for overall viability of GSCM initiatives. The final condition rating is obtained in the form of degree of believe measures on the evaluation grades that were previously set. Table 2.5 illustrates the condition rating for each general factor, whereas Table 2.6 is used to present the overall condition rating as well as comparing it with the final result obtained through Yager's rule of combination.

Table 2.5 Overall condition ratings using DST

General factors	Poor	Fairly poor	Average	Good	Excellent	P_H
Internal factors	0.004	0.041	0.204	0.591	0.140	0.006
External factors	0.000	0.073	0.237	0.597	0.093	0.007
Overall	0.002	0.044	0.192	0.647	0.095	0.083

Table 2.6 Overall condition ratings using Yager's rule

Factors	Poor	Fairly poor	Average	Good	Excellent	P_H
Internal factors	0.005	0.043	0.209	0.594	0.149	0.000
External factors	0.000	0.073	0.238	0.588	0.101	0.001
Overall	0.002	0.048	0.204	0.638	0.108	0.031

2.6.2.2 Interpretive Evaluation of Viability Index for GSCM Initiative Using Yager's Rule

Assessment of the Internal Factors Using Yager's Rule

The normalization in the data combination is what that sets Yager's rule apart from the DS rule of combination. There is no multiplication of factor K needed in Yager's rule as the factor K is already shifted to ignorance or uncertainty while it is calculated. The calculation for the factor K in Yager's recursive rule is performed according to the following equation:

$$K_2 = \sum_{n=1}^{4} \sum_{p=1,p\neq n}^{4} M_1^n m_2^p$$
$$= \begin{pmatrix} 0 + \ldots + m_1^{H_2} m_2^{H_1} + m_1^{H_2} m_2^{H_3} + m_1^{H_3} m_2^{H_4} + \\ m_1^{H_2} m_2^{H_5} + m_1^{H_3} m_2^{H_1} + m_1^{H_3} m_2^{H_2} + \ldots + 0 + \ldots + 0 \end{pmatrix}$$
$$= \begin{pmatrix} 0.010 \times 0.013 + 0.010 \times 0.052 + 0.010 \times 0.065 + 0.030 \times 0.013 + 0.030 \times \\ 0.052 + 0.030 \times 0.065 + 0.010 \times 0.013 + 0.010 \times 0.052 + 0.010 \times 0.065 \end{pmatrix}$$
$$= 1.004177 \tag{2.17}$$

The combination according to Yager's rule for the first two factors of GSCM initiatives is as follows:

$$M_2^{H_1} = \left(m_1^{H_1} m_2^{H_1} + m_1^{H_1} m_2^{H} + m_1^{H} m_2^{H_1}\right)$$
$$= (0.000 \times 0.000 + 0.000 \times 0.870 + 0.950 \times 0.000) = 0.000$$
$$M_2^{H_2} = \left(m_1^{H_2} m_2^{H_2} + m_1^{H_2} m_2^{H} + m_1^{H} m_2^{H_2}\right)$$
$$= (0.108 \times 0.020 + 0.108 \times 0.870 + 0.950 \times 0.658) = 0.02118$$
$$M_2^{H_3} = \left(m_1^{H_3} m_2^{H_3} + m_1^{H_3} m_2^{H} + m_1^{H} m_2^{H_3}\right)$$
$$= (0.018 \times 0.030 + 0.018 \times 0.870 + 0.950 \times 0.658) = 0.07706$$
$$M_2^{H_4} = \left(m_1^{H_4} m_2^{H_4} + m_1^{H_4} m_2^{H} + m_1^{H} m_2^{H_4}\right)$$
$$= (0.054 \times 0.130 + 0.054 \times 0.870 + 0.950 \times 0.658) = 0.0711$$
$$M_2^{H_5} = \left(m_1^{H_5} m_2^{H_5} + m_1^{H_5} m_2^{H} + m_1^{H} m_2^{H_5}\right)$$
$$= (0.000 \times 0.020 + 0.000 \times 0.870 + 0.950 \times 0.658) = 0.000$$
$$M_2^{H} = K_2 + \left(m_1^{H} m_2^{H}\right)$$
$$= 1.004177 + (0.870 \times 0.950) = 0.83066$$

Overall Assessment for GSCM Initiative Factors Using Yager's Rule

The same process is then sequentially repeated for every other internal factor, as it was done in the case of DS rule of combination. Following the combination, the final condition states in terms of degrees of belief on evaluation grades are determined and represented as $M_8^{H_1} = 0.003$, $M_8^{H_2} = 0.023$, $M_8^{H_3} = 0.111$, $M_8^{H_4} = 0.315$, $M_8^{H_5} = 0.079$, and $M_8^{H} = 0.470$. Finally, the condition ratings of the viability index are determined to be {(poor, 0.005), (fairly poor, 0.043), (average, 0.209), (good, 0.594), and (excellent, 0.149)}.

2.6.2.3 Utility Perspective Overall Assessment of Viability Index for GSCM Initiatives

Utility-Based Calculation of the Viability Index

Utilizing this condition rating, a single numerical value for the viability index is computed as the utility perspective value. A utility perspective value, as a single numerical rating between 0 and 1, is developed in order to make it possible to assess the viability on a scale of 0–1. Having a scale of 0–1 in terms of viability can facilitate the decision-making by clearly representing the condition state. This also creates the opportunity to make relative comparisons with set standards. For this purpose, the maximum, minimum, and average expected utility in terms of viability of GSCM initiatives are sequentially determined as described in the following equations:

$$u_{\max}(VI) = \sum_{n=1}^{4} P_n u(H_n) + (P_n + P_H) u(H_N) \tag{2.18}$$

$$\begin{aligned} u_{\max}(VI) &= P_1 u(H_1) + P_2 u(H_2) + P_3 u(H_3) + P_4 u(H_4) + (P_N + P_H) u(H_N) \\ &= 0.002 \times 0 + 0.044 \times 0.25 + 0.192 \times 0.5 + 0.647 \times 0.75 + (0.095 + 0.083) \times 1 \\ &= 0.770341 \end{aligned}$$

$$u_{\min}(VI) = (P_1 + P_H) u(H_1) + \sum_{n=2}^{N} P_n u(H_n) \tag{2.19}$$

$$\begin{aligned} u_{\min}(VI) &= (P_1 + P_H) u(H_1) + P_2 u(H_2) + P_3 u(H_3) + P_4 u(H_4) + P_5 u(H_5) \\ &= (0.002 + 0.083) \times 0 + 0.044 \times 0.25 + 0.192 \times 0.5 + 0.647 \times 0.75 + 0.095 \times 1 \\ &= 0.687639 \end{aligned}$$

$$\begin{aligned} u_{\text{avg}}(VI) &= \frac{u_{\max}(VI) + u_{\min}(VI)}{2} \\ &= \frac{0.770341 + 0.687639}{2} \\ &= 0.728990 \end{aligned} \tag{2.20}$$

The average utility perspective measure for the viability index for GSCM initiatives in the supply chain is 0.728990, which defines that the supply chain is in "good" condition to introduce successful GSCM initiatives.

Similarly, the utility perspective viability index value is computed for the condition ratings from Yager's rule. Both of these utility values are then illustrated in Table 2.7.

2.6.2.4 Overall Result Interpretation and Comparative Study Between DST and Yager's Rule

The result from Yager's rule indicates the condition state "good" is the state that contains the highest degree of belief in the viability index with a 59.4% degree of belief, which is fairly consistent with the result obtained from the DST as the DST yielded a 59.1% degree of belief in favor of "good" condition state. Hence, from these values, it can reasonably be said that there is no significant variation between the results obtained through different data combination techniques. Thus, the Yager's rule supports the result of DST claiming the condition state to be "good." Having a "good" condition state demonstrates that the supply chain in question, at the time of the assessment, is in a favorable or good state to adopt GSCM initiatives. Table 2.8 illustrates the results for overall condition ratings.

The final condition ratings are presented in Fig. 2.3, which illustrate the values of the viability index for introducing GSCM initiatives from two data combination approaches, namely, DST and Yager's rule. Figure 2.3 also illustrates the finding that the condition of the organization is in a "good" state for introducing GSCM initiative with a very similar degree of confidence for both approaches. The term P_H is used to denote the uncertainty in the calculation process. On both occasions, the viability index denotes the overall condition of the textile company to introduce GSCM initiatives in the supply chain to be above average, i.e., "good" condition state. Additionally, the utility perspective value gives the assessment result on a scale from 0 to 1. While 0 indicates the poorest condition, 1 represents the best possible state of the organization that is suitable to introduce GSCM initiatives and sustain them. The resultant value of final condition rating, as indicated by Fig. 2.3, in addition to the utility perspective value *0.728990* as shown in Table 2.7, demonstrates that the textile company has a "good" organizational condition state with a

Table 2.7 Utility perspective values obtained from DST and Yager's rule for GSCM

Computational approach	U_{avg}
DST	0.728990
Yager's rule	0.716036

Table 2.8 Overall condition ratings based on DST and Yager's rule

Approach	Poor	Fairly poor	Average	Good	Excellent	P_H
DST	0.002	0.044	0.192	0.647	0.095	0.083
Yager's rule	0.002	0.048	0.204	0.638	0.108	0.031

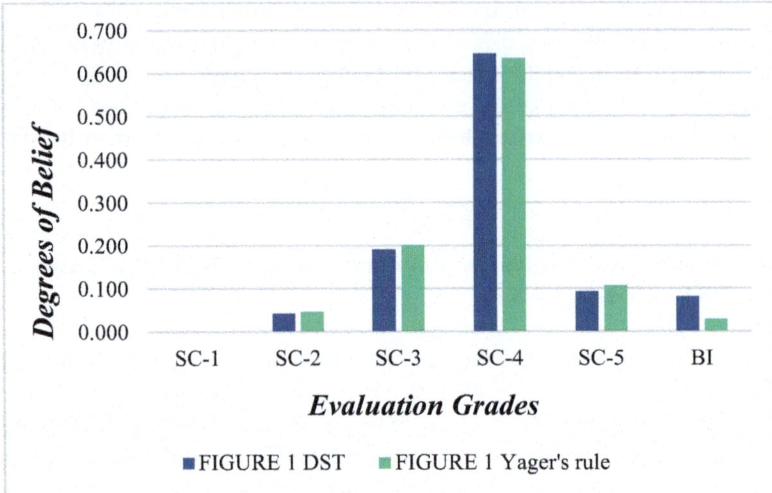

Fig. 2.3 Degree of beliefs for each condition state in the viability index

59.4% degree of belief for having the capability to introduce successful GSCM initiatives. This could be a possible way of comparing the state of an organization at any given moment with the ideal realistic scenarios for assessing the prospect of any supply chain to succeed in introducing GSCM initiatives. This also facilitates the scope of setting a benchmark to improve conditions for introducing GSCM initiatives.

2.7 Sensitivity Analysis

This section discusses the sensitivity analysis performed using the weight assignments on the factors of GSCM initiatives. This sensitivity analysis is done to verify the accuracy of the computational process by testing under different conditions. These conditions are taken as different weight assignments. In this study, 36 different weight assignments are considered for the computation. The results obtained through both the combination rules are compared to see if there is any significant deviation. Since the utility perspective value from Yager's rule fairly matches that of DST, as illustrated in Fig. 2.4, it is concluded that the Yager's rule supports the result of DST. The weight combinations used in this sensitivity analysis are described as follows:

1. *Scenarios 1–16*: In this scenario, one factor is attributed the weight of 0.5, and all the remaining factors are assigned a weight of 0.0333.
2. *Scenarios 17–32*: In this scenario, one factor is considered to have the weight of 0.05, and the remaining factors are given a weight of 0.0633.
3. *Scenario 33*: A weight of 0.5 is assigned to the factors with the highest and lowest utility value, and the remaining 14 factors are assigned the weight of 0.

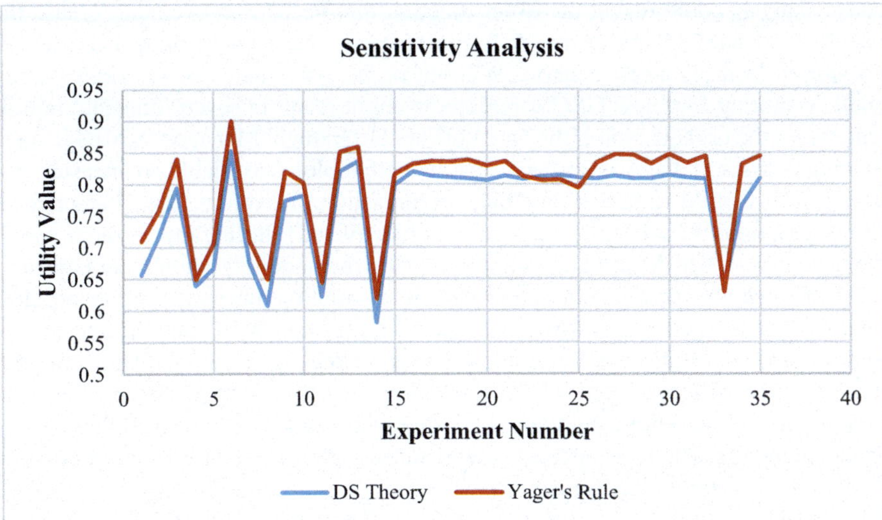

Fig. 2.4 Results of sensitivity analysis

4. *Scenario 34*: A weight of 0 is attributed to the factors with the highest and lowest utility value, and all other remaining factors are assigned the weight of 0.0714.
5. *Scenario 35*: In this final scenario, an equal weight of 0.0625 is assigned to each factor.

From Fig. 2.4, the fluctuations in the utility index value with the change of weight combinations are clearly identifiable. These fluctuations follow the same pattern for both the computational processes, and the differences between these two computational methods are minimal. Thus, the sensitivity analysis served its purpose by concluding that the Yager's rule fairly matches with the DST in their results.

2.8 Discussion

The utility value, presented in Table 2.7, in addition to Fig. 2.3, clearly gives an understanding of the condition state of the organization in terms of their capability of successfully introducing GSCM initiatives throughout their supply chain and sustaining such initiatives to get a viable outcome from those initiatives. This proposed methodology creates an opportunity for managers to pinpoint the exact state of their organization from a holistic point of view of their supply chain to know if their GSCM initiatives are really viable or not. This methodology also provides managers with a measure of degree of belief to which the organization is capable of ensuring a successful GSCM initiative. This would allow the managers to consider what changes need to be made in order to ensure successful GSCM initiative adoption. In case the condition state is found to be below average or in poor condition, a root cause analysis could reveal the exact function that needs improvement to ease the obstacles for GSCM initiatives. A future research scope could be to explore this possibility in a practical scenario. Without proper knowledge of the organization itself, venturing into new GSCM initiatives might result in severe financial ramifications for the organization if they are not capable enough. Managers can avoid such financial losses if there is an assessment methodology available to measure the viability of GSCM initiatives for their organization. Therefore, this characteristic of our proposed model satisfies our initial objective of fulfilling the research gap in the viability of GSCM initiative adoption. This methodology allows the consideration of subjective data without having to suffer from the uncertain and incomplete nature of the subjective data (Silva et al. 2019; Kang et al. 2019), and the uncertainty gets dissipated during the calculation process (Wu et al. 2005). This provides managers the benefit of performing the assessment by taking subjective data and making the assessment methodology less time-consuming. Moreover, if there is any other factor identified for special circumstances, then such factors could also be easily incorporated in this framework.

2.8.1 *Significant Implications for Managerial Decision-Making*

This study has some major implications for managers in terms of their decision-making process. First, the study allows managers to have a decision-making tool to assess the condition of their own organization. The complex nature of supply chain, stemming from the involvement of different stakeholders, makes any GSCM initiative to face certain obstacles. A proper holistic assessment of the supply chain and an organization's condition could reveal those obstacles to the managers, reducing the probability of failure in such GSCM initiatives. Second, the framework proposed in this article will allow managers to make relative comparisons with other organizations with some ideal set conditions. This makes benchmarking and goal-setting for organizations a lot more effective. Third, as this methodology permits the use of subjective data by dealing with the uncertainties through the data fusion technique, managers could utilize surveys in order to assess their organization's condition regarding the viability of introducing GSCM initiatives.

2.9 Conclusion

The environmental legislations create pressure for organizations to make their supply chain as green as possible, while consumer awareness is increasing this pressure on the supply chain stakeholders continuously. To keep up with the pressure, organizations and businesses are opting for several GSCM initiatives.

But before going down this road of often massive financial investment that could potentially disturb the structure of the supply chain, the organizations need to properly analyze their situation. That is why it is paramount to assess the viability of venturing into such GSCM initiatives. According to the assessment, managers need to devise their strategies and the timing of introducing GSCM initiatives.

The viability of GSCM initiative might not depend only on the financial aspects of the initiative itself but also the structure of the supply chain and the organization itself. So, a holistic assessment approach in terms of the supply chain needs to be developed. The proposed assessment methodology served this purpose through this study. As such, a holistic assessment might have to deal with subjective data more often than not, and a computational approach is needed for the assessment framework that can deal with the uncertainties and incompleteness of the subjective data efficiently. The Dempster-Shafer theory is used to serve this purpose in our study. The combination of AHP, DST, and Yager's rule of combination formulates the overall assessment framework's computational approach. This integrated framework, in addition with the identification of the factors affecting GSCM initiatives, *addresses* all the complexities of having an assessment framework for GSCM initiative adoption.

Appendix: Assessing the Evaluation Grades for External Factors Using DST

$$\text{External factors index} = S(e_1) \times W(e_1) \oplus S(e_2) \times W(e_2) \oplus S(e_3) \times W(e_3) \oplus \ldots\ldots\ldots \oplus S(e_8) \times W(e_8) \tag{2.21}$$

BPAs determined through the multiplication of the conditional ratings and respective weights are:

$$m(e_1) = \{0.000, 0.000, 0.012, 0.048, 0.000, 0.940\}$$

$$m(e_2) = \{0.000, 0.000, 0.048, 0.272, 0.000, 0.680\}$$

$$m(e_3) = \{0.000, 0.006, 0.024, 0.000, 0.000, 0.970\}$$

$$m(e_4) = \{0.000, 0.000, 0.006, 0.015, 0.009, 0.970\}$$

$$m(e_5) = \{0.000, 0.000, 0.020, 0.040, 0.040, 0.900\}$$

$$m(e_6) = \{0.000, 0.085, 0.085, 0.000, 0.000, 0.830\}$$

$$m(e_7) = \{0.000, 0.000, 0.018, 0.063, 0.000, 0.919\}$$

$$m(e_8) = \{0.000, 0.000, 0.042, 0.105, 0.063, 0.790\}$$

The BPA obtained from the aggregation of BPAs of two different factors is denoted as M. The initial aggregation is taken as to be $M_1 = m_1$. The aggregation of the remaining factors is then continued as the following approach:

$$
\begin{aligned}
K_2 &= \left(1 - \sum_{n=1}^{4} \sum_{p=1, p \neq n}^{4} M_1^n m_2^p\right)^{-1} \\
&= \left[1 - \begin{pmatrix} 0 + \ldots + m_1^{H_2} m_2^{H_1} + m_1^{H_2} m_2^{H_3} + m_1^{H_2} m_2^{H_4} + \\ m_1^{H_2} m_2^{H_5} + m_1^{H_3} m_2^{H_2} + m_1^{H_3} m_2^{H_4} + \ldots + 0 + \ldots + 0 \end{pmatrix}\right]^{-1} \\
&= [1 - (0.012 \times 0.048 + 0.012 \times 0.272 + 0.048 \times 0.048 + 0.048 \times 0.272)]^{-1} \\
&= 1.005599
\end{aligned}
\tag{2.22}
$$

After calculating the factor K, the combined BPA for the first two factors can be computed as follows:

$$
\begin{aligned}
M_2^{H_1} &= K_2 \times \left(m_1^{H_1} m_2^{H_1} + m_1^{H_1} m_2^{H} + m_1^{H} m_2^{H_1}\right) \\
&= 1.005599 \times (0.000 \times 0.000 + 0.000 \times 0.680 + 0.940 \times 0.000) = 0.000 \\
M_2^{H_2} &= K_2 \times \left(m_1^{H_2} m_2^{H_2} + m_1^{H_2} m_2^{H} + m_1^{H} m_2^{H_2}\right) \\
&= 1.005599 \times (0.000 \times 0.000 + 0.0000 \times 0.680 + 0.940 \times 0.0000) = 0.000 \\
M_2^{H_3} &= K_2 \times \left(m_1^{H_3} m_2^{H_3} + m_1^{H_3} m_2^{H} + m_1^{H} m_2^{H_3}\right) \\
&= 1.005599 \times (0.012 \times 0.048 + 0.012 \times 0.680 + 0.940 \times 0.048) = 0.054158 \\
M_2^{H_4} &= K_2 \times \left(m_1^{H_4} m_2^{H_4} + m_1^{H_4} m_2^{H} + m_1^{H} m_2^{H_4}\right) \\
&= 1.005599 \times (0.048 \times 0.272 + 0.048 \times 0.680 + 0.940 \times 0.272) = 0.303063 \\
M_2^{H_5} &= K_2 \times \left(m_1^{H_5} m_2^{H_5} + m_1^{H_5} m_2^{H} + m_1^{H} m_2^{H_5}\right) \\
&= 1.005599 \times (0.000 \times 0.000 + 0.000 \times 0.680 + 0.940 \times 0.000) = 0.000 \\
M_2^{H} &= K_2 \times \left(m_1^{H} m_2^{H}\right) \\
&= 1.005599 \times (0.680 \times 0.940) = 0.642779
\end{aligned}
$$

The remaining external factors are aggregated following the same process as the aggregation of first two factors:

$$M_8 = \{0.000, 0.044, 0.142, 0.357, 0.056, 0.401684\}$$

The normalization is done according to the following equation:

$$\{H_n\} : P_n = \frac{M_8^{H_n}}{1 - M_8^{H}}; n = 1, 2, .., N \tag{2.23}$$

Thus, $P_1 = 0.000$, $P_2 = 0.073$, $P_3 = 0.237$, $P_4 = 0.597$, $P_5 = 0.093$.

Hence, the final condition ratings found through DS theory for the external factors are *{(poor, 0.000), (fairly poor, 0.073), (average, 0.237), (good, 0.597)*, and *(excellent, 0.093)}*.

References

Anisul Huq F, Stevenson M, Zorzini M (2014) Social sustainability in developing country suppliers: an exploratory study in the ready-made garments industry of Bangladesh. Int J Oper Prod Manag 34(5):610–638. https://doi.org/10.1108/IJOPM-10-2012-0467

Asgari B, Hoque MA (2013) A system dynamics approach to supply chain performance analysis of the ready-made-garment industry in Bangladesh. Retsumeikan J Asia Pac Stud 32(1):51–61

Bappy MM, Ali SM, Kabir G, Paul SK (2019) Supply chain sustainability assessment with Dempster-Shafer evidence theory: implications in cleaner production. J Clean Prod 237:117771

Barua U, Ansary MA (2017) Workplace safety in Bangladesh ready-made garment sector: 3 years after the Rana Plaza collapse. Int J Occup Saf Ergon 23(4):578–583. https://doi.org/10.1080/10803548.2016.1251150

Beynon M, Curry B, Morgan P (2000) The Dempster–Shafer theory of evidence: an alternative approach to multi-criteria decision modeling. Omega 28(1):37–50

Carter CR, Rogers DS (2008) A framework of sustainable supply chain management: moving toward new theory. Int J Phys Distrib Logist Manag 38(5):360–387

Chin TA, Tat HH, Sulaiman Z (2015) Green supply chain management, environmental collaboration and sustainability performance. Procedia CIRP 26:695–699

Chowdhury NA, Ali SM, Mahtab Z, Rahman T, Kabir G, Paul SK (2019) A structural model for investigating the driving and dependence power of supply chain risks in the readymade garment industry. J Retail Consum Serv 51:102–113. https://doi.org/10.1016/j.jretconser.2019.05.024

Dalkir K (2005) Knowledge management in theory and practice. The MIT Press, Cambridge, MA

Darnall N, Jolley GJ, Handfield R (2008) Environmental management systems and green supply chain management: complements for sustainability? Bus Strateg Environ 17(1):30–45

Dempster AP (1967) Upper and lower probabilities induced by a multivalued mapping. Ann Math Stat 38(2):325–339

Deng Y, Chan FTS (2011) A new fuzzy dempster MCDM method and its application in supplier selection. Expert Syst Appl 38(8):9854–9861

Dhull S, Narwal M (2016) Drivers and barriers in green supply chain management adaptation: a state-of-art review. Uncertain Supply Chain Manag 4(1):61–76

Diabat A, Govindan K (2011) An analysis of the drivers affecting the implementation of green supply chain management. Resour Conserv Recycl 55(6):659–667

Fahimnia B, Sarkis J, Davarzani H (2015) Green supply chain management: a review and bibliometric analysis. Int J Prod Econ 162:101–114

Govindan K, Kaliyan M, Kannan D, Haq AN (2014) Barriers analysis for green supply chain management implementation in Indian industries using analytic hierarchy process. Int J Prod Econ 147:555–568

Green KW, Zelbst PJ, Meacham J, Bhadauria VS (2012) Green supply chain management practices: impact on performance. Supply Chain Manag Int J 17(3):290–305

Huynh P (2013) Outbound transportation and its environmental impact: case company: drilling mud corporation. Bachelors thesis, HAAGA-HELIA University of Applied Science

Islam A, Maideen MBH, Abdullah ARB (2019) The role of employee knowledge and behavior towards sustainable development: an investigative study based on readymade garments industry of Chittagong, Bangladesh. Int J Recent Technol Eng 7:359–362

Jumadi H, Zailani S (2010) Integrating green innovations in logistics services towards logistics service sustainability: a conceptual paper. Environ Res J 4(4):261–271

Kang B, Zhang P, Gao Z, Chhipi-Shrestha G, Hewage K, Sadiq R (2019) Environmental assessment under uncertainty using Dempster–Shafer theory and Z-numbers. J Ambient Intell Humaniz Comput 11:2041–2060

Kassinis G, Vafeas N (2006) Stakeholder pressures and environmental performance. Acad Manag J 49(1):145–159

Khiewnavawongsa S, Schmidt EK (2013) Barriers to green supply chain implementation in the electronics industry. In: Industrial engineering and engineering management (IEEM), 2013 IEEE international conference on, pp 226–230
Kong G, Xu DL, Yang JB, Ma X (2015) Combined medical quality assessment using the evidential reasoning approach. Expert Syst Appl 42(13):5522–5530
Kumar R, Chandrakar R (2012) Overview of green supply chain management: operation and environmental impact at different stages of the supply chain. Int J Eng Adv Technol 1:1–6
Lee SY, Klassen RD (2008) Drivers and enablers that foster environmental management capabilities in small-and medium-sized suppliers in supply chains. Prod Oper Manag 17(6):573–586
Lin RJ, Tan KH, Geng Y (2013) Market demand, green product innovation, and firm performance: evidence from Vietnam motorcycle industry. J Clean Prod 40:101–107
Mathiyazhagan K, Govindan K, NoorulHaq A, Geng Y (2013) An ISM approach for the barrier analysis in implementing green supply chain management. J Clean Prod 47:283–297
Naciti V (2019) Corporate governance and board of directors: the effect of a board composition on firm sustainability performance. J Clean Prod 237:117727
Narayana AS, Elias A, Pati KR (2014) Reverse logistics in the pharmaceuticals industry: a systemic analysis. Int J Logist Manag 25(2):379–398
Ninlawan C (2010) The implementation of green supply chain management practices in electronics industry. In: Proceedings of the international multi-conference of engineers and computer scientists
Qi Y, Boyer KK, Zhao X (2009) Supply chain strategy, product characteristics, and performance impact: evidence from Chinese manufacturers. Decis Sci 40(4):667–695
Rivera-Camino J (2007) Re-evaluating green marketing strategy: a stakeholder perspective. Eur J Mark 41(11/12):1328–1358
Saaty TL (1990) How to make a decision: the analytic hierarchy process. Eur J Oper Res 48(1):9–26
Sarkis J, Zhu Q, Lai K (2011) An organizational theoretic review of green supply chain management literature. Int J Prod Econ 130(1):1–15
Sellitto MA, Bittencourt SA, Reckziegel BI (2015) Evaluating the implementation of GSCM in industrial supply chains: two cases in the automotive industry. Chem Eng 43:1315–1320. Brazil
Shafer G (1976) A mathematical theory of evidence, vol 42. Princeton University Press
Shang K-C, Lu C-S, Li S (2010) A taxonomy of green supply chain management capability among electronics-related manufacturing firms in Taiwan. J Environ Manag 91(5):1218–1226
Silva MM, Hipel KW, Kilgour DM, Costa APCS (2019) Strategic analysis of a regulatory conflict using Dempster-Shafer theory and AHP for preference elicitation. J Syst Sci Syst Eng 28(4): 415–433
Testa F, Iraldo F (2010) Shadows and lights of GSCM (Green Supply Chain Management): determinants and effects of these practices based on a multi-national study. J Clean Prod 18(10–11):953–962
Van Rensburg SLJ (2015) A framework in green logistics for companies in South Africa. MBA thesis, University of South Africa
Walker H, Brammer S (2009) Sustainable procurement in the United Kingdom public sector. Supply Chain Manag Int J 14(2):128–137
Wilkerson T (2005) Best practices in implementing green supply chains. In: North America supply chain world conference and exposition, April, vol 5
Wood S, Reynolds J (2013) Knowledge management, organisational learning and memory in UK retail network planning. Serv Ind J 33(2):150–170
Wu HJ, Dunn SC (1995) Environmentally responsible logistics systems. Int J Phys Distrib Logist Manag 25(2):20–38
Wu W-Z, Zhang M, Li H-Z, Mi J-S (2005) Knowledge reduction in random information systems via Dempster–Shafer theory of evidence. Inf Sci 174(3–4):143–164
Yang F, Wei H, Feng P (2020) A hierarchical Dempster-Shafer evidence combination framework for urban area land cover classification. Measurement 151

Zhang H, Yang F (2016) On the drivers and performance outcomes of green practices adoption. Ind Manag Data Syst 116(9):2011–2034
Zhu Q, Sarkis J, Geng Y (2005) Green supply chain management in China: pressures, practices and performance. Int J Oper Prod Manag 25(5):449–468
Zhu Q, Sarkis J, Lai K (2012) Examining the effects of green supply chain management practices and their mediations on performance improvements. Int J Prod Res 50(5):1377–1394
Zsidisin GA, Siferd SP (2001) Environmental purchasing: a framework for theory development. Eur J Purch Supply Manag 7(1):61–73

Chapter 3
Supply Chain Sustainability and Supply Chain Resilience: A Performance Measurement Framework with Empirical Validation

Marta Negri, Enrico Cagno, Claudia Colicchia, Leonardo Coppellotti, and Caterina Baldassarri

Abstract Sustainability and resilience are both fundamental to ensure the long-term survival of supply chains. Considering that synergies and trade-offs exist, ensuring that their implementation results in positive and non-detrimental applications is fundamental. Having an adequate performance measurement framework, able to include all relevant aspects of sustainability and resilience, in a comprehensive yet concise way, exploiting synergies and links between the topics, is of paramount importance. A literature review allowed retrieving that there is a lack of such a framework as well as the lack of operativity and adaptability to the changing needs of firms. Building on this, a new performance measurement framework was built to simultaneously include sustainability and resilience indicators, integrating the two concepts via the capabilities and the performances. Besides, it was designed to be operative and to be scalable, adapting to different firms and needs. Thus, it can be adapted to the needs of small and medium enterprises or firms at the beginning of their measurement journey. The framework was validated with three supply chain case studies that confirmed its usefulness, comprehensiveness, and ease of use. These results contribute to the discussion about sustainability and resilience performance measurement and provide guidance for practitioners and regulators.

Keywords Supply chain · Sustainability · Resilience · Performance measurement system · Indicators · Capabilities · Empirical evidence · Scalability · Small and medium enterprises · New adopters

---

M. Negri (✉) · E. Cagno · C. Colicchia · L. Coppellotti · C. Baldassarri
Politecnico di Milano – Department of Management, Economics and Industrial Engineering, Milan, Italy
e-mail: marta.negri@polimi.it

© The Author(s), under exclusive license to Springer Nature Switzerland AG 2023
N. U. Ibne Hossain (ed.), *Data Analytics for Supply Chain Networks*, Greening of Industry Networks Studies 11, https://doi.org/10.1007/978-3-031-29823-3_3

3.1 Introduction

Performance measurement (PM) is fundamental for several reasons. For instance, it allows an improved understanding of processes, spotting issues or success opportunities, taking informed decisions, monitoring progress, and communicating effectively (Akyuz and Erkan 2010). In light of the current trends affecting supply chains (SCs), performance measurement has been receiving increased attention. In particular, some scholars have initiated a conversation on how the PM should mirror the changing needs of SCs (Elgazzar et al. 2019), suggesting that a balanced set of traditional economic performance indicators and non-cost-related ones should be adopted (Kaplan and Norton 2005), possibly looking not only to past performance but also providing predictions on the future (Mishra et al. 2018). It has also been argued that there is still a lack of adequate PM that is able to grasp the complexity of the SC (Simão et al. 2021).

In the context of PM, indicators are "a quantitative or qualitative factor or variable that provides a simple and reliable means to measure achievement, to reflect changes connected to an intervention, or to help assess the performance of a development actor" (Medini et al. 2015; Saidani et al. 2019); a PM framework, instead, has been defined as "a collection of indicators that conveys a broader purpose and significance to the individual indicator and provides a comprehensive picture of some entities" (Saidani et al. 2019). It is thus important to carefully select and organize indicators to provide a broader meaning to practitioners, who will then be using them to make informed decisions.

A nourished debate concerns the key, unavoidable characteristics of a good PM framework (Elgazzar et al. 2019) and the indicators contained in it. PM frameworks should include all relevant aspects related to the SC performance in an integrated way (Simão et al. 2021). The relevant aspects included in PM frameworks should, in turn, be related to the trends experienced at the supply chain level. Sustainability—intended as the balance of economic, environmental, and social aspects (Cagno et al. 2019)—and resilience have been identified as some of the fastest-growing research streams (Swanson et al. 2018), and their mutual influence has been recognized (Fahimnia et al. 2019), calling for a more synergic evaluation of performance. Considering that synergies and trade-offs exist (Negri et al. 2022), this has translated into a call for an integrated PM framework, able to monitor the intersection of sustainability and resilience in order not to have detrimental applications (Negri et al. 2021a). Indeed, initial but strong evidence is present on the influence sustainability and resilience have on each other. For instance, a corporate strategy that builds on the overexploitation of resources will reduce the SC's resilience (Perrings 2006); disruptions may sometimes open room for a sustainability turn (Sarkis et al. 2020); at a more operational level, improving efficiency increases sustainability, but it will reduce effectiveness and resilience (Fahimnia et al. 2019). The first study of the expected impacts of sustainability and resilience practices was carried out by Negri et al. (2022). This study highlights that although most of the practices develop synergies between sustainability and resilience, some might end up as trade-offs. It is the case, for instance, of eco-design, which might have initial detrimental effects on the cost performance of firms (Green et al. 2012). Additionally, certifications have positive environmental, social, and resilience impacts but entail higher initial costs (Negri et al. 2022). All these elements should be accounted for in an integrated PM framework, so as to ensure that firms and SCs are heading towards improved sustainability and resilience.

Such a framework is still absent from the literature, but there have been attempts to broaden the scope of traditional PM to include these relevant aspects. For instance, in recent developments, PM frameworks evolved to include environmental and social aspects (Simão et al. 2021). Nonetheless, there still needs to be a structured approach to integrate sustainability and resilience PM.

Finally, there is the need to address a practical problem in firms. As mentioned, firms should measure all relevant aspects of their performance to track progress correctly, benchmark results, and communicate effectively. However, firms are often time- and resource-constrained and are often not willing or capable of monitoring too many indicators (Neri 2021). Thus, they often call for synthetic yet complete frameworks that allow them to have a limited number of indicators while maximizing the information contained (Cagno et al. 2019). This is exacerbated in small and medium enterprises (SMEs), which usually are time-, staff-, money-, and competency-constrained, and new adopters (NAs), which are at the beginning of their PM journey and may lack adequate awareness to use sophisticated tools (Negri et al. 2021b). It has therefore been argued that PM should adapt to the changing firms' requirements (Bititci et al. 2015).

In the following sections, an overview of the sustainability and resilience PM frameworks is provided (Sects. 3.1.1, 3.1.2, and 3.1.3), and the main gaps found in the extant literature are highlighted in Sect. 3.1.4. Section 3.2 proposes a methodology for the development of a new, integrated, sustainability, and resilience performance measurement framework. Section 3.3 presents the methodology used to test the developed framework, and the results are outlined in Sect. 3.4. Finally, Sects. 3.5 and 3.6 report, respectively, the discussion of results and final conclusions, with some suggestions for interesting future research paths.

3.1.1 Supply Chain Sustainability Performance Measurement

The contributions related to sustainability PM have been growing, although some major issues concerning the lack of triple bottom line (TBL) balance and supply chain consideration are still raised (Saeed and Kersten 2020).

Ahi and Searcy (2015) reviewed the literature and provided a methodology to measure performance in sustainable and green SCs. The authors stressed the general lack of agreement on how to measure the sustainability performance, although the number of indicators has been proliferating. This resulted in the need to standardize metrics and terminology to facilitate benchmarking activities and avoid confusion.

The authors highlighted that PM frameworks should address multiple characteristics of the SCs.

Neri et al. (2021) comprehensively evaluated the past literature on SC PM and built a TBL-balanced framework of indicators. The authors stressed the fact that PM frameworks should have a long-term perspective; should balance economic, environmental, and social aspects; and should focus on the overall SC instead of on local optimization at the firm level. Most importantly, the authors stressed the fact that PM frameworks should adapt to the context and firms' requirements, suggesting that a scalable framework might initially guide firms in their PM journey, turning them into a more complex and complete tool as firms' awareness and resources increase.

This aspect has been underlined several times in the past literature, especially in relation to SMEs. SMEs make up a significant part of the European economy, accounting for 99% of the firms in the private sector (Mura et al. 2020), whose collective environmental impact accounts for 60–70% (Witjes et al. 2017). As traditional tools may not be suitable for the characteristics of SMEs (Johnson 2015), it is necessary to provide adequate guidance for these firms. A similar reasoning could be valid for larger firms that approach sustainability for the first time, i.e. new adopters (Negri et al. 2021b), that might have similar requirements in terms of tools and guidance as SMEs.

In light of the trend that sees sustainability PM expands its scope to include deeper meaning, there are interesting avenues of research that combine sustainability and circular economy (CE) indicators, merging them as complementary and partially overlapping concepts (Rossi et al. 2020; Lee et al. 2021), which combine sustainability and industrial symbiosis (IS) indicators, considering the complementarity between the concepts (Fraccascia and Giannoccaro 2020), or which provide integration for sustainability, CE, and IS indicators (Cagno et al. 2023). Increasing attention has also been given to developing and standardizing indicators for the social dimension, which has been somewhat neglected in the past literature (Mies and Gold 2021).

Subramanian and Gunasekaran (2015) proposed a classification of performance indicators based on an extensive literature review and found that some areas are not covered by proper sustainability indicators, such as product design and IT.

In general terms, while the economic dimension is well-established, there still needs to be consensus on the environmental dimension, which has seen a proliferation of new indicators, and the social one that is usually limited to health and safety (Cagno et al. 2019). The economic dimension usually consists of costs, profits, and investments. The environmental dimension usually concentrates on resource use, emissions, waste, and pollution. Finally, the social dimension generally consists of employees, customers, health and safety, and community (Mengistu and Panizzolo 2022). Table 3.1 reports the PM framework formulated by Neri et al. (2021), which represents one of the most recent, complete, and TBL-balanced proposed frameworks in the literature.

Some studies have also started analysing the role of capabilities in the outcome obtained by firms and SCs. Ordinary capabilities "involve the performance of administrative, operational, and governance-related functions that are (technically)

Table 3.1 Performance measurement framework and indicators

Category	Indicator	Performance areas	Performance
Financial	Return on investment	Economic and financial, cost	Financial performance (profitability)
	Return on sales	Economic and financial, cost	Financial performance (profitability)
	Return on assets	Economic and financial, cost	Financial performance (profitability)
	Total SC cost	Cost	SC cost, direct cost, indirect cost
	Inventory costs	Cost, inventory	Direct cost, inventory level, inventory performance
	Cash-to-cash cycle time	SC cycle times: Economic and financial, cost	Cycle time: Financial performance (profitability)
Internal process	Capacity utilization	Production, flexibility	Performance, operation flexibility
	Recycling	Environment, cost, product	Reuse and recycling, direct cost, responsibility
	Certification	Management, quality, environment, social	Awareness, procedure, ethical conduct, quality management, environmental management, environmental ethical conduct, environmental cost, social management, social-related cost, social ethical conduct
	SC responsiveness	SC cycle times, flexibility, management, information, suppliers, customers	Cycle time, production flexibility, motivational effort, extent of sharing information, characteristics of information, collaboration, characteristics (suppliers), service (customers)
	SC cycle time	SC cycle times, performance, information, suppliers	Cycle time; schedule, lead time; time, flexibility; extent of sharing information; characteristics of information; collaboration; characteristics (suppliers)
	Process cycle time	SC cycle times, performance, flexibility	Cycle time, lead time, production, process, schedule, efficiency, throughput, operation flexibility
Learning and growth	Labour efficiency	Production, flexibility	Performance, operation flexibility
	New product development time	Performance, product, flexibility, SC cycle time	R & D, innovation, design, responsibility, production flexibility, cycle time, lead time
	Investments	Financial and economic, performance	Economic performance, R & D, process
	Integration with SC partners	Suppliers, information	Dependency, performance (suppliers), collaboration, characteristics (suppliers); extent of sharing information, characteristics of information, reverse SC
	Use of new technology	Performance	IT

(continued)

Table 3.1 (continued)

Category	Indicator	Performance areas	Performance
Customer	Market share	Financial and economic; customers	Economic performance, customers' characteristics
	Customer satisfaction	Customer, quality	Customer satisfaction, customer service, return service
	Product quality	Product, quality	Defectiveness, responsibility, product quality
	Product/service variety	Product, flexibility; customers	Product characteristics, production flexibility, customer service
	Order fulfilment	Order procedures and delivery	Order performance, invoice, delivery performance
	Delivery reliability	Order procedures and delivery, cost, product, inventory, SC cycle times	SC cost, direct cost, indirect cost, lead time
Environment	Energy use	Environment	Resource consumption
	Water use	Environment	Resource consumption, reuse and recycling
	Material use	Environment, performance, cost	Resource consumption, direct costs, production, reuse and recycling
	Environmental impacts	Environment	Emission, environmental management, environmental ethical conduct
	Waste	Environment, cost	Waste, reuse and recycling, direct cost
Social	Community relationships	Social	External stakeholders, community
	Philanthropic investments	Social	Ethical conduct, community
	OHS performance	Social	Employees; social-related cost
	Labour turnover	Social, costs	Employees, direct cost
	Employee satisfaction	Management	Employees, motivational effort, indirect cost

Neri et al. (2021)

necessary to accomplish tasks", while dynamic capabilities "involve higher-level activities that can enable an enterprise to direct its ordinary activities towards high-payoff endeavours" (Teece 2014). Considering that SCs are evaluated not only according to their financial performance but also on their environmental and social ones (Beske 2012), there is initial evidence that the firms' sustainability performance is driven by the ability of firms to monitor their performance and promote the creation of the necessary operational routines, metrics and behaviours (Gelhard and von Delft 2016). It has also been pointed out that the relation between practices and firms' performance is mediated by the capabilities (Gelhard and von Delft 2016). More often, sustainability is studied according to the dynamic capabilities theory, which requires firms to reconfigure their ordinary capabilities to obtain a long-term competitive advantage (Beske 2012; Felsberger et al. 2022).

3.1.2 Supply Chain Resilience Performance Measurement

The PM concerning supply chain resilience (SCRes) is relatively new, and only recently have some studies attempted to standardize the topic (Karl et al. 2018). Common SCRes indicators are the time to recovery, recovery level, and lost performance because of the disruption (Behzadi et al. 2020). However, there has been a proliferation of indicators and different methodologies to assess resilience. Rajesh (2016) proposed five indicators classified into flexibility, responsiveness, quality, productivity, and accessibility. Chowdhury and Quaddus (2017) elaborated a scale based on the dynamic capability theory, proposing a set of 63 indicators.

Much of the attention of the literature is now on assessing the level of resilience of SCs and networks (Sahu et al. 2017), often computing an index of resilience (Wang et al. 2016). While this can be useful for communication purposes, how this information should be used by decision-makers should be made clearer.

SCRes is often linked to the concept of capabilities (Chowdhury and Quaddus 2017), as SCs are required to have a continuous process of adaptation and reconfiguration of their resources and capabilities to respond to disruptions. The role of capabilities is quite established in the SCRes literature, as SCRes itself is defined as an operational capability that allows firms to survive, adapt, and grow in turbulent environments (Brusset and Teller 2017). Some authors argue that the vulnerabilities present in all SCs should be balanced with elements that form capabilities (Pettit et al. 2013). More recently, the SCRes capabilities have turned to be used as a proxy for SCRes performance (Kaviani et al. 2020), as the implementation of capabilities is often linked with an improved SCRes performance (Carvalho et al. 2012b). This study confirms what was established in his research by Birkie et al. (2017), who stated that the practices implemented regularly by SCs may help develop capabilities and ultimately ensure SCRes performance.

Fiksel et al. (2015) identified 16 capabilities that allow SCs to counterbalance vulnerabilities: "flexibility in sourcing", "flexibility in manufacturing", "flexibility in order fulfilment", "production capacity", "efficiency", "visibility", "adaptability", "anticipation", "recovery", "dispersion", "collaboration", "organization", "market position", "security", "financial strength", and "product stewardship".

Han et al. (2020) performed a study that systematically reviewed the recent literature on SCRes performance (Table 3.2). This represents one of the first comprehensive frameworks found in the extant literature, which links the PM of resilience with capabilities. As commonly done (Ali et al. 2017), the authors divided capabilities and performance into three phases of resilience, namely, readiness, response, and recovery.

Table 3.2 Framework for SCRes measurement

Phase	Capabilities	Performance
Readiness	Situation awareness	Performance of discerning possible disruptions
	Visibility	Performance of overseeing the supply chain situation
	Redundancy	Performance of production and inventory
Response	Agility	Efficiency of completing supply chain processes
	Flexibility	Efficiency of responding the disruptions
	Collaboration	Performance of relationship management
Recovery	Contingency planning	Reconstruction of the supply chain
		Efficiency of recovery to normality
	Market position	Performance of maintaining customer satisfaction
		Damage from disruption
		Financial performance

Adapted from Fiksel et al. (2015), Chowdhury and Quaddus (2016), Ali et al. (2017), and Han et al. (2020)

3.1.3 Supply Chain Sustainability and Resilience Performance Measurement

Initial evidence suggests that integrating sustainability and resilience into strategic decision-making will improve the performance of SCs (Ruiz-Benitez et al. 2019; Negri et al. 2022).

Ramezankhani et al. (2018) developed a model for SC sustainability and resilience evaluation, by developing a dynamic network DEA model to identify inefficiencies and propose improvements to managers. However, this model tends to be quantitative and complex to use for practitioners, who would need the guidance of experts, and it only considers a few factors in the evaluation, namely, costs, number of employees, average inventory, employees' satisfaction, profits, and recyclable waste. Arguably, there are many more aspects that need to be assessed to grasp the level of sustainability and resilience in SCs.

A quite nourished field of study is the development of LARG indexes, to measure the leanness, agility, resilience, and greenness of supply chains (Cabral et al. 2012; Azevedo et al. 2016). Azevedo et al. (2013) proposed an Ecosilient index computed starting from the practices and performance of single firms in the SC, constituting two separate resilience and greenness indicators. These two indicators are then merged into an Ecosilient index. Also in this case, the TBL balance is not included, as only the environmental aspect is considered.

Kaur et al. (2020) developed a model to assess and rank suppliers and production modes that consider environmental criteria, such as carbon emissions during ordering, holding, and transportation, under an uncertainty scenario. However, the simultaneous PM of sustainability and resilience is not considered.

Kazemi Matin et al. (2021) proposed a model to measure sustainability and resilience in blood supply chains, although focusing on mainly economic factors for their analysis. Ruiz-Benitez et al. (2019) performed a study in the aerospace sector to measure sustainability and resilience performance. The authors included a mix of operational and economic performance indicators to assess, i.e. costs, on-time deliveries, product quality, and capacity utilization. Owida et al. (2022) analysed the COVID-19 pandemic as a case study, to provide insights on newly developed PM for linking sustainability and resilience. The authors found that some new indicators were applied, mainly measuring economic sustainability and robustness (i.e. products sold, number of weeks of the horizon, percentage of changeover loss in the overall equipment effectiveness (OEE) loss tree, days on hand, percentage of reduced errors, and cost reduction) and measuring social sustainability and robustness (number of workers per line per zone, the social distance between workers, percentage of virtual teams, number of subject 1 cases, and further virus transmission). The study concluded that there needs to be better integration of the concepts and more research on proper indicators to measure the integration of sustainability and resilience.

Hervani et al. (2022) provided a PM framework to evaluate the social pillar in sustainable and resilient supply chains, which is fundamental to allow firms to thrive in uncertain times. The authors argued that during disruptions, sustainable firms tend to be more sustainable from a social and environmental viewpoint, by allowing the right degree of redundancies in terms of collaboration, coordination, and flexibility . The dimensions of socially sustainable capabilities considered in the article are "internal human resources", "external population", "stakeholder and participation", and "macro-social issues and concerns" (Hervani et al. 2022). The authors also stated that firms may use market-based indicators—"avoided cost method", "cost-of-illness method", "expenditure approach method", "damage assessment method", and "productivity method"—to evaluate the impact of their social practices on building more resilience.

3.1.4 Research Gaps

The literature review presented shows that there are several gaps concerning SC sustainability and resilience PM.

First, there is a lack of an adequate framework able to simultaneously measure sustainability and resilience performance, also because of the complexity of developing adequate considerations for the purpose (Negri et al. 2021a). Indeed, several authors still focus on green and resilient SCs (Sen et al. 2018; Ruiz-Benitez et al. 2019; Mohammed 2020), overlooking a TBL-balanced sustainability. Some authors focus on sustainability and risk assessment (Xu et al. 2019; Abdel-Basset and Mohamed 2020), on sustainability assessment under risk conditions (Almeida et al. 2016), or on resilient assessment of green supply chains (Mohammed et al. 2020). A comprehensive study that delivers a balanced PM framework to measure sustainability and resilience is still absent.

Second, the frameworks present in the extant literature lack operativity (Um and Han 2020), as they often do not provide guidance or metrics for practitioners to use. In other cases, the indicators and indexes are way too complex (Suryawanshi et al. 2021), so their use and interpretation may be too difficult for practitioners.

Finally, none of the frameworks present in the literature provide scalability considerations. As mentioned, firms are often resource-, skills-, and time-constrained and may require a simpler tool to use, while larger, more aware, or competent firms may benefit from a more sophisticated one (Cagno et al. 2019). This encourages the development of different versions of the framework, with different levels of complexity, so firms can select the one that best suits their needs and possibly allow to scale it up as firms' requirements evolve (Bititci et al. 2015).

3.2 Developing a New, Integrated Framework for Sustainability and Resilience Performance Measurement

This section aims at presenting a new, integrated SC sustainability and resilience PM framework, developed to provide an answer to the gaps identified. Considering that there are very few contributions exploring the PM of sustainability and resilience, the first step consisted of integrating the existing considerations on sustainability PM and resilience PM to solve the first research gap identified. The starting point was therefore an analysis of the existing separate frameworks to measure SC sustainability and SCRes. The two concepts, performance and indicators, have been elaborated and merged to obtain a first version of the framework (full framework, as reported in step 1 of Fig. 3.1). To address the second research gap, the framework has been intentionally built to be operative (Cagno et al. 2019). Thus, the indicators including

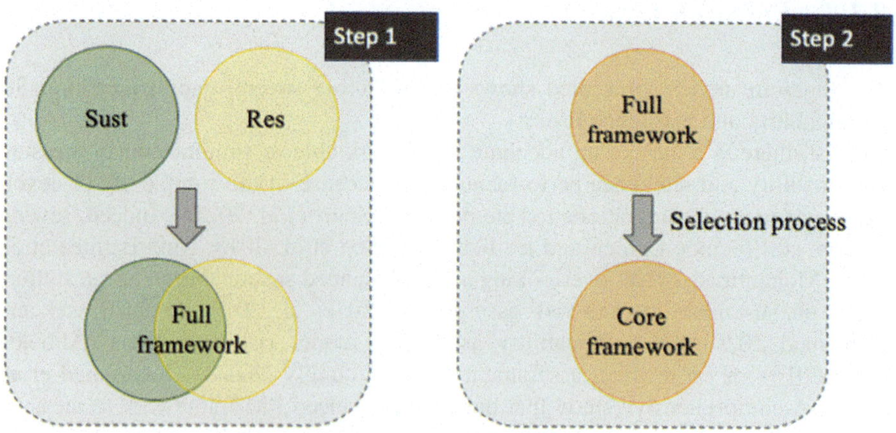

Fig. 3.1 Development process of the full and core frameworks

a description and metrics are provided. Besides, the link to the capabilities of sustainability and resilience allows to have further operativity, by highlighting the relation between the wanted performance and the practices to implement to obtain it (Han et al. 2020).

Subsequently, in response to the third research gap identified, the framework has been refined to come up with a simpler, more synthetic version (core framework, as reported in step 2 of Fig. 3.1). Taking inspiration from Cagno et al. (2019), the indicators were analysed to eliminate repetitions and better exploit synergies, so as to come up with a lower number of indicators but keeping a sufficient amount of information, as reported in Fig. 3.2.

Details of the passages and assumptions made are presented in Sects. 3.2.1 and 3.2.2.

As mentioned, the first highlighted literature gap is the lack of an adequate framework to measure SC sustainability and resilience simultaneously, so as to exploit their synergies and avoid detrimental applications.

As the literature on sustainability and resilience PM, taken separately, is quite developed, two recent and complete PM frameworks have been taken from the literature. These have been re-elaborated to fit the purpose of creating a unique framework for sustainability and resilience PM.

Concerning sustainability, the study by Neri et al. (2021) was taken as a reference for the development for the following reasons: Firstly, the paper performs a comprehensive literature review on the topic of SC sustainability PM, contributing to establish standards to measure sustainability. Secondly, it provides balance among the TBL, which was often highlighted by authors as a still existing gap. Thirdly, the framework is tested empirically with firms and supply chains, which makes it more robust.

Concerning SCRes, the study by Han et al. (2020) was selected. Once again, this study represents one of the most recent literature reviews on the topic, thus creating

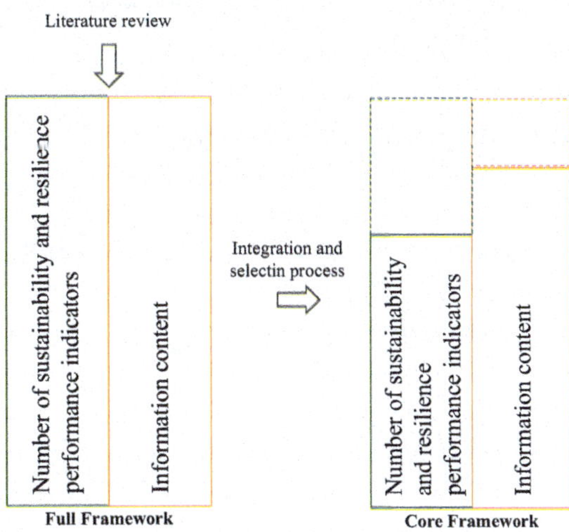

Fig. 3.2 Content of information of the full and core frameworks

established concepts on which to build further considerations. Second, this model allows connecting performance with the capabilities of resilience. This is particularly useful to make the framework operative, as performance could be easily connected to the practices and bring higher levels of implementation. The practices are here intended as the operational interventions that allow to reach a certain performance (Um and Han 2020).

Building on the two frameworks retrieved in the past contributions, the final structure of the proposed PM framework is presented below.

To begin with, a common structure between sustainability and resilience had to be established.

Four main levels have been used (Fig. 3.3): First, it is the dimensions of sustainability and resilience, as the widest categorization possible. The concept of dimensions is quite established in the SCRes literature, and they correspond to the phases of resilience, namely, readiness, response, and recovery (Ali et al. 2017). The same concept is not as present in the extant sustainability literature, but we linked it to the pillars of the TBL, namely, economic, environmental, and social (Carter and Rogers 2008) for symmetry.

The second level consists of capabilities. As mentioned, the concept of capabilities is quite established in the SCRes literature, but it is also present for sustainability. In the context of sustainability, the proposed framework refers to ordinary capabilities, defined as those capabilities that allow firms to operate and obtain certain performances (Teece 2014; Kalubanga and Gudergan 2022). In this proposed framework, only operational capabilities are included since the framework aims at assessing the capabilities in a specific moment in time and not their evolution over time. The sustainability capabilities considered are "financial", "internal process", "learning and growth", "customer", "environment", and "social" (Neri et al. 2021), which have been elaborated considering various authors. The resilience capabilities have been re-elaborated based on the studies by Han et al. (2020) and other authors

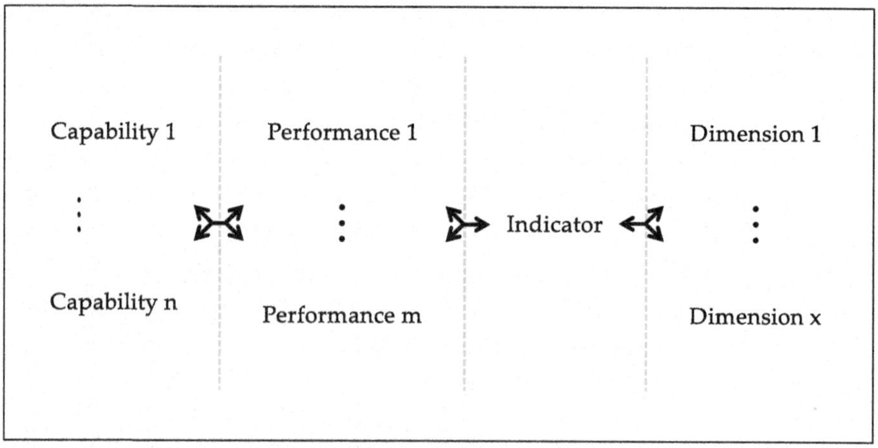

Fig. 3.3 Relationship among the level of the framework

(Karl et al. 2018). The capabilities allow to give operativity to the framework, as they can be connected to the practices to implement to improve specific performance (Han et al. 2020). By adding this level, practitioners can focus on the most relevant practices to obtain the wanted outcome. Besides, both in the sustainability (Gelhard and von Delft 2016) and in the resilience literature (Birkie et al. 2017), the link between practices and performance is mediated by the capabilities, so much so that capabilities often are considered a proxy of performance (Kaviani et al. 2020).

The third level consists of the performance. The sustainability performance has been derived from the indicators proposed by Neri et al. (2021) and is present in Table 3.3. The resilience performance has been derived by Han et al. (2020), with the addition of the knowledge management performance in correspondence of the respective capability (Karl et al. 2018).

Finally, the fourth level consists of the indicators. Both sustainability and resilience indicators have been collected from various contributions in the past literature. Some indicators have been added, for instance, "Speed of information-sharing" and "Accuracy of information-sharing", as they are present in the literature as important aspects of sustainability (Charkha and Jaju 2014; Narimissa et al. 2020) and they are also widely considered in resilience (Fiksel et al. 2015; Chowdhury and Quaddus 2017). Therefore, it is interesting to highlight this synergy by adding these two indicators. More specifically, resilience indicators have been retrieved by Cabral et al. (2012), Azevedo et al. (2013), and Behzadi et al. (2020), as the number and typology of indicators are relatively less standardized.

There are several links among the levels described so far: First, one indicator can measure one or more performances and can belong to one or more dimensions. Second, one performance can be linked to one or more capabilities, and one capability can refer to one or more performances (n:n) (Table 3.4).

3.2.1 Full Framework

The selection process was initiated by the indicator selection, which was done based on the following criteria:

1. Overlapping definitions, sufficiently similar to eliminate indicators without losing key information. Indicators that present the same definition but have different names or indicators that can be derived from each other have been incorporated (Cagno et al. 2019). In this way, the model is streamlined in the number of indicators (thus becoming more usable by firms) without losing important information and thus completeness.
2. Measurability, as some indicators have clearer formulas or are more easily computable with data already available for firms. Indicators with clear and simpler formulas have been prioritized (Tangen 2005).
3. Ratios rather than absolute numbers, as they are easier to understand (Tangen 2005).

Table 3.3 Sustainability structure

Capabilities	Performance	Indicators	Dimensions
Financial	Performance in having a profitable business	Return on investment	Economic
		Return on sales	
		Return on assets	
	Performance in controlling SC-related costs	Total SC cost	
		Inventory cost	
	Performance in converting cash outflows in inflows	Cash-to-cash cycle time	
Growth	Performance in having a productive and flexible workforce	Labour efficiency	
	Performance in improving the quality of business over time	New product development time	
		Use of new technology	
		Investments in innovation	
Customer-related	Performance in winning the customers' choice	Market share	
		Customer satisfaction	
	Performance in providing the product/service and the correlated services	Product quality	
		Product/service variety	
		Order fulfilment	
		Delivery reliability	
Process-related	Performance in having a set of efficient production activities	SC cycle time	
		Process cycle time	
		SC responsiveness	
		Capacity utilization	
Efficient consumption of resources	Performance in efficiently managing natural/non-natural resources across the processes	Energy use	Environmental
		Water use	
		Key material use	
	Performance in having a proof of environmentally sustainable operations	Certifications	
Effective waste management	Performance in virtuously managing processes to efficiently control waste creation	Environmental impact	
		Waste	
		Recycling	

(continued)

Table 3.3 (continued)

Capabilities	Performance	Indicators	Dimensions
Positive public relations	Performance in managing relations outside the firm with other firms and single people	Community	Social
		Relationships	
		Philanthropic investments	
	Efficient supply chain visibility	Speed of information-sharing	
		Accuracy of information-sharing	
Positive employee-firm relations	Performance in managing relations within the firm, with existing workers	OHS performance	
		Labour turnover	
		Employee satisfaction	

Adapted from Neri et al. (2021)

Table 3.4 Resilience structure

Capabilities	Performance	Indicators	Dimensions
Situation awareness	Performance of discerning possible disruptions	Quality of forecast	Readiness
		Supply chain alertness	
		Disruption probability	
Visibility	Performance of overseeing the supply chain situation	Order accuracy	
		Visibility	
		Monitoring and maintenance	
		Reliability	
		Disaster preparation	
Redundancy	Performance of production and inventory	Production changeover/production run length	
		Total average inventory across distribution centres	
		Stock-out rate	
		Reserve capacity	
Knowledge management and culture	Development of supply chain knowledge management	Investment in knowledge-sharing	
		Preparation of contingency plans	
Agility	Efficiency of completing supply chain processes	Lead time ratio	Response
		Lead time reduction	
		Speed of critical activities	
Flexibility	Efficiency of responding the disruptions	Responsiveness	
		Internal flexibility	
		Alternative options to ensure production	
		External flexibility	

(continued)

Table 3.4 (continued)

Capabilities	Performance	Indicators	Dimensions
Collaboration	Performance of relationship management	Collaboration, information-sharing, trust, and risk- and revenue-sharing	
		SC relationship	
		Risk management infrastructure	
		Supply network resilience	
		Cooperation	
Contingency planning	Reconstruction of the supply chain	Resource reconfiguration scale	Recovery
		Contingency plan	
	Efficiency of recovery to normality	Profile length	
		Performance loss	
		Recovery rate	
Market position	Performance of maintaining customer satisfaction	Customer service level	
		Customer satisfaction	
		Percentage of unfulfilled demand	
		Post-disruption mitigation capabilities	
	Damage from disruption	Supply chain disruption scale	
		Disruption impact	
		Ripple effect	
	Financial performance	Total SC cost	
		Efficiency	
		Financial perspective	

Adapted from Han et al. (2020)

4. Relevance. Indicators that are particularly relevant for the sustainability or resilient PM have been kept (Cagno et al. 2019).
5. Balance among the dimensions, considering both sustainability and resilience broadly, and the single dimensions (i.e. economic, environmental, social, readiness, response, recovery). In doing so, we have preferred comprehensiveness to precision over certain aspects, considering that having a comprehensive view allows to have a better understanding and take better decisions.

It is important to note that when an indicator was discarded from the framework, its information was not lost but absorbed into the remaining indicator, as done in previous literature (Cagno et al. 2019). In particular, the remaining indicator would acquire the capabilities, performance, and dimensions of the discarded indicator. In this way, the framework keeps as much information as possible but has a more manageable number of indicators (Negri et al. 2021b).

The complete list from the two single sustainability and resilience frameworks consisted of 76 indicators, 26 performance, 17 capabilities, and 6 dimensions.

A first screening of the lists of indicators allowed to remove duplicates. For instance, the "Customer satisfaction" indicator was included twice, as a measure of sustainability and as a measure of resilience. In the first case, it represents a customer-related capability and measures the performance of winning the customers' choice, therefore the economic dimension of sustainability. In the second case, it belongs to the recovery dimension. It is therefore more focused on the ability of the organization to recover performance after a disruption and thus keep the customer satisfied. In both cases, however, the indicator is defined as the degree of customer satisfaction, i.e. the number of satisfied customers out of the total (Aramyan et al. 2007; Cabral et al. 2012; Han et al. 2020). As the two indicators have the same formula, only one was kept in the framework. To clarify, the "Customer satisfaction" indicator left contains both the recovery and the economic dimensions, including the performance of maintaining customer satisfaction and winning the customers' choice.

As a second example, "Total SC cost" is present in both resilience and sustainability lists. It measures the total cost of fulfilment related to the company's operations (economic sustainability) and at the same time the economic losses in the recovery phase (Carvalho et al. 2012b; Xu et al. 2016). Since the indicators have the same formula, only one of them was kept in the framework, and the dimensions, capabilities, and performance have been integrated into a single indicator.

Finally, the last couple of duplicate indicators were "Process cycle time" in the sustainability part and "Production changeover/production run length" in the resilience part. The first one is defined as "the time required by the SC from the time the product begins its manufacture to the time it is completely processed" (Neri et al. 2021). Likewise, the second one is a measure of the time to produce and the time to change production (Han et al. 2020). Therefore, since they have the same formula, only "Process cycle time" was kept in the framework.

Subsequently, the indicators were further analysed in order to understand whether the indicators presented as a measure of resilience can equally measure sustainability and vice versa, in order to have a successful integration of the two concepts in a single framework. Each indicator was classified in terms of dimensions covered. For example, "Capacity utilization" was present in the sustainability list, under the economic dimension. Indeed, it represents the efficiency of resource use, i.e. the intensity with which resources are used in production (Gunasekaran et al. 2001, Neri et al. 2021). At the same time, it can also be seen as a measure of the environmental dimension, as it accounts for the oversizing of production means which generate environmental harm without an adequate value added to the process (Du et al. 2020). Moreover, "Capacity utilization" is also capable of measuring resilience, since it quantifies the flexibility and the ability of production systems to quickly respond to a disruption (Ivanov 2021). Therefore, based on the literature, it is possible to state that the "Capacity utilization" indicator represents three different dimensions: economic, environmental, and response. Hence, it is capable of measuring both resilience and sustainability. Another interesting example is "Cooperation", listed as a measure of the agility of the company in responding to a disruption. However, in the literature, this indicator also includes aspects of social and environmental sustainability.

Indeed, by having closer collaboration with suppliers, firms may have higher visibility of their regulations and certifications and can influence them to adopt more sustainable practices (Gualandris and Kalchschmidt 2015; Hannibal and Kauppi 2019). To summarize, three dimensions can be attributed to the "Cooperation" indicator: response, social, and environmental.

This process of the intersection was repeated for all indicators, ultimately arriving at a set of 73 indicators, as presented in Table 3.5.

3.2.2 Core Framework

The last highlighted literature gap is the lack of scalability of existing frameworks, which are therefore not capable of adapting to the different and evolving requirements of firms. The importance of scalability in PM has already been recognized in past contributions (Bititci et al. 2015), especially for some firms. Indeed, small and medium enterprises (SMEs) (Johnson 2015) or new adopters (NAs), i.e. firms at the beginning of their measurement journey (Negri et al. 2021b), may not have the resources or skills to use the same tools larger and more aware firms adopt. For these reasons, the core framework was developed. In order to make the framework usable by firms with less resources and awareness, reducing the number of indicators is essential (Cagno et al. 2019).

Using the full framework as a starting point, the level of performance was further analysed to find similarities and overlaps. As mentioned, there are several similarities and links between sustainability and resilience, and this is reflected in the level of performance. The sustainability and resilience performance were then grouped based on similarity, as shown in Table 3.6.

As an example, in group A, it is apparent that the two performances are connected. They both focus on process efficiency and organizational agility. Consequently, the indicators measuring these two performances are connected and can thus be easily incorporated. Out of the 26 performances in the framework, 22 were placed in a group and assessed as similar to other performances. This further reinforces the concept that resilience and sustainability are strongly linked. Only four performances were not included in a group (one sustainability performance and three resilience performance) because of their direct and close reference to resilience or sustainability and therefore hardly similar to any other performance. These are "Reconstruction of the supply chain", "Performance of discerning possible disruption", "Performance in managing relations within the firm with existing workers", and "Efficiency to recovery to normality".

Inside each identified performance group, the indicators were analysed and merged following the criteria listed in Sect. 3.2. As an example of the rationale behind the selection, in group A, out of six indicators ("Capacity utilization", "Lead time ratio", "Lead time reduction", "SC cycle time", "SC responsiveness", "Speed of critical activities"), four are selected. First, "SC cycle time" and "SC lead time" have an overlapping definition. Indeed, the first is defined as "Time required for the SC to

Table 3.5 Full framework

Capabilities	Performance	Indicator	Dimension
Positive public relations	Efficient supply chain visibility	*Accuracy of information-sharing*	Readiness Social
Flexibility	Efficiency of responding the disruptions	*Alternative options to ensure production*	Response
Process-related	Performance in having a set of efficient production activities	*Capacity utilization*	Economic Environmental Response
Financial	Performance in converting cash outflows in inflows	*Cash-to-cash cycle time*	Economic Readiness
Efficient consumption of resources	Performance in having a proof of environmentally sustainable operations	*Certifications*	Environmental Readiness
Collaboration	Performance of relationship management	*Collaboration between supply chain agents*	Environmental Response Social
Positive public relations	Performance in managing relations outside the firm with other firms and single people	*Community*	Social
Contingency planning	Reconstruction of the supply chain	*Contingency plan*	Recovery
Collaboration	Performance of relationship management	*Cooperation between supply chain agents*	Environmental Response Social
Market position	Performance in winning the customers' choice		Economic
Customer-related	Performance of maintaining customer satisfaction	*Customer satisfaction*	Recovery Social
Market position	Performance of maintaining customer satisfaction	*Customer service level*	Recovery
Customer-related			Social
Customer-related	Performance in providing the product/service and the correlated services	*Delivery reliability*	Economic
Visibility	Performance of overseeing the supply chain situation	*Disaster preparation*	Readiness
Market position	Damage from disruption	*Disruption impact*	Recovery
Situation awareness	Performance of discerning possible disruptions	*Disruption probability*	Readiness
Market position	Financial performance	*Financial efficiency*	Recovery
Positive employee-firm relations	Performance in managing relations within the firm, with existing workers	*Employee satisfaction*	Social

(continued)

Table 3.5 (continued)

Capabilities	Performance	Indicator	Dimension
Efficient consumption of resources	Performance in efficiently managing natural/non-natural resources across the processes	*Energy consumption*	Environmental
Effective waste management	Performance in virtuously managing processes to efficiently control waste creation	*Environmental impact*	Environmental
Flexibility	Efficiency of responding the disruptions	*External flexibility*	Response
Market position	Financial performance	*Financial perspective*	Economic Recovery
Flexibility	Efficiency of responding the disruptions	*Internal flexibility*	Response
Financial	Performance in controlling SC-related costs	*Inventory cost*	Economic
Knowledge management and culture	Development of supply chain knowledge management	*Investment in knowledge-sharing*	Readiness
Growth	Performance in improving the quality of business over time	*Investments in innovation*	Economic Readiness
Efficient consumption of resources	Performance in efficiently managing natural/non-natural resources across the processes	*Key material*	Environmental Readiness Social
Growth	Performance in having a productive and flexible workforce	*Labour efficiency*	Economic Readiness
Positive employee-firm relations	Performance in managing relations within the firm, with existing workers	*Labour turnover*	Social
Agility	Efficiency of completing supply chain processes	*Lead time ratio*	Response Agility
Agility	Efficiency of completing supply chain processes	*Lead time reduction*	Response Agility
Customer-related	Performance in winning the customers' choice	*Market share*	Economic
Visibility	Performance of overseeing the supply chain situation	*Monitoring and maintenance*	Readiness
Growth	Performance in improving the quality of business over time	*New product development time*	Economic
Positive employee-firm relations	Performance in managing relations within the firm, with existing workers	*OHS performance*	Social
Visibility	Performance of overseeing the supply chain situation	*Order accuracy*	Readiness
Customer-related	Performance in providing the product/service and the correlated services	*Order fulfilment*	Economic

(continued)

Table 3.5 (continued)

Capabilities	Performance	Indicator	Dimension
Market position	Performance of maintaining customer satisfaction	*Percentage of unfulfilled demand*	Recovery
Contingency planning	Efficiency of recovery to normality	*Performance loss*	Economic Recovery
Positive public relations	Performance in managing relations outside the firm with other firms and single people	*Philanthropic investments*	Social
Market position	Performance of maintaining customer satisfaction	*Post-disruption mitigation capabilities*	Recovery
Knowledge management and culture	Development of supply chain knowledge management	*Preparation of contingency plans*	Readiness
Customer-related	Performance in providing the product/service and the correlated services	*Product quality*	Economic
Customer-related	Performance in providing the product/service and the correlated services	*Product/service variety*	Economic
Process-related	Performance in having a set of efficient production activities	*Process cycle time*	Economic
Redundancy	Performance of production and inventory		Readiness
Contingency planning	Efficiency of recovery to normality	*Profile length*	Recovery
Situation awareness	Performance of discerning possible disruptions	*Quality of forecast*	Readiness
Contingency planning	Efficiency of recovery to normality	*Recovery rate*	Recovery
Effective waste management	Performance in virtuously managing processes to efficiently control waste creation	*Recycling*	Environmental
Positive public relations	Performance in managing relations outside the firm with other firms and single people	*Number of relevant relationships*	Response Social
Visibility	Performance of overseeing the supply chain situation	*Reliability*	Readiness
Redundancy	Performance of production and inventory	*Reserve capacity*	Readiness
Contingency planning	Reconstruction of the supply chain	*Resource reconfiguration scale*	Recovery
Flexibility	Efficiency of responding the disruptions	*Responsiveness*	Economic Response
Financial	Performance in having a profitable business	*Return on assets*	Economic Recovery
Financial	Performance in having a profitable business	*Return on investment*	Economic Recovery

(continued)

Table 3.5 (continued)

Capabilities	Performance	Indicator	Dimension
Financial	Performance in having a profitable business	*Return on sales*	Economic Recovery
Market position	Damage from disruption	*Ripple effect*	Recovery
Collaboration	Performance of relationship management	*Risk management infrastructure*	Response
Process-related	Performance in having a set of efficient production activities	*SC cycle time*	Economic
Process-related	Performance in having a set of efficient production activities	*SC responsiveness*	Economic Response
Agility	Efficiency of completing supply chain processes	*Speed of critical activities*	Response
Positive public relations	Efficient supply chain visibility	*Speed of information-sharing*	Readiness Social
Redundancy	Performance of production and inventory	*Stock-out rate*	Economic Readiness
Collaboration	Performance of relationship management	*SC relationship*	Response
Situation awareness	Performance of discerning possible disruptions	*Supply chain alertness*	Readiness
Market position	Damage from disruption	*Supply chain disruption scale*	Recovery
Collaboration	Performance of relationship management	*Supply network resilience*	Response
Redundancy	Performance of production and inventory	*Total average inventory across distribution centres*	Readiness
Market position	Financial performance	*Total SC cost*	Economic Recovery
Growth	Performance in improving the quality of business over time	*Use of new technology*	Economic
Visibility	Performance of overseeing the supply chain situation	*Visibility*	Economic Readiness Social
Effective waste management	Performance in virtuously managing processes to efficiently control waste creation	*Waste generation*	Economic Environmental Social
Efficient consumption of resources	Performance in efficiently managing natural/nonnatural resources across the processes	*Water use*	Environmental

fulfil an order" (Neri et al. 2021) and the latter as "the elapsed period from receipt of customer order to delivery" (Christopher and Peck 2004). As a consequence, "SC cycle time" and "Lead time ratio" are strongly connected, since their formulas are

Table 3.6 Performance groups

	Sustainability performance	Resilience performance
A	Performance in having a set of efficient production activities	Efficiency of completing supply chain processes
B	Performance in managing relations outside the firm with other firms and single people	Performance of relationship management
C	Efficient supply chain visibility	Performance of overseeing the supply chain situation
D	Performance in improving the quality of business over time	Development of supply chain knowledge management
E	Performance in winning the customers' choice	Performance of maintaining customer satisfaction
	Performance in providing the product/service and the correlated services	
F	Performance in controlling supply chain-related costs	Damage from disruption
	Performance in converting cash outflows in inflows	Financial performance
	Performance in having a profitable business	
G	Performance in having a productive and flexible workforce	Efficiency of responding the disruptions
H	Performance in having a proof of environmentally sustainable operations	Performance of production and inventory
	Performance in efficiently managing natural/nonnatural resources across the process	
	Performance in virtuously managing processes to efficiently control waste creation	

based on "SC lead time". As mentioned in the selection criteria (Sect. 3.2.1), having two indicators with equal definition, a ratio is preferred. Therefore, "Lead time ratio" was selected (Mason-Jones and Towill 1999). The other indicators considered similar are "Lead time reduction" and "SC responsiveness". Indeed, the "Lead time reduction" measures both cost and time performance, which means that it can measure the agility of a supply chain, i.e. "SC responsiveness" (Carvalho et al. 2012a). Between the two, "Lead time reduction" was selected as it is easier to measure. "Speed of critical activities" and "Capacity utilization" are kept in the framework since no overlapping definitions are found with the other indicators in group A. Figure 3.4 reports the selections made for each performance group.

After this selection process, the performances not included in any groups were analysed. For the indicators included in "Performance of discerning possible disruption", there are "Disruption probability", "Quality of forecast", and "Supply chain alertness". The latter is defined as "Capability of a supply chain to detect changes either from the external business environment or from the internal supply chain network, in a timely manner" (Han et al. 2020; Li et al. 2021). "Quality of forecast" is defined as "the percentage of difference between forecasts previously made for a year and the actual consumption or other operational aspects for that year" (Rajesh 2018). Being a ratio and more measurable, "quality of forecast" was selected.

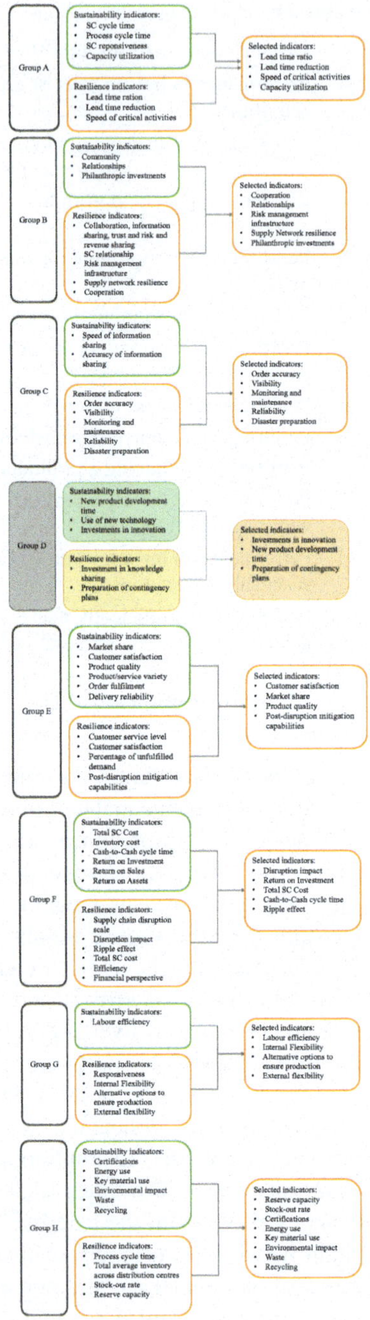

Fig. 3.4 Indicators' selection for each performance group

In the performance "Reconstruction of the supply chain", no overlapping definitions are detected, and so both indicators are kept ("Contingency plan" and "Resource reconfiguration scale").

In the "Efficiency of recovery to normality" performance, an analogy between "Performance loss" and "Recovery rate" is highlighted. Indeed, the formulas contain the same elements, and in the first case, it measures the percentage of performance that the organization is able to recover, while the other measures the opposite, i.e. the percentage of performance which the organization is not able to recover (Munoz and Dunbar 2015; Zeng and Yen 2017).

Lastly, the "Performance in managing relations within the firm, with existing workers" contains three indicators ("Employee satisfaction", "Labour turnover", and "OHS performance"). "Employee satisfaction" and "Labour turnover" were merged because of their overlapping definition. The latter, indeed, measures the percentage of employees leaving the organization (Demartini et al. 2018), and therefore it is an indirect measure of employee satisfaction (Cotton and Tuttle 1986). Therefore, "Employee satisfaction" was selected.

At the end of this selection process, the number of selected indicators is 47. The indicators were further analysed to merge the ones that could be considered similar or that could be derived from another one. For instance, "Consumption of key resources" and "Labour efficiency" present some similarities. Indeed, the latter is an indirect measure of the consumption of the labour resource, which is indeed a key resource (Stocker et al. 2015). Therefore, "Consumption of key resources" was kept in the framework. Similarly, "Alternative options to ensure production" is merged into "External flexibility", as both measure the agility of the supplier to change capacity and quantity (Wicher et al. 2015). "Lead time ratio" and "Lead time reduction" both have formulas based on lead time, and they have therefore been incorporated without losing essential information (Tersine and Hummingbird 1995). "Quality of forecast" and "Disaster preparation" are merged because these two concepts are strictly connected. Indeed, if the forecast demonstrates a strong ability to anticipate reality, it is more likely that the disturbances are predicted (Chowdhury and Quaddus 2016). "Recovery rate", "Disruption impact", and "Resource reconfiguration scale" are incorporated into one indicator. The latter is defined as the "ability to realign, reconfigure, restructure, and renew the resource" (Han et al. 2020), and therefore it is indirectly connected to the performance recovered by firms and to the impact of the disruption (Ambulkar et al. 2015). Among the three, "Recovery rate" was selected since it is a ratio.

"Reliability" and "Post-disruption mitigation capabilities" are merged; the first measures the "ability to satisfy immediate demand before any risk-mitigating actions, preventative or post-disruption, are taken" (Chen et al. 2017). Hence, it also includes the capability to recover after a disruption occurs (Ambulkar et al. 2015). "Contingency plan" and "Preparation of contingency plan" are quite similar by definition, although they refer to the readiness dimension and to the response one. However, both indicators measure the ability of firms to prepare response actions (Lam and Bai 2016) and are therefore merged. Finally, "Capacity utilization" and "Reserve capacity" both indicate the presence of idle capacity to use in case of emergencies (Kamalahmadi and Parast 2016). "Reserve capacity" was selected due to measurability.

This process allowed defining a final core framework of 36 indicators (Table 3.7), which can be considered synthetic enough to be used by firms with limited resources and awareness (Cagno et al. 2019; Negri et al. 2021b).

3.2.3 *Final Remarks*

The concept of balance for PM framework has already been discussed and is deemed central (Neri et al. 2021). In particular, the balance among the sustainability and resilience dimensions is essential for the purpose of this study. The final balance among the indicators is reported in Table 3.8. This was calculated taking into account that each selected indicator would maintain the dimension, performance, and capability of the discarded ones. As shown, both versions of the framework can be considered balanced. However, the elaborations and selections carried out to build the core framework allowed to have more control on its balance.

The operativity of both the full and the core framework is ensured by connecting the capabilities to the performance, which in turn allows connecting the performance to the practices. This allows firms to know where to act to improve one or more performance, making the framework more useful. Furthermore, it is operative as the indicators present metrics for calculation, providing guidance to firms on how to measure them. Finally, the use of this framework is twofold. On the one hand, firms can individuate the specific indicators they should use to measure the desired performance. For example, if firms need to improve their "performance of discerning possible disruptions", they should start monitoring the "Disruption probability" and "Quality of forecast" indicators.

On the other hand, they are also able to know what capabilities, performance, and dimensions are related to a single indicator and thus what impacts should be expected by improving or worsening an indicator. For instance, "Customer satisfaction" is linked to the "Performance in providing the product/service and the correlated services", the "Performance in winning the customers' choice", and the "Performance of maintaining customer satisfaction" performance. Thus, it contains information related to economic and recovery dimensions. This will help firms make informed decisions on sustainability and resilience.

3.3 Framework Testing: Case Studies

3.3.1 *Methodology*

The framework has been tested empirically with manufacturing SCs. The methodology used is the case study research, as the most appropriate tool to carry out this research (Yin 2003). Indeed, qualitative methodologies can serve as forms of theory

Table 3.7 Core framework

Capabilities	Performance	Indicator	Dimension
Financial	Financial performance	*Cash-to-cash cycle time*	Economic
Market position	Performance in converting cash outflows in inflows		Readiness
			Recovery
Efficient consumption of resources	Performance in having a proof of environmentally sustainable operations	*Certifications*	Environmental
			Readiness
Efficient consumption of resources	Performance in efficiently managing natural/non-natural resources across the processes	*Consumption of energy*	Environmental
Efficient consumption of resources	Performance in efficiently managing natural/non-natural resources across the processes	*Consumption of key resources*	Economic
Growth	Performance in having a productive and flexible workforce		Environmental
			Readiness
			Social
Collaboration	Performance of relationship management	*Cooperation between supply chain agents*	Environmental
			Response
			Social
Customer-related	Performance in providing the product/service and the correlated services	*Customer satisfaction*	Economic
Market position	Performance in winning the customers' choice		Recovery
	Performance of maintaining customer satisfaction		Social
Situation awareness	Performance of discerning possible disruptions	*Disruption probability*	Readiness
Positive employee-firm relations	Performance in managing relations within the firm, with existing workers	*Employee satisfaction*	Social
Effective waste management	Performance in virtuously managing processes to efficiently control waste creation	*Environmental impact*	Environmental
Flexibility	Efficiency of responding the disruptions	*External flexibility*	Economic
			Response
Flexibility	Efficiency of responding the disruptions	*Internal flexibility*	Response
Growth	Development of supply chain knowledge management	*Investments in innovation*	Economic
Knowledge management and culture	Performance in improving the quality of business over time		Readiness
Agility	Efficiency of completing supply chain processes	*Lead time ratio*	Economic
Process-related	Performance in having a set of efficient production activities		Response

(continued)

Table 3.7 (continued)

Capabilities	Performance	Indicator	Dimension
Customer-related	Performance in winning the customers' choice	*Market share*	Economic
Visibility	Performance of overseeing the supply chain situation	*Monitoring and maintenance*	Readiness
			Environmental
Growth	Performance in improving the quality of business over time	*New product development time*	Economic
Positive employee-firm relations	Performance in managing relations within the firm, with existing workers	*OHS performance*	Social
Positive public relations	Efficient supply chain visibility	*Order accuracy*	Readiness
Visibility	Performance of overseeing the supply chain situation		Social
Positive public relations	Performance in managing relations outside the firm with other firms and single people	*Philanthropic investments*	Social
Customer-related	Performance in providing the product/service and the correlated services	*Product quality*	Economic
			Social
Contingency planning	Efficiency of recovery to normality	*Profile length*	Recovery
Situation awareness	Performance of discerning possible disruptions	*Quality of forecast*	Readiness
Contingency planning	Efficiency of recovery to normality	*Recovery rate*	Economic
Market position	Reconstruction of the supply chain		Recovery
Effective waste management	Performance in virtuously managing processes to efficiently control waste creation	*Recycling*	Environmental
Collaboration	Performance in managing relations outside the firm with other firms and single people	*Number of relevant relationships*	Response
Positive public relations	Performance of relationship management		Social
Market position	Performance of maintaining customer satisfaction	*Reliability*	Readiness
Visibility	Performance of overseeing the supply chain situation		Recovery
Contingency planning	Development of supply chain knowledge management	*Reserve capacity*	Economic
Knowledge management and culture	Performance in having a set of efficient production activities		Environmental
Process-related	Performance of production and inventory		Readiness
Redundancy	Reconstruction of the supply chain		Response
			Recovery

(continued)

Table 3.7 (continued)

Capabilities	Performance	Indicator	Dimension
Financial	Financial performance	*Return on investment*	Economic
Market position	Performance in having a profitable business		Recovery
Market position	Damage from disruption	*Ripple effect*	Recovery
Collaboration	Performance of relationship management	*Risk management infrastructure*	Response
Agility	Efficiency of completing supply chain processes	Speed of critical activities	Response
Process-related	Performance of production and inventory		Economic
Redundancy	Performance in having a set of efficient production activities		Readiness
Redundancy	Performance of production and inventory	*Stock-out rate*	Economic
			Readiness
Collaboration	Performance of relationship management	*Supply network resilience*	Response
Financial	Financial performance	*Total SC cost*	Economic
Market position	Performance in controlling SC-related costs		Recovery
Positive public relations	Efficient supply chain visibility	*Visibility*	Economic
Visibility	Performance of overseeing the supply chain situation		Readiness
			Social
Effective waste management	Performance in virtuously managing processes to efficiently control waste creation	*Waste generation*	Environmental
			Social
			Economic

testing, when the comparison does not rely on single measures between groups of observations, but rather the aim is to compare a pattern of observations with expected outcomes from theory (Bitektine 2008). The case study methodology allowed retrieving substantial information on the contextual factors, which helped explain the outcomes obtained and provide insights on the sustainability and resilience PM, looking for analytical generalization of results rather than statistical (Barratt et al. 2011; Baškarada 2014).

From a practical standpoint, the focal firms of the different supply chains selected were interviewed, namely, the one that "rules or governs the SC, provides the direct contact to the customer, and designs the product" (Masi et al. 2018). This was done with the assumption that the behaviour of the focal firm could influence that of the entire SC (Marshall et al. 2015), as assumed in past studies (Govindan et al. 2020; Negri et al. 2022). Besides, a focal firm usually monitors the performance of its suppliers and, more in general, the entire chain (Beske 2012). In all three cases, the key informant(s) (DiCicco-Bloom and Crabtree 2006) was the people managing the SC in their firms, and when possible the SC manager was interviewed. In this way, we were able to collect information on the SC and validate the framework.

Table 3.8 Balance among the dimensions

	Full framework						Core framework					
	Sustainability			Resilience			Sustainability			Resilience		
	Eco	Env	Social	Readiness	Response	Recovery	Eco	Env	Social	Readiness	Response	Recovery
Relative weight	21.7%	13.0%	19.3%	9.7%	14.9%	21.5%	17.9%	15.8%	13.6%	15.7%	18.1%	18.9%
Relative weight	54%			46%			47.3%			52.7%		

3.3.1.1 Sample Selection

Multiple case studies have been conducted as a way to increase robustness (Miles et al. 2014). Diversity among SCs was ensured by selecting SCs that differ in terms of size, market coverage, products, and awareness level as a way to expand the range of characteristics in the sample (Løkke and Sørensen 2014) to improve the external validity of results beyond the sample itself (Voss et al. 2002). The manufacturing sector has been selected as the most appropriate for this research. Indeed, it is responsible for substantial environmental impact (Acerbi and Taisch 2020; International Energy Agency 2021), therefore having a strong influence on sustainability. It has also been receiving strong pressure from various stakeholders to improve this impact, and therefore it has developed higher awareness and attention towards sustainability implementation and measurement (Neri et al. 2018). Besides, the economic relevance of the manufacturing sector in Europe is widely recognized (Eurostat 2020), and the several challenges experienced in these last few years—the COVID-19 pandemic, supply shortages, and the energy crisis—have demonstrated that firms need to improve their resilience to cope with unexpected disruptions (European Commission 2022).

Purposive sampling has been used (Acharya et al. 2013), intentionally designing a diverse sample to increase robustness. The international database Orbis has been used, and three SCs have been selected.

The case studies from supply chain A and supply chain C have been complemented by close observation and collaboration over a period of 6 months. This has allowed to become acquainted with the practices, assumptions, and general behaviour of the firms and to collect a substantial amount of information.

The respondents who have been involved in the structured interviews are all SC and operations experts, who have been selected for their privileged visibility across the SC of each of the firms. Furthermore, when contacting the respondents, priority has been given to personnel with a specific focus on either supply chain sustainability or supply chain resilience (Table 3.9).

3.3.1.2 Data Collection and Analysis

The primary source of data was semi-structured interviews. An interview protocol was used to reduce the variability of the questions, allowing at the same time interviewees to freely add comments and insights (Cooper et al. 2006; McIntosh and Morse 2015). Several interviews have been conducted for each case study, with more than one informant to reduce bias and increase robustness (Voss et al. 2002). Secondary sources served for triangulation to corroborate the insights that emerged during the interviews (Woodside and Wilson 2003; Zainal 2007).

The interviewees started with information about the firm's operations and characteristics of SC, which have been used to interpret the outcomes of the interviews, followed by a general part about sustainability and resilience and the attitude of the SC towards the two.

Table 3.9 Characteristics of the selected sample

	Supply chain A	Supply chain B	Supply chain C
Activity	Manufacture of beauty care products, make-up, and perfumes	Manufacture of heating systems and thermal pumps	Manufacturing of rubber-based tires for all types of vehicles
Contacts	Supply chain director Director of operations	Demand planning manager Production planning manager	Sustainable and digital procurement manager Sustainability manager
Upstream	Quasi-monopolies with high contractual power	High degree of vertical integration, plus third parties. High power of suppliers, also due to the scarcity of raw materials	Very limited number of suppliers, mainly natural rubber, that have high power due to concentration and scarcity of resources. Other suppliers have low power. No partnerships
Downstream	Mostly B2B to large retailers, but the e-commerce is growing (and variable in the different markets worldwide)	70% B2B, 30% B2C; 150 distribution countries, worldwide coverage	Higher degree of collaboration; often products are developed together
Production	Main production activities are carried out close to the country of distribution	Either produced internally by the focal firm or acquired by third parties	Production located in 12 different countries. Production is very flexible and can be moved easily

Subsequently, the interviewees were asked to assess the novel PM framework. In particular, the aim was to understand how much the framework was able to answer the needs of firms. Three variables were tested (Trianni et al. 2017):

- Capacity of representation: whether the framework is able to include all relevant aspects related to sustainability and resilience PM.
- Ease of use: the framework's user-friendliness.
- Usefulness: whether the framework facilitates decision-making.

The interviewees were presented at first with the core framework, as it emerged from a more substantial refinement process and contained more assumptions than the full one. Once the firms had validated the core, the full framework was presented, and the interviewees were asked to comment on the additional indicators, with the aim of investigating the value of having a more complete tool. Finally, the interviewees were asked to evaluate the single indicators and provide feedback.

The information collected through the interviews and secondary material have then been analysed. The analysis in case study research consists of searching and interpreting patterns in data (Kohlbacher 2006). Interviews were transcribed (Rowley 2012) and analysed by three different researchers to reduce bias (Miles et al. 2014). Section 3.4 presents the within-case analysis and cross-case analysis. The results are then discussed and compared to the extant literature.

3.4 Results

3.4.1 Supply Chain A

3.4.1.1 General Information

Supply chain A's focal firm is a multinational group competing in the beauty care industry, manufacturing and selling beauty products in 150 countries around the world. The company is among the most relevant worldwide in its reference sector, both for its size and its turnover.

Upstream in the chain, suppliers are located worldwide and operate in quasi-monopolies, which ensure that any negotiation leaves the power in their hands. Thus, the sustainability and resilience initiatives are often confined in the boundaries of single firms, with little coordination or collaboration. For instance, the focal firm evaluates its suppliers and only selects those that meet its standards.

The products are processed in plants that are owned by the firms and distributed in the markets served, i.e. all products that are sold in Europe are produced in the continent and the same happens for other zones. The SC operates mostly as a B2B business, selling products to distributors downstream, that then reach the final customer; still, a percentage of sales happen through a B2C channel via e-commerce, which is raising in importance, although with high variability in the different markets worldwide, due to the characteristics of local consumers and the different products sold. For example, Chinese consumers are much more used to purchasing through e-commerce, while it is less common for Europeans, even if the percentage is increasing sharply. The product portfolio of SC A is divided into five product categories which are skincare, make-up, haircare, fragrances, and hair colouring, with most of sales coming from the categories of skincare and make-up.

3.4.1.2 Approach to Sustainability and Resilience

As mentioned, the sustainability and resilience initiatives tend to happen within the perimeter of single firms of the SC. The focal firm's approach to sustainability is particularly mature, with actions in place since 2013 to improve performance in this field and in line with the three pillars of the TBL. Indeed, the interviewees stated: "it has been many years actually that the firm operates on sustainability with really demanding commitments, challenging and clearly defined at the corporate level. [...] I am referring to creating a business that can adapt to a fast-changing context, with structured skills, backup capacity, organizational capabilities that allow to face and survive in such a system. Obviously, being financially sustainable allows firms to be also environmentally and socially sustainable, but the environmental and social parts are core here".

Despite little collaboration, several initiatives involve the group and all local subsidiaries. The initiatives cover all pillars of the TBL, for example, with topics such as the efficient consumption of water, the efficient consumption of plastics and reduction of waste sent to landfills for the environmental pillar, programs for employment in underprivileged communities, gender equality in the roles inside the firm, and the measurement of gender stereotypes present in communications for the social pillar. For the financial pillar, the measurement of performance is more standard, with all traditional measures in place.

In this sense, both interviewees seem to intend sustainability and resilience as a means to achieve long-term growth.

For what concerns SCRes, the respondents stated that it "is the ability of a firm to thrive in a context of extended volatility and uncertainty. [...] between zero and one hundred, the focus of the firm is one hundred. Indeed, you can see this by looking at the performance of the firm in the context of the Covid-19 pandemic [...] a large context of economic, political, social crisis".

3.4.1.3 Framework Validation

Regarding the clarity of the model, the first respondent stated that it is very clear, also considering that most of the indicators are already in use in the firm. The second respondent agreed and affirmed that "It is fluid and I found only a few indicators that were not clear to me at first glance. It would be interesting to see if there are priorities among the indicators, if there are targets". Therefore, for both respondents, the model was clear. Concerning the capacity of representation, the two interviewees were asked whether the framework was able to well represent both sustainability and resilience while highlighting their synergies. The first respondent stated: "Absolutely yes, because we discussed interactions, waste, performance, risk, ROI, [...] in the indicators that you included I find all that I deem necessary". The second respondent commented that "Yes of course, it can represent very well, 100% these two concepts".

Ease of use was also recognized, although the second respondent stated that the number of indicators might be too high, even in the core framework. However, commenting on its usefulness, both interviewees acknowledged that the framework allows to effectively and efficiently measure sustainability and resilience performance.

Finally, after presenting the full framework, both respondents did not find any essential indicators that were not contained in the core framework. In fact, the second respondent also added that the number of indicators in the core framework is already more than he would have expected and would not add any further indicators to represent supply chain resilience and sustainability, which are perceived as already present in their completeness.

Concerning more specifically the single indicators, Table 3.10 reports the main comments from the two interviewees. The firm already uses most of the indicators proposed, which is a proxy of its advancement in sustainability and resilience measurement. All proposed indicators appeared to be useful to measure both the firm's and the SC's performance. The only indicator that appeared as too complex and not used was "Ripple effect". Both interviewees stated that it would be very complex to measure, although they recognized its value.

Table 3.10 Comments on the single indicators of the core framework for firm A

Indicators	Valuable to measure SC's performance	Valuable to measure firm's performance	Currently in use + comments
Cash-to-cash cycle time	Yes/yes	Yes/yes	Not specifically, but similar time measures
Certifications	Yes/yes	Yes/yes	Yes, it is a proof of excellence for the supply chain
Consumption of energy	Yes/yes	Yes (plants)/yes	Yes, important for energy-intensive steps; fundamental
Consumption of key resources	Yes/yes	Yes/yes	Yes, valuable for actors upstream and drive stock quality; fundamental
Cooperation between supply chain agents	Yes/yes (very much)	Yes/yes	Yes, not in a single indicator but as a scattered information; not easy to measure
Customer satisfaction	Yes/yes	Yes/yes	Yes
Disruption probability	Yes/yes	Yes/yes	Yes, locally and at group level
Employee satisfaction	Yes/yes	Yes/yes	Yes, similar to cooperation
Environmental impact	Yes/yes	Yes/yes	Yes, CO_2 emissions and energy
External flexibility	Yes/yes	Yes/yes	Yes, absolutely, important in a complex context. Third parties are involved in production
Internal flexibility	Yes/yes	Yes/yes	Yes
Investments in innovation	Yes/yes	Yes/yes	Yes, innovation is a business driver
Lead time ratio	Yes/yes	Yes/yes	Yes, compared to an objective lead time
Market share	Yes/yes	Yes/yes	Yes, crucial
Monitoring and maintenance	Yes/yes	Yes/yes	Yes, monitoring is locally fundamental. There are people working in a continuous improving business
New product development time	Yes/yes	Yes/yes	Yes, time measures are crucial
OHS performance	Yes/yes	Yes/yes	Yes
Order accuracy	Yes/yes	Yes/yes	Yes
Philanthropic investments	Yes/yes	Yes/yes	Not known
Product quality	Yes/yes	Yes/yes	Yes, for different levels (product, categories, components, etc.)
Profile length	Yes/yes	Yes/yes	Yes
Quality of forecast	Yes/yes	Yes/yes	Yes, crucial for anticipation

(continued)

Table 3.10 (continued)

Indicators	Valuable to measure SC's performance	Valuable to measure firm's performance	Currently in use + comments
Recovery rate	Yes/yes	Yes/yes	Yes, similar to profile length
Recycling	Yes/yes	Yes/yes	No, more focus on preventing waste
Number of relevant relationships	Yes (both for suppliers and clients)/yes	Yes (crucial for information)/yes	Yes
Reliability	Yes/yes	Yes (also for data)/yes	Yes
Reserve capacity	Yes/yes	Yes/yes	Yes
Return on investment	Yes/yes	Yes/yes	Yes
Ripple effect	No/no	No/no	No, too complex
Risk management infrastructure	Yes/yes	Yes/yes	Yes, locally and at group level
Speed of critical activities	Yes/yes	Yes/yes	Yes, critical in terms of value that they provide to the customer
Stock-out rate	Yes/yes	Yes/yes	Yes
Supply network resilience	Yes/yes	Yes/yes	Yes
Total SC cost	Yes/yes	Yes/yes	Yes, but also compared to the quality it can drive
Visibility	Yes/yes	Yes/yes	Yes
Waste generation	Yes/yes	Yes/yes	Yes

Finally, after presenting the full framework, both respondents did not find any essential indicators that were not contained in the core framework. In fact, the second respondent also added that the number of indicators in the core framework is already more than he would have expected and would not add any additional ones.

3.4.2 *Supply Chain B*

3.4.2.1 General Information

Supply chain B operates in the thermal comfort industry and reaches a worldwide market of consumers. Its focal firm offers a large range of products for water and space heating, components, and burners. Thanks to its 90 years of experience, it is one of the market leaders with 1.66 billion € turnover in 2020.

The SC stretches from the raw material suppliers to the distributors. Starting from the upstream supply chain, two different flows can be distinguished: the internal production, where the vendors are raw material suppliers, and outsourcing, although this latter is marginal. In addition, the focal firm is vertically integrated, as many components required for production are made by other companies of the group. Therefore, the suppliers can be divided into direct and indirect material, consumable vendors, and suppliers of finished products. As raw materials are becoming increasingly scarce in the last few years, the supply is particularly critical and strategic. Producers of finished products are also extremely important and particularly considered in risk mitigation plans. In terms of production, the focal firm has 26 production sites in 15 different countries. Despite the strong international footprint, about 90% of the products sold in Italy are manufactured locally. The same applies to other countries, which are served mostly by Italian production and, where present, by local production. Downstream, there are distributors, with some exceptions of sales to the final consumer. Indeed, 70% of the products are sold in B2B sales. The distributors cover over 150 countries.

The focal firm operates in several layers of the supply chain, spanning from manufacturing of finished goods to distribution. Indeed, it owns both production plants and distribution centres across multiple regions of the world, with the goal to serve multiple markets. However, the control of operations remains in the Italian headquarters.

As mentioned, this supply chain can be split into two different flows, which are different mainly in the upstream section. In regard to this, the respondents stated: "There is a double structure [...] on one side what we make is what we sell, while on the other we have the products that we purchased as finished goods and put our brands on them, this is what we sell". So the focal firm is in a constant "make-or-buy" decision when evaluating its alternatives for procurement.

3.4.2.2 Approach to Sustainability and Resilience

The interviewees defined sustainability as core and integrated it in the SC's strategy. Even if their sustainability journey started relatively recently, in 2017, they have already achieved many interesting goals. In 2020, 29% of the R & D investments were dedicated to renewable products, also achieving high energy efficiency from plants to products. Thanks to the technological efficiency of the firm, 1.3 million tons of CO_2 were saved from 2018 to 2020. On the other side, the firm is not one of the most mature companies in the field of sustainability, as it disregards the three pillars of the TBL and mostly considers economic and, to a lesser degree, environmental aspects. One of the respondents has defined the attitude of the SC to sustainability a: "Sustainability is a core item for the SC; indeed it is also present in our mission statement. The concept is very relevant for the field of thermal comfort [...] the environmental sustainability is the most important as consuming the material is crucial. [...] Social sustainability is not particularly considered in relation to the supply chain". The two SCs that are differentiated by the make-or-buy decision are similar in the approach to sustainability, but the presence of massive companies providing finished goods in the buy side makes it more complex for the focal firm to have the adequate contractual power to push sustainability goals to its suppliers. The firm has anyway the possibility to carefully assess suppliers before activating a procurement channel, which is also done through certifications.

As regards resilience, SC B has a more readiness-oriented attitude. Indeed, it aims at being proactive and to be ready for disruptions with contingency plans. In order to achieve this, the focal firm conducted several studies and analyses on possible future risks and their probability of occurrence. Moreover, the overall SC's approach to disruption is not to return to its previous performance levels but to try to perform better than pre-disruption. The aim is to learn from shocks and adapt to changes by anticipating future disruptions. The approach is defined by respondents as more proactive than reactive. Indeed, one of the respondents told us: "The idea here is to adapt to a new normal post-shock [. . .] particularly in the last relevant disruption that we saw, I think that will never return to the previous environment". And also "The idea in the whole business is to centralize, [. . .] some firms have tried to delocalize before due to costs but now it is riskier to do anything, such as transportation [. . .]".

In spite of this, the interviewees presented a quite interesting approach and awareness about resilience. The presence of make-or-buy decisions partially influences the attitude towards SC resilience. Indeed, when assessing its suppliers' characteristics and performance before the activation of the contract and during the actual supply phase, the focal firm monitors sources of disruptions and potential risks that derive from each specific supplier and, if necessary, modifies its suppliers' base accordingly. However, given the nature of contractual agreements, full-fledged flexibility is not always possible.

3.4.2.3 Framework Validation

Concerning the capacity of representation, the answer of the first respondent was "Yes, I feel both are represented but I felt that resilience is a little more represented [. . .] I would say this is correct for a supply chain [function], where is crucial and sustainability, especially social, is often not the focus. But this is not wrong in any way. [. . .] I would say that the balance is 60% [for supply chain resilience] and 40% [for supply chain sustainability]". The first interviewee also recognized the value of the framework: "I see the value of this framework, especially in the case that this becomes a standard, in the sense that this can become a way to perform a benchmark [. . .] it can create a guideline for other firms too". The second respondent of firm B commented that "In my opinion it represents both concepts very well, it deepens both resilience and sustainability".

Concerning the ease of use, the first interviewee noted a certain complexity of the framework: "[It is clear] but with a correct explanation, I feel that the structure itself is valuable, but it requires to be correlated with the explanation of the structure and the description of indicators". The second interviewee highlighted that SC workers might have more facility in using the framework, as they are already acquainted with many of the concepts mentioned, i.e. "For people working in the supply chain, the indicators are very explanatory. So, it is easy to understand how the model should be used". Both interviewees acknowledged the value of the framework. Table 3.11 reports the main comments from the two interviewees.

Table 3.11 Comments on the single indicators of the core framework for firm B

Indicators	Valuable to measure SC's performance	Valuable to measure firm's performance	Currently in use + comments
Cash-to-cash cycle time	Yes (more for the supply chain)/yes	Yes/yes	Yes, linked to statutory reporting
Certifications	Not clear/no	Yes/yes	Yes, mostly international standards; law requirements
Consumption of energy	Yes/yes (more for the plants)	Yes (building on law regulations)	Yes
Consumption of key resources	Yes (not for assets leased)/yes	Yes/yes	Yes, for sustainability but also for financial (costs)
Cooperation between supply chain agents	Yes (but undervalued now)/yes	Yes (very)/yes	No, but would be very useful
Customer satisfaction	Yes/yes	Yes/yes	Yes
Disruption probability	Yes (more for the supply chain)/yes	Yes/yes	Yes, qualitative
Employee satisfaction	Yes (as a consequence)/yes	Yes/yes	Yes, qualitative
Environmental impact	Yes (more for the supply chain)/yes	Yes/yes	Yes, only through certifications and CO_2. Could be very valuable to further explore
External flexibility	Yes/yes	Yes (broad in scope)/yes	Yes, qualitative but would be valuable as quantitative
Internal flexibility	Yes/yes	Yes (very)/yes	Yes, qualitative
Investments in innovation	Yes/yes	Yes/yes	Yes
Lead time ratio	Yes (good for logistics and production)/yes	Yes/yes	Yes, once per year
Market share	Not really/yes	Yes (more for the firm)/yes	Yes
Monitoring and maintenance	Yes (more for the supply chain)/yes	Yes/yes	Yes, qualitative
New product development time	Not really/not sure	Yes/yes	Yes
OHS performance	Yes/yes	Yes (good for reputation)/yes	Yes, mandatory by law
Order accuracy	Yes (very)/yes	Yes/yes	Yes
Philanthropic investments	Yes/yes	Yes/yes (less relevant)	Yes
Product quality	Yes/yes	Yes/yes	Yes, for the firms and suppliers
Profile length	Yes/yes	Yes/yes	Yes, qualitative but very interesting

(continued)

Table 3.11 (continued)

Indicators	Valuable to measure SC's performance	Valuable to measure firm's performance	Currently in use + comments
Quality of forecast	Yes/yes	Yes/yes	Yes, as SOE (sales order error)
Recovery rate	Yes/yes	Yes (very)/yes	Yes, through target values to achieve under particular contexts
Recycling	Yes/yes	Yes/yes	No, could be useful to evaluate third-party providers of this service
Number of relevant relationships	Yes/yes	Yes/yes	Yes, for suppliers
Reliability	Yes/yes	Yes/yes	Yes, through safety stocks
Reserve capacity	Yes (more for the supply chain)/yes	Yes/yes	Yes, through machinery saturation
Return on investment	No/yes	Yes/yes	Yes
Ripple effect	Yes/yes	Yes/yes	Yes, through the size of buffers in the areas of the chains
Risk management infrastructure	Yes/yes	Yes/yes	No, but would be useful
Speed of critical activities	Yes (more for the supply chain)/yes	Yes/yes	Yes
Stock-out rate	Yes/yes	Yes/yes	Yes, but in total number not as a rate
Supply network resilience	Yes/yes	Yes/yes	No, just an evaluation of suppliers initially
Total SC cost	Yes/yes	Yes/yes	Yes
Visibility	Yes/yes (very)	Yes/yes	No, but not particularly valuable
Waste generation	Yes/yes	Yes/yes	Yes, as scrap and reworks amount

Finally, both respondents commented on the full framework. The first respondent stated that "Definitely [the full framework] is much more complete, but I think the core is still comprehensive. I didn't find an indicator that was particularly needed but not present in the core". The second respondent was also quite in line with this comment, also stating that since the core is already complete, she/he would not add further indicators in order to not complicate the model.

3.4.3 Supply Chain C

3.4.3.1 General Information

Supply chain C operates in the automotive industry; in particular, it is involved in the production of tyres for cars, motorcycles, and machinery. This focal firm is one of the leaders in the sector, with a 4.3 billion € turnover, over 30,000 employees, and a worldwide presence, with 19 production sites in 12 different countries.

The supply chain is composed of several tiers, starting from the extraction of natural rubber and ending with the customers, i.e. car, motor, and bike manufacturer. Upstream, there are more than 14,000 suppliers, divided mainly into suppliers of raw materials, indirect materials, and capital goods and services. Most of the spending is dedicated to raw materials, of which natural rubber is one of the most important. The number of suppliers in this case is very limited, being the most strategic component of the company's purchasing. The shortage of natural resources, such as natural rubber, gives higher bargaining power to the suppliers. Machinery suppliers, on the other hand, are not easily replaceable because their products are company-specific and customized. However, the spending dedicated to such suppliers is relatively low. The focal firm's expenditures for services are high but divided among a relatively high number of suppliers with mostly spot contracts, thus with little collaboration. Finally, for indirect materials, there are a low average expenditure and a high average number of suppliers. Also in this case, the relationship with suppliers is not particularly strong. To summarize, for most of its suppliers, the focal firm has higher bargaining power, while it has more difficulties for suppliers of raw materials. Downstream, there are 19 production sites—located around the world—where most products are manufactured, with only a small percentage produced by third parties. Among these production sites, the approach is very flexible and interconnected, with the possibility of moving production from one plant to another as needed. On the downstream side, there are more than 17,000 points of sales worldwide, showing good market coverage. The portfolio of products offered by the SC, as mentioned above, consists of three main categories: car, motorcycle, and bicycle tires. Partnerships allow the focal firm to develop tyres that are tailored to meet the specific needs of their customers through collaborative projects. Such collaborations are obviously made for high-value products that generate 70% of the revenue. These products consist of innovative tyres (car, motorcycle, and bicycle tyres) designed to achieve higher levels of performance and safety. Considering the three end product categories, there are no fundamental differences in their management. The only difference is in the high-value products which are managed with a stronger focus on performance and produced in collaboration with customers.

3.4.3.2 Approach to Sustainability and Resilience

The focal firm has been working to improve its sustainability for 15 years, both at strategic and operational levels. As an example of the firm's commitment, it is certified ISO 20400, which attests that the company procurement prioritizes sustainability in its activities. However, both interviewees confirmed that it is not just a matter of compliance with regulations but also an effort to minimize the firm's environmental and human negative impact, in all steps of the supply chain.

Several practices have been implemented in order to have end-to-end visibility, also involving suppliers, for instance, the land cultivators. In this context, control over resource consumption is particularly critical. Consequently, the focal firm has enforced high standards that suppliers have to comply with, starting from the adoption of various certifications. In addition, since the firm's main raw material is natural rubber, it is a priority for them to have visibility on its extraction practices and on suppliers' deforestation impact. Suppliers are generally not selected if they do not meet the minimum sustainability criteria. In addition, the focal firm requires compliance with their Code of Conduct and Sustainability Clause.

Therefore, the focal firm finds integration, communication of core values, and training of suppliers fundamental.

The SC's approach to resilience is cross-functional and well-integrated in the strategy and operations, once again guided by the focal firm. Indeed, in every function of the firm, a risk management team is present with the aim of promptly tackling disruptions. The firm's strategy is more focused on the readiness phase of resilience, aiming to be proactive. For this reason, there are contingency plans in place in order to be prepared in case of several possible disruptions, with a natural higher attention to raw materials and natural rubber and logistics.

3.4.3.3 Framework Validation

Concerning the capacity of representation, both interviewees agreed that the model represents both sustainability and resilience very well, treating the two concepts in a comprehensive manner. The first respondent indeed said that "At this company I don't see sustainability and resilience as a single concept because sustainability is flat, it's at the base. But yes, seeing this model I can see them integrated". The comment of the second respondent also confirms that the framework provides a good representation of sustainability and resilience: "I see sustainability and resilience as quite complementary. I see resilience as an important pillar of sustainability [. . .]. On the indicators chosen, I would say yes [. . .] I think they are all elements that we maybe sometimes take for granted, but it actually makes sense to see them in one model. Going through them one by one has certainly helped to give a more global view of the concepts".

Concerning the ease of use, both respondents agreed that the framework was intuitive and easy to use. The first interviewee stated that "the model seems very simple to me. Even if I were alone, I would have had no difficulty in understanding the indicators in the list with the definitions attached. Perhaps I would have clustered the indicators to make it more user-friendly". The answers of the second interviewee are very much in line with the former, i.e. the model is "very easy to understand. Without a definition, I would have struggled, also because some are very transversal".

Finally, both interviewees agreed that the model is useful to measure sustainability and resilience performance (Table 3.12).

Table 3.12 Comments on the single indicators of the core framework for firm C

Indicators	Valuable to measure SC's performance	Valuable to measure firm's performance	Currently in use + comments
Cash-to-cash cycle time	Yes/yes	Yes/yes	Yes, fundamental for the purchasing function
Certifications	Yes/yes	Yes/yes	Yes, in addition to the legal requirements
Consumption of energy	Yes/yes	Yes/yes	Yes, calculated for each Tyre typology; dedicated function to monitor energy
Consumption of key resources	Yes/yes	Yes/yes	Yes, especially for raw materials
Cooperation between supply chain agents	Yes/yes	Yes/yes	Yes, mostly with strategic suppliers
Customer satisfaction	Yes/yes (but it is difficult)	Yes/yes	Yes
Disruption probability	Yes/yes	Yes/yes	Yes, risk mapping is done especially for raw materials
Employee satisfaction	Yes/yes	Yes/yes	Yes, through questionnaires and additional follow-up
Environmental impact	Yes/yes	Yes/yes	Yes, water consumption, CO_2 emission, and more are monitored
External flexibility	Yes/yes	Yes/yes	Yes, through collaboration with suppliers
Internal flexibility	Yes/yes	Yes/yes	Yes, interconnection between plants
Investments in innovation	Yes/yes	Yes/yes	Yes
Lead time ratio	Yes/yes	Yes/yes	Yes
Market share	Yes/yes	Yes/yes	Yes
Monitoring and maintenance	Yes/yes	Yes/yes	Yes, the company has a dedicated function
New product development time	Yes/yes	Yes/yes	Yes
OHS performance	Yes/yes	Yes/yes	Yes
Order accuracy	Yes/yes	Yes/yes	Yes
Philanthropic investments	Yes/yes	Yes/yes	Yes
Product quality	Yes/yes	Yes/yes	Yes
Profile length	Yes/yes	Yes/yes	Yes, after each disruption, a comparison is made
Quality of forecast	Yes/yes	Yes/yes	Yes
Recovery rate	Yes/yes	Yes/yes	Yes
Recycling	Yes/yes	Yes/yes	Yes

(continued)

Table 3.12 (continued)

Indicators	Valuable to measure SC's performance	Valuable to measure firm's performance	Currently in use + comments
Number of relevant relationships	Yes/yes	Yes/yes	Yes
Reliability	Yes/yes	Yes/yes	Yes
Reserve capacity	Yes/yes	Yes/yes	Yes, for each plant
Return on investment	Yes/yes	Yes/yes	Yes
Ripple effect	Yes/yes	Yes/yes	Yes
Risk management infrastructure	Yes/yes	Yes/yes	Yes, in each function, there is a dedicated team
Speed of critical activities	Yes/yes	Yes/yes	Yes
Stock-out rate	Yes/yes	Yes/yes	Yes
Supply network resilience	Yes/yes	Yes/yes	Yes, especially for raw material suppliers
Total SC cost	Yes/yes	Yes/yes	Yes
Visibility	Yes/yes	Yes/yes	Yes, it is monitored for tier one, but the company wants to achieve tier 2
Waste generation	Yes/yes	Yes/yes	Yes

Regarding the full framework, the first respondent indicated that he would add the indicator "Community" to the core framework. Indeed, she/he commented that "I think the one thing that is missing [from the core framework] is stakeholder engagement in sustainability. I think this indicator could be useful to understand engagement". On the other hand, the second respondent stated that the core framework seems very complete and that she would not add any indicators included in the full framework.

3.5 Discussion

The three firms found the framework comprehensive, easy to use, and useful to measure both the SC's and the firm's performance. The framework resulted in being easier to handle for firms with a more structured approach to sustainability and resilience already in place and also with advanced methods to performance measurement. Indeed, SC A confirmed that it is already monitoring most of the indicators present in the framework, while the focal firm of SC B stated that the framework might require an explanation before use.

All the interviewees confirmed that the framework allows to have a complete overview of sustainability and resilience, which contributes to filling a research gap (Negri et al. 2021a). Concerning the framework usefulness, interviewees from SC B stated that it would be valuable to have it as a standard tool among companies, in order to have a benchmark on fundamental topics such as sustainability and resilience, and have best practices that can improve the SC performance. This is an interesting point that emerged during the case studies, that is, firms with lower power in the SC, such as the focal firm in SC B, may encounter difficulties in measuring the SC performance. On the other hand, firms with high brand awareness and power in the SC, as the focal firm in SC A, have higher interest in monitoring the performance of all actors in the SC.

These points are expansions on the findings from available literature (Beamon 1999; Gunasekaran et al. 2004; Chia et al. 2009), in which researchers started postulating about the fundamental relevance of providing a complete set of measurement tools to represent the complexities of the chain without exceeding the complexity of the tool itself. Indeed, it could be argued that the present work is advancing the research towards the construction of a shared set of indicators that can foster benchmarking and provide a shared understanding of SC sustainability and SCRes with a balanced set of indicators (Ivanov 2018; Fahimnia et al. 2019).

The framework was intentionally built to have a balance between sustainability and resilience perspectives. The cases confirmed that both are equally represented in the framework, which allows an adequate measurement of performance. The interviewees, especially from SC C, also confirmed that sustainability and resilience are tightly linked and the success of SCs is only possible if both are considered equally important. This is in line with what was found in previous literature (Fahimnia and Jabbarzadeh 2016). As an example, Ivanov (2020) defined the "viable supply chain" as the one that is able to both sustain disruptions and increase its sustainability, which has become even more evident after recent disruptions. As a further proof of how much sustainability and resilience are connected, SC A's first respondent provided a definition of sustainability that largely overlaps with resilience. Indeed, she/he defined sustainability as a way to build "a business that can adapt to a fast-changing context, with structured skills, backup capacity, organizational capabilities that allow to face and survive in such a system". In this way, sustainability is seen as an enabler of long-term resilience, in line with the findings of Gouda and Saranga (2018) and Bag et al. (2019).

The structure of the framework allowed to easily grasp the synergies between sustainability and resilience by referring to the dimensions as indicators that belong to more than one dimension. For instance, in the core framework, "consumption of key resources" is an economic, environmental, and readiness indicator, while "customer satisfaction" is an economic, social, and recovery indicator. In a similar way, the indicators belong to more than one performance and more than one capability. This allows, on the one hand, to understand what the indicator is measuring; on the other, it allows to select the indicators to monitor, should firms decide to improve one specific performance or capability. The appendix presents a practical visualization of the core framework.

As regards the individual indicators, almost all indicators were assessed as relevant and valuable for measuring both the SC and the firm's performance, by all the three SCs. The "Ripple effect" indicator was rated from SC A and SC C as the most difficult to use and to measure, although its importance within the supply chain is recognized. Indeed, the "ripple effect", which is measuring the likelihood of having the effects from a disruption spreading among actors in the network, is a qualitative indicator that may require a periodic evaluation by SCRes experts, such as consultants. Generally, however, all indicators are considered valid for measuring a firm's and SC's performance in terms of sustainability and resilience.

The interviewees, mainly from SC A and SC B, reported some overlaps between the indicators presented in the core framework. Starting from these reported overlaps, the indicators were once again analysed in order to see whether there might actually be overlaps that enable to choose one indicator over the other. After a careful evaluation, also based on the literature, no indicators were removed from the core framework. Indeed, all the indicators reported are different in definition and pertain to different performances. In particular, the overlaps reported are listed as follows:

- "Waste generation" and "Environmental impact", although obviously linked, do not overlap. The first measures the amount of waste produced, in terms of materials and resources, while the second represents the impact in terms of CO_2 emissions, energy consumption, and sustainable procurement (Neri et al. 2021).
- "Lead time ratio" measures the time ratio between the actual lead time and the promised one (Carvalho et al. 2011), while "Speed of critical activities" measures the time it takes to the company to complete the most critical steps (Han et al. 2020). Indeed, it is argued that the two indicators are measuring different timespans in the supply chains, as the first is measuring the whole lead time, while the second is focusing on "critical" activities, which may be constituted, for example, by operational bottlenecks.
- "Profile length" measures the time between a disruption and the recovery of performance (Han et al. 2020), while "Recovery rate" measures the percentage of performance recovered after a specific period of time that is defined in advance. Both relate to the profile that describes the ability of a firm to recover its performance after a disruption, but as the first measures the speed of recovery, the latter measures the amount of recovery achieved after a defined period of time. It is important to have both measures in order to have a complete view of post-disruption performance (Han et al. 2020).

- "Reliability" measures the ability to satisfy the demand immediately after a disruption, while "Ripple effect" measures the probability of a disruption propagation (Han et al. 2020). The first is considering the ability to satisfy the market demand, while the latter is focusing on the risk of having a disruption that propagates its effects in the network, hindering the ability of some nodes in the chain to perform. Thus, the latter does not provide information on the demand served, although, in some cases, the impact can also reach the market.
- "Number of relevant relationships" and "Cooperation between supply chain agents". The former focuses on the number of relevant relationships (Han et al. 2020), as a means to quantify the number of value-adding relationships in the network, while the latter measures the quality of the relationships with the stakeholders (Gualandris et al. 2015), including a qualitative assessment that considers information about joint investments with other actors in the chain and investments in a shared ERP solution with supplier and/or distributors, among others.

It can be concluded that there are no significant overlaps among indicators, as each of them contributes to adding important information for decision-making. The ones mentioned by interviewees may be linked to their specific understanding of the concepts, which may be partial and dependent upon their experience.

Considering the comments by interviewees and the indicators already in use by firms, the interviewees showed a disproportionate focus on risk management and the readiness level rather than on the response to disruption. This can be found in the extant literature as well. However, as not all disruptions are foreseeable, avoidable, or reliably anticipated, this approach tends to be limited and partial (Pettit et al. 2013; Fiksel et al. 2015). Concerning sustainability, the cases allowed to retrieve that there is high attention to environmental sustainability. The social pillar is relatively less monitored (Hervani et al. 2022), while the economic one is so established that it is sometimes difficult for firms to see that sustainability and resilience may have economic benefits too (Dabhilkar et al. 2016; Gouda and Saranga 2018; Silva et al. 2019; Malesios et al. 2020). Some interviewees, as for firm A, stated that economic sustainability may be an enabler of environmental and social initiatives. In their opinion, indeed, an economically stable business is able to invest in environmental protection and social projects. However, the literature suggests that the contrary might also be true, meaning that environmental and social practices do strengthen the business and result in economic sustainability (Choi et al. 2017; Islam et al. 2017; Miroshnychenko et al. 2017).

One of the respondents from SC A believes it might be interesting to prioritize the indicators in the framework to also give a roadmap on how to start measuring sustainability and resilience. This could indeed help firms, especially SMEs, to increase the adoption of performance measurement frameworks and guide them towards improved awareness and implementation of sustainability and resilience (Bititci et al. 2015). It could be argued that the versions of the core and full framework do provide guidance to firms, by providing a first set of fundamental indicators (core framework) and then scaling up to include more indicators and information (full framework), thus adapting to the changing needs of firms.

3.6 Conclusions and Further Research

This chapter provides insights that advance knowledge on performance measurement, by providing a framework to simultaneously measure the sustainability and resilience performance.

This framework aims at filling the first research gap identified, namely, the lack of a framework to measure sustainability and resilience. The framework proposed exploits the synergies between the concepts to arrive at a limited number of indicators while keeping a comprehensive view on the concepts. The second research gap, namely, the lack of operativity, is addressed by providing clear definitions and metrics attached to the indicators. Besides, the link with the capabilities and performance allows to clearly identify the areas that need improvement and, reversely, to select adequate indicators to improve a specific capability or performance. Finally, the third research gap identified, namely, the lack of scalability, is addressed by providing two versions of the framework. The core framework is particularly indicated for SMEs, for new adopters, and for day-to-day decision-making.

This study provides insights for policy-making activities, to better design policies to suit SCs' needs and increase sustainability and resilience.

Some limitations may also be pointed out. First, the case studies conducted are limited in number and only served as a first empirical validation. More extensive empirical analyses may need to be conducted in the future, to have a better understanding of the dynamics shaping firms and SCs' performance measurement. Second, only the focal firm in the SC was interviewed, thus limiting the perspective on the entire SC. Multi-tier case studies may be conducted to better grasp the differences among the tiers of the SC. This would end up becoming useful in collecting insights on different perceptions of performance measurement at different levels of the SC. Finally, further research should be developed to apply the framework to SCs, measuring the performance of firms at the SC level and at the single firm level.

Appendix: Core Framework

	Dimensions						Capabilities																	Performance																									
indicator	Economic	Environmental	Readiness	Recovery	Response	Social	Agility	Collaboration	Contingency planning	Customer-related	Effective waste management	Efficient consumption of resources	Financial	Flexibility	Growth	Knowledge management and culture	Market position	Positive employee-firm-relations	Positive public relations	Process-related	Redundancy	Situation awareness	Visibility	Damage from disruption	Development of supply chain knowledge management	Efficiency of completing supply chain processes	Efficiency of recovery to normality	Efficiency of responding the disruptions	Efficient supply chain visibility	Financial performance	Performance in controlling SC-related costs	Performance in converting cash outflows in inflows	Performance in efficiently managing natural/non-natural resources across the processes	Performance in having a productive and flexible workforce	Performance in having a profitable business	Performance in having a proof of environmentally sustainable operations	Performance in having a set of efficient production activities `	Performance in improving the quality of business over time	Performance in managing relations outside the firm with other firms and single people	Performance in managing relations within the firm, with exisiting workers	Performance in providing the product/service and the correlated services	Performance in virtuously managing processes to efficiently control waste-creation	Performance in winning the customers' choice	Performance of discerning possible disruptions	Performance of maintaining customer satisfaction	Performance of overseeing the supply chain situation	Performance of production and inventory	Performance of relationship management	Reconstruction of the supply chain
Cash-to-cash cycle time	1	0	1	1	0	0	0	0	0	0	0	0	1	0	0	0	1	0	0	0	0	0	0	0	0	0	0	0	0	1	0	1	0	0	0	0	0	0	0	0	0	0	0	0	0	0	0	0	0
Certifications	0	1	1	0	0	0	0	0	0	0	0	1	0	0	0	0	0	0	0	0	0	0	0	0	0	0	0	0	0	0	0	0	0	0	0	1	0	0	0	0	0	0	0	0	0	0	0	0	0
Consumption of energy	0	1	0	0	0	0	0	0	0	0	0	1	0	0	0	0	0	0	0	0	0	0	0	0	0	0	0	0	0	0	0	0	1	0	0	0	0	0	0	0	0	0	0	0	0	0	0	0	0
Consumption of key resources	1	1	1	0	0	1	0	0	0	0	0	1	0	0	1	0	0	0	0	0	0	0	0	0	0	0	0	0	0	0	0	0	1	1	0	0	0	0	0	0	0	0	0	0	0	0	0	0	0
Cooperation between supply chain agents	0	1	0	0	1	1	0	1	0	0	0	0	0	0	0	0	0	0	0	0	0	0	0	0	0	0	0	0	0	0	0	0	0	0	0	0	0	0	0	0	0	0	0	0	0	0	0	1	0
Customer satisfaction	1	0	0	1	0	1	0	0	0	1	0	0	0	0	0	0	1	0	0	0	0	0	0	0	0	0	0	0	0	0	0	0	0	0	0	0	0	0	0	0	1	0	1	0	1	0	0	0	0
Disruption probability	0	0	1	0	0	0	0	0	0	0	0	0	0	0	0	0	0	0	0	0	0	1	0	0	0	0	0	0	0	0	0	0	0	0	0	0	0	0	0	0	0	0	0	1	0	0	0	0	0
Employee satisfaction	0	0	0	0	0	1	0	0	0	0	0	0	0	0	0	0	0	1	0	0	0	0	0	0	0	0	0	0	0	0	0	0	0	0	0	0	0	0	0	1	0	0	0	0	0	0	0	0	0
Environmental impact	0	1	0	0	0	0	0	0	0	0	1	0	0	0	0	0	0	0	0	0	0	0	0	0	0	0	0	0	0	'0	0	0	0	0	0	0	0	0	0	0	0	1	0	0	0	0	0	0	0
External flexibility	0	0	0	0	1	0	0	0	0	0	0	0	0	1	0	0	0	0	0	0	0	0	0	0	0	0	0	1	0	0	0	0	0	0	0	0	0	0	0	0	0	0	0	0	0	0	0	0	0
Internal flexibility	0	0	0	0	1	0	0	0	0	0	0	0	0	1	0	0	0	0	0	0	0	0	0	0	0	0	0	1	0	0	0	0	0	0	0	0	0	0	0	0	0	0	0	0	0	0	0	0	0
Investments in innovation	1	0	1	0	0	0	0	0	0	0	0	0	0	0	1	1	0	0	0	0	0	0	0	0	1	0	0	0	0	0	0	0	0	0	0	0	0	1	0	0	0	0	0	0	0	0	0	0	0

(continued)

Lead time ratio	1	0	0	0	1	0	1	0	0	0	0	0	0	0	0	0	0	0	0	1	0	0	0	0	0	1	0	0	0	0	0	0	0	0	0	0	1	0	0	0	0	0	0	0	0	0	0	0	0
Market share	1	0	0	0	0	0	0	0	0	1	0	0	0	0	0	0	0	0	0	0	0	0	0	0	0	0	0	0	0	0	0	0	0	0	0	0	0	0	0	0	0	0	1	0	0	0	0	0	0
Monitoring and maintenance	0	0	1	0	0	0	0	0	0	0	0	0	0	0	0	0	0	0	0	0	0	0	1	0	0	0	0	0	0	0	0	0	0	0	0	0	0	0	0	0	0	0	0	0	0	1	0	0	0
New product development time	1	0	0	0	0	0	0	0	0	0	0	0	0	0	1	0	0	0	0	0	0	0	0	0	0	0	0	0	0	0	0	0	0	0	0	0	1	0	0	0	0	0	0	0	0	0	0	0	0
OHS performance	0	0	0	0	0	1	0	0	0	0	0	0	0	0	0	0	0	1	0	0	0	0	0	0	0	0	0	0	0	0	0	0	0	0	0	0	0	0	0	1	0	0	0	0	0	0	0	0	0
Order accuracy	0	0	1	0	0	1	0	0	0	0	0	0	0	0	0	0	0	0	1	0	0	0	1	0	0	0	0	0	1	0	0	0	0	0	0	0	0	0	0	0	0	0	0	0	0	1	0	0	0
Philanthropic investments	0	0	0	0	0	1	0	0	0	0	0	0	0	0	0	0	0	0	1	0	0	0	0	0	0	0	0	0	0	0	0	0	0	0	0	0	0	0	1	0	0	0	0	0	0	0	0	0	0
Product quality	1	0	0	0	0	1	0	0	0	1	0	0	0	0	0	0	0	0	0	0	0	0	0	0	0	0	0	0	0	0	0	0	0	0	0	0	0	0	0	0	1	0	0	0	0	0	0	0	0
Profile length	0	0	0	1	0	0	0	0	1	0	0	0	0	0	0	0	0	0	0	0	0	0	0	0	0	0	1	0	0	0	0	0	0	0	0	0	0	0	0	0	0	0	0	0	0	0	0	0	0
Quality of forecast	0	0	1	0	0	0	0	0	0	0	0	0	0	0	0	0	0	0	0	0	0	1	0	0	0	0	0	0	0	0	0	0	0	0	0	0	0	0	0	0	0	0	0	1	0	0	0	0	0
Recovery rate	0	0	0	1	0	0	0	0	1	0	0	0	0	0	0	0	1	0	0	0	0	0	0	1	0	0	1	0	0	0	0	0	0	0	0	0	0	0	0	0	0	0	0	0	0	0	0	0	1
Recycling	0	1	0	0	0	0	0	0	0	0	1	0	0	0	0	0	0	0	0	0	0	0	0	0	0	0	0	0	0	0	0	0	0	0	0	0	0	0	0	0	0	1	0	0	0	0	0	0	0
Number of relevant relationships	0	0	0	0	1	1	0	1	0	0	0	0	0	0	0	0	0	0	1	0	0	0	0	0	0	0	0	0	0	0	0	0	0	0	0	0	0	0	1	0	0	0	0	0	0	0	0	1	0
Reliability	0	0	1	1	0	0	0	0	0	0	0	0	0	0	0	0	1	0	0	0	0	0	1	0	0	0	0	0	0	0	0	0	0	0	0	0	0	0	0	0	0	0	0	0	1	1	0	0	0
Reserve capacity	1	0	1	1	0	0	0	0	1	0	0	0	0	0	0	1	0	0	0	1	1	0	0	0	1	0	0	0	0	0	0	0	0	0	0	0	1	0	0	0	0	0	0	0	0	0	1	0	1
Return on investment	1	0	0	1	0	0	0	0	0	0	0	0	1	0	0	0	1	0	0	0	0	0	0	0	0	0	0	0	0	1	0	0	0	0	1	0	0	0	0	0	0	0	0	0	0	0	0	0	0
Ripple effect / disruption propagation	0	0	0	1	0	0	0	0	0	0	0	0	0	0	0	0	1	0	0	0	0	0	0	1	0	0	0	0	0	0	0	0	0	0	0	0	0	0	0	0	0	0	0	0	0	0	0	0	0
Risk management infrastructure	0	0	0	0	1	0	0	1	0	0	0	0	0	0	0	0	0	0	0	0	0	0	0	0	0	0	0	0	0	0	0	0	0	0	0	0	0	0	0	0	0	0	0	0	0	0	0	1	0
Speed of critical activities	0	0	0	0	1	0	1	0	0	0	0	0	0	0	0	0	0	0	0	0	0	0	0	0	0	1	0	0	0	0	0	0	0	0	0	0	0	0	0	0	0	0	0	0	0	0	0	0	0
Stock-out rate	1	0	1	0	0	0	0	0	0	0	0	0	0	0	0	0	0	0	0	0	1	0	0	0	0	0	0	0	0	0	0	0	0	0	0	0	0	0	0	0	0	0	0	0	0	0	1	0	0
Supply network resilience	0	0	0	0	1	0	0	1	0	0	0	0	0	0	0	0	0	0	0	0	0	0	0	0	0	0	0	0	0	0	0	0	0	0	0	0	0	0	0	0	0	0	0	0	0	0	0	1	0
Total SC cost	1	0	0	1	0	0	0	0	0	0	0	0	1	0	0	0	1	0	0	0	0	0	0	0	0	0	0	0	0	1	1	0	0	0	0	0	0	0	0	0	0	0	0	0	0	0	0	0	0
Visibility	1	0	1	0	0	1	0	0	0	0	0	0	0	0	0	0	0	0	1	0	0	0	1	0	0	0	0	0	1	0	0	0	0	0	0	0	0	0	0	0	0	0	0	0	0	1	0	0	0
Number of relevant relationships	0	1	0	0	0	0	0	0	0	0	1	0	0	0	0	0	0	0	0	0	0	0	0	0	0	0	0	0	0	0	0	0	0	0	0	0	0	0	0	0	0	1	0	0	0	0	0	0	0

References

Abdel-Basset M, Mohamed R (2020) A novel plithogenic TOPSIS- CRITIC model for sustainable supply chain risk management. J Clean Prod 247:119586. https://doi.org/10.1016/j.jclepro.2019.119586

Acerbi F, Taisch M (2020) A literature review on circular economy adoption in the manufacturing sector. J Clean Prod 273:123086. https://doi.org/10.1016/j.jclepro.2020.123086

Acharya AS, Prakash A, Saxena P, Nigam A (2013) Sampling: why and how of it? Indian J Med Spec 4:3–7. https://doi.org/10.7713/ijms.2013.0032

Ahi P, Searcy C (2015) An analysis of metrics used to measure performance in green and sustainable supply chains. J Clean Prod 86:360–377. https://doi.org/10.1016/j.jclepro.2014.08.005

Akyuz GA, Erkan TE (2010) Supply chain performance measurement: a literature review. Int J Prod Res 48:5137–5155. https://doi.org/10.1080/00207540903089536

Ali A, Mahfouz A, Arisha A (2017) Analysing supply chain resilience: integrating the constructs in a concept mapping framework via a systematic literature review. Supply Chain Manag 22:16–39. https://doi.org/10.1108/SCM-06-2016-0197

Almeida A, Bastos J, Francisco RDP et al (2016) Sustainability assessment framework for proactive supply chain management. Int J Ind Syst Eng 24:198–222

Ambulkar S, Blackhurst J, Grawe S (2015) Firm's resilience to supply chain disruptions: scale development and empirical examination. J Oper Manag 33–34:111–122. https://doi.org/10.1016/j.jom.2014.11.002

Aramyan LH, Lansink AGJMO, Van Der Vorst JGAJ, Van Kooten O (2007) Performance measurement in agri-food supply chains: a case study. Supply Chain Manag 12:304–315. https://doi.org/10.1108/13598540710759826

Azevedo SG, Govindan K, Carvalho H, Cruz-Machado V (2013) Ecosilient Index to assess the greenness and resilience of the upstream automotive supply chain. J Clean Prod 56:131–146. https://doi.org/10.1016/j.jclepro.2012.04.011

Azevedo SG, Carvalho H, Cruz-Machado V (2016) LARG index: a benchmarking tool for improving the leanness, agility, resilience and greenness of the automotive supply chain. Benchmarking 23:1472–1499. https://doi.org/10.1108/BIJ-07-2014-0072

Bag S, Gupta S, Foropon C (2019) Examining the role of dynamic remanufacturing capability on supply chain resilience in circular economy. Manag Decis 57:863–885. https://doi.org/10.1108/MD-07-2018-0724

Barratt M, Choi TY, Li M (2011) Qualitative case studies in operations management: trends, research outcomes, and future research implications. J Oper Manag 29:329–342. https://doi.org/10.1016/j.jom.2010.06.002

Baškarada S (2014) Qualitative case study guidelines. Qual Rep 19:1–18

Beamon BM (1999) Measuring supply chain performance. Int J Oper Prod Manag 19:275–292. https://doi.org/10.1108/01443579910249714

Behzadi G, O'Sullivan MJ, Olsen TL (2020) On metrics for supply chain resilience. Eur J Oper Res 287:145–158. https://doi.org/10.1016/j.ejor.2020.04.040

Beske P (2012) Dynamic capabilities and sustainable supply chain management. Int J Phys Distrib Logist Manag 42:372–387. https://doi.org/10.1108/09600031211231344

Birkie SE, Trucco P, Fernandez Campos P (2017) Effectiveness of resilience capabilities in mitigating disruptions: leveraging on supply chain structural complexity. Supply Chain Manag 22:506–521. https://doi.org/10.1108/SCM-01-2017-0009

Bitektine A (2008) Prospective case study design. Organ Res Methods 11:160–180. https://doi.org/10.1177/1094428106292900

Bititci US, Garengo P, Ates A, Nudurupati SS (2015) Value of maturity models in performance measurement. Int J Prod Res 53:3062–3085. https://doi.org/10.1080/00207543.2014.970709

Brusset X, Teller C (2017) Supply chain capabilities, risks, and resilience. Int J Prod Econ 184:59–68. https://doi.org/10.1016/j.ijpe.2016.09.008

Cabral I, Grilo A, Cruz-Machado V (2012) A decision-making model for Lean, Agile, Resilient and Green supply chain management. Int J Prod Res 50:4830–4845. https://doi.org/10.1080/00207543.2012.657970

Cagno E, Neri A, Howard M et al (2019) Industrial sustainability performance measurement systems: a novel framework. J Clean Prod 230:1354–1375. https://doi.org/10.1016/j.jclepro.2019.05.021

Cagno E, Negri M, Neri A, Giambone M (2023) One framework to rule them all: an integrated, multi-level and scalable performance measurement framework of sustainability, circular economy and industrial symbiosis. Sustain Prod Consum 35:55–71. https://doi.org/10.1016/j.spc.2022.10.016

Carter CR, Rogers DS (2008) A framework of sustainable supply chain management: moving toward new theory. Int J Phys Distrib Logist Manag 38:360–387. https://doi.org/10.1108/09600030810882816

Carvalho H, Duarte S, Machado VC (2011) Lean, agile, resilient and green: Divergencies and synergies. Int J Lean Six Sigma 2:151–179. https://doi.org/10.1108/20401461111135037

Carvalho H, Azevedo SG, Cruz-Machado V (2012a) Agile and resilient approaches to supply chain management: influence on performance and competitiveness. Logist Res 4:49–62. https://doi.org/10.1007/s12159-012-0064-2

Carvalho H, Barroso AP, MacHado VH et al (2012b) Supply chain redesign for resilience using simulation. Comput Ind Eng 62:329–341. https://doi.org/10.1016/j.cie.2011.10.003

Charkha PG, Jaju SB (2014) Designing innovative framework for supply chain performance measurement in textile industry. Int J Logist Syst Manag 18:216–230. https://doi.org/10.1504/IJLSM.2014.062327

Chen X, Xi Z, Jing P (2017) A unified framework for evaluating supply chain reliability and resilience. IEEE Trans Reliab 66:1144–1156. https://doi.org/10.1109/TR.2017.2737822

Chia A, Goh M, Hum SH (2009) Performance measurement in supply chain entities: balanced scorecard perspective. Benchmarking 16:605–620. https://doi.org/10.1108/14635770910987832

Choi B, Hwang S, Lee SH (2017) What drives construction workers' acceptance of wearable technologies in the workplace? Indoor localization and wearable health devices for occupational safety and health. Autom Constr 84:31–41. https://doi.org/10.1016/j.autcon.2017.08.005

Chowdhury MMH, Quaddus M (2016) Supply chain readiness, response and recovery for resilience. Supply Chain Manag 21:709–731. https://doi.org/10.1108/SCM-12-2015-0463

Chowdhury MMH, Quaddus M (2017) Supply chain resilience: conceptualization and scale development using dynamic capability theory. Int J Prod Econ 188:185–204. https://doi.org/10.1016/j.ijpe.2017.03.020

Christopher M, Peck H (2004) Building the resilient supply chain. Int J Logist Manag 15:1–13. https://doi.org/10.1108/09574090410700275

Cooper DR, Schindler PS, Sun J (2006) Business research methods. McGraw-Hill, New York

Cotton JL, Tuttle JM (1986) Linked references are available on JSTOR for this article: employee turnover: a meta-analysis and review with implications for research. Acad Manag 11:55–70

Dabhilkar M, Birkie SE, Kaulio M (2016) Supply-side resilience as practice bundles: a critical incident study. Int J Oper Prod Manag 36:948–970. https://doi.org/10.1108/IJOPM-12-2014-0614

Demartini M, Pinna C, Aliakbarian B et al (2018) Soft drink supply chain sustainability: a case based approach to identify and explain best practices and key performance indicators. Sustainability (Switzerland) 10:1–24. https://doi.org/10.3390/su10103540

DiCicco-Bloom B, Crabtree BF (2006) The qualitative research interview. Med Educ 40:314–321. https://doi.org/10.1111/j.1365-2929.2006.02418.x

Du W, Wang F, Li M (2020) Effects of environmental regulation on capacity utilization: evidence from energy enterprises in China. Ecol Indic 113:106217. https://doi.org/10.1016/j.ecolind.2020.106217

Elgazzar S, Tipi N, Jones G (2019) Key characteristics for designing a supply chain performance measurement system. Int J Product Perform Manag 68:296–318. https://doi.org/10.1108/IJPPM-04-2018-0147
European Commission (2022) European economic forecast spring 2022
Eurostat (2020) Manufacturing statistics. https://ec.europa.eu/eurostat/statistics-explained/index.php/Manufacturing_statistics_-_NACE_Rev._2#Structural_profile
Fahimnia B, Jabbarzadeh A (2016) Marrying supply chain sustainability and resilience: a match made in heaven. Transport Res E Log Transp Rev 91:306–324. https://doi.org/10.1016/j.tre.2016.02.007
Fahimnia B, Sarkis J, Talluri S (2019) Editorial design and management of sustainable and resilient supply chains. IEEE Trans Eng Manag 66:2–7. https://doi.org/10.1109/TEM.2018.2870924
Felsberger A, Qaiser FH, Choudhary A, Reiner G (2022) The impact of Industry 4.0 on the reconciliation of dynamic capabilities: evidence from the European manufacturing industries. Prod Plan Control 33:277–300. https://doi.org/10.1080/09537287.2020.1810765
Fiksel J, Polyviou M, Croxton KL, Pettit TJ (2015) From risk to resilience: learning to deal with disruption. MIT Sloan Manag Rev 56:79–86
Fraccascia L, Giannoccaro I (2020) What, where, and how measuring industrial symbiosis: a reasoned taxonomy of relevant indicators. Resour Conserv Recycl 157:104799. https://doi.org/10.1016/j.resconrec.2020.104799
Gelhard C, von Delft S (2016) The role of organizational capabilities in achieving superior sustainability performance. J Bus Res 69:4632–4642. https://doi.org/10.1016/j.jbusres.2016.03.053
Gouda S-K, Saranga H (2018) Sustainable supply chains for supply chain sustainability: impact of sustainability efforts on supply chain risk. Int J Prod Res 56:5820–5835. https://doi.org/10.1080/00207543.2018.1456695
Govindan K, Rajeev A, Padhi SS, Pati RK (2020) Supply chain sustainability and performance of firms: a meta-analysis of the literature. Transport Res E Logist Transp Rev 137:101923. https://doi.org/10.1016/j.tre.2020.101923
Green KW, Zelbst PJ, Meacham J, Bhadauria VS (2012) Green supply chain management practices: impact on performance. Supply Chain Manag 17:290–305. https://doi.org/10.1108/13598541211227126
Gualandris J, Kalchschmidt M (2015) Supply risk management and competitive advantage: a misfit model. Int J Log Manag 26:459–478
Gualandris J, Klassen RD, Vachon S, Kalchschmidt M (2015) Sustainable evaluation and verification in supply chains: aligning and leveraging accountability to stakeholders. J Oper Manag 38:1–13. https://doi.org/10.1016/j.jom.2015.06.002
Gunasekaran A, Patel C, Tirtiroglu E (2001) Performance measures and metrics in a supply chain environment. Int J Oper Prod Manag 21:71–87. https://doi.org/10.1108/01443570110358468
Gunasekaran A, Patel C, McGaughey RE (2004) A framework for supply chain performance measurement. Int J Prod Econ 87:333–347. https://doi.org/10.1016/j.ijpe.2003.08.003
Han Y, Chong WK, Li D (2020) A systematic literature review of the capabilities and performance metrics of supply chain resilience. Int J Prod Res 58:1–26. https://doi.org/10.1080/00207543.2020.1785034
Hannibal C, Kauppi K (2019) Third party social sustainability assessment: is it a multi-tier supply chain solution? Int J Prod Econ 217:78–87. https://doi.org/10.1016/j.ijpe.2018.08.030
Hervani AA, Nandi S, Helms MM, Sarkis J (2022) A performance measurement framework for socially sustainable and resilient supply chains using environmental goods valuation methods. Sustain Prod Consum 30:31–52. https://doi.org/10.1016/j.spc.2021.11.026
International Energy Agency (2021) Tracking industry 2021. IEA. https://www.iea.org/reports/tracking-industry-2021
Islam S, Karia N, Fauzi FBA, Soliman MSM (2017) A review on green supply chain aspects and practices. Manag Mark 12:12–36. https://doi.org/10.1515/mmcks-2017-0002

Ivanov D (2018) Revealing interfaces of supply chain resilience and sustainability: a simulation study. Int J Prod Res 56:3507–3523. https://doi.org/10.1080/00207543.2017.1343507

Ivanov D (2020) Viable supply chain model: integrating agility, resilience and sustainability perspectives—lessons from and thinking beyond the COVID-19 pandemic. Ann Oper Res 319:1411. https://doi.org/10.1007/s10479-020-03640-6

Ivanov D (2021) Lean resilience: AURA (Active Usage of Resilience Assets) framework for post-COVID-19 supply chain management. Int J Logist Manag 33:1196. https://doi.org/10.1108/IJLM-11-2020-0448

Johnson MP (2015) Sustainability management and small and medium-sized enterprises: managers' awareness and implementation of innovative tools. Corp Soc Responsib Environ Manag 22:271–285. https://doi.org/10.1002/csr.1343

Kalubanga M, Gudergan S (2022) The impact of dynamic capabilities in disrupted supply chains—the role of turbulence and dependence. Ind Mark Manag 103:154–169. https://doi.org/10.1016/j.indmarman.2022.03.005

Kamalahmadi M, Parast MM (2016) A review of the literature on the principles of enterprise and supply chain resilience: major findings and directions for future research. Int J Prod Econ 171:116–133. https://doi.org/10.1016/j.ijpe.2015.10.023

Kaplan RS, Norton DP (2005) The balanced scorecard: measures that drive performance. Harv Bus Rev 83(7):172

Karl AA, Micheluzzi J, Leite LR, Pereira CR (2018) Supply chain resilience and key performance indicators: a systematic literature review. Production 28. https://doi.org/10.1590/0103-6513.20180020

Kaur H, Singh SP, Garza-Reyes JA, Mishra N (2020) Sustainable stochastic production and procurement problem for resilient supply chain. Comput Ind Eng 139:105560. https://doi.org/10.1016/j.cie.2018.12.007

Kaviani MA, Tavana M, Kowsari F, Rezapour R (2020) Supply chain resilience: a benchmarking model for vulnerability and capability assessment in the automotive industry. Benchmarking 27:1929–1949. https://doi.org/10.1108/BIJ-01-2020-0049

Kazemi Matin R, Azadi M, Saen RF (2021) Measuring the sustainability and resilience of blood supply chains. Decis Support Syst 161:113629. https://doi.org/10.1016/j.dss.2021.113629

Kohlbacher F (2006) The use of qualitative content analysis in case study research. Forum Qual Soc Res 7:1–30

Lam JSL, Bai X (2016) A quality function deployment approach to improve maritime supply chain resilience. Transport Res E Log Transp Rev 92:16–27. https://doi.org/10.1016/j.tre.2016.01.012

Lee SY, Hu J, Lim MK (2021) Maximising the circular economy and sustainability outcomes: an end-of-life tyre recycling outlets selection model. Int J Prod Econ 232:107965. https://doi.org/10.1016/j.ijpe.2020.107965

Li Y, Chen K, Collignon S, Ivanov D (2021) Ripple effect in the supply chain network: forward and backward disruption propagation, network health and firm vulnerability. Eur J Oper Res 291:1117–1131. https://doi.org/10.1016/j.ejor.2020.09.053

Løkke A, Sørensen PD (2014) Theory testing using case studies interdisciplinary Centre for Organizational Architecture (ICOA), School of Business and Social. Electron J Bus Res Methods 12:66–74

Malesios C, Dey PK, Abdelaziz FB (2020) Supply chain sustainability performance measurement of small and medium sized enterprises using structural equation modeling. Ann Oper Res 294:623–653. https://doi.org/10.1007/s10479-018-3080-z

Marshall D, McCarthy L, Heavey C, McGrath P (2015) Environmental and social supply chain management sustainability practices: construct development and measurement. Prod Plan Control 26:673–690. https://doi.org/10.1080/09537287.2014.963726

Masi D, Kumar V, Garza-Reyes JA, Godsell J (2018) Towards a more circular economy: exploring the awareness, practices, and barriers from a focal firm perspective. Prod Plan Control 29:539–550. https://doi.org/10.1080/09537287.2018.1449246

Mason-Jones R, Towill DR (1999) Total cycle time compression and the agile supply chain. Int J Prod Econ 62:61–73. https://doi.org/10.1016/S0925-5273(98)00221-7
McIntosh MJ, Morse JM (2015) Situating and constructing diversity in semi-structured interviews. Glob Qual Nurs Res 2. https://doi.org/10.1177/2333393615597674
Medini K, Da Cunha C, Bernard A (2015) Tailoring performance evaluation to specific industrial contexts – application to sustainable mass customisation enterprises. Int J Prod Res 53:2439–2456. https://doi.org/10.1080/00207543.2014.974844
Mengistu AT, Panizzolo R (2022) Analysis of indicators used for measuring industrial sustainability: a systematic review. Environ Dev Sustain 25:1979. https://doi.org/10.1007/s10668-021-02053-0
Mies A, Gold S (2021) Mapping the social dimension of the circular economy. J Clean Prod 321: 128960. https://doi.org/10.1016/j.jclepro.2021.128960
Miles MB, Huberman AM, Saldaña J (2014) Qualitative data analysis – a methods sourcebook, 3rd edn. Sage, London
Miroshnychenko I, Barontini R, Testa F (2017) Green practices and financial performance: a global outlook. J Clean Prod 147:340–351. https://doi.org/10.1016/j.jclepro.2017.01.058
Mishra D, Gunasekaran A, Papadopoulos T, Dubey R (2018) Supply chain performance measures and metrics: a bibliometric study. Benchmarking: An Int J, 25(3):932–967. https://doi.org/10.1108/BIJ-08-2017-0224
Mohammed A (2020) Towards 'gresilient' supply chain management: a quantitative study. Resour Conserv Recycl 155:104641. https://doi.org/10.1016/j.resconrec.2019.104641
Mohammed A, Harris I, Soroka A et al (2020) Gresilient supplier assessment and order allocation planning. Ann Oper Res 296:335. https://doi.org/10.1007/s10479-020-03611-x
Munoz A, Dunbar M (2015) On the quantification of operational supply chain resilience. Int J Prod Res 53:6736–6751. https://doi.org/10.1080/00207543.2015.1057296
Mura M, Longo M, Zanni S (2020) Circular economy in Italian SMEs: a multi-method study. J Clean Prod 245:118821. https://doi.org/10.1016/j.jclepro.2019.118821
Narimissa O, Kangarani-Farahani A, Molla-Alizadeh-Zavardehi S (2020) Evaluation of sustainable supply chain management performance: indicators. Sustain Dev 28:118–131. https://doi.org/10.1002/sd.1976
Negri M, Cagno E, Colicchia C, Sarkis J (2021a) Integrating sustainability and resilience in the supply chain: a systematic literature review and a research agenda. Bus Strateg Environ 30:1–29. https://doi.org/10.1002/bse.2776
Negri M, Neri A, Cagno E, Monfardini G (2021b) Circular economy performance measurement in manufacturing firms: a systematic literature review with insights for small and medium enterprises and new adopters. Sustainability (Switzerland) 13. https://doi.org/10.3390/su13169049
Negri M, Cagno E, Colicchia C (2022) Building sustainable and resilient supply chains: a framework and empirical evidence on trade-offs and synergies in implementation of practices. Prod Plan Control 0:1–24. https://doi.org/10.1080/09537287.2022.2053758
Neri A (2021) Industrial sustainability performance measurement system—challenges for the development. In: Methods in sustainability science. Elsevier. pp. 87–104
Neri A, Cagno E, Di Sebastiano G, Trianni A (2018) Industrial sustainability: modelling drivers and mechanisms with barriers. J Clean Prod 194:452–472. https://doi.org/10.1016/j.jclepro.2018.05.140
Neri A, Cagno E, Lepri M, Trianni A (2021) A triple bottom line balanced set of key performance indicators to measure the sustainability performance of industrial supply chains. Sustain Prod Consum 26:648–691. https://doi.org/10.1016/j.spc.2020.12.018
Owida A, Galal NM, Elrafie A (2022) Decision-making framework for a resilient sustainable production system during COVID-19: an evidence-based research. Comput Ind Eng 164: 107905. https://doi.org/10.1016/j.cie.2021.107905
Perrings C (2006) Resilience and sustainable development. Environ Dev Econ 11:417–427. https://doi.org/10.1017/S1355770X06003020

Pettit TJ, Croxton KL, Fiksel J (2013) Ensuring supply chain resilience: development and implementation of an assessment tool. J Bus Logist 34:46–76. https://doi.org/10.1111/jbl.12009
Rajesh R (2016) Forecasting supply chain resilience performance using grey prediction. Electron Commer Res Appl 20:42–58. https://doi.org/10.1016/j.elerap.2016.09.006
Rajesh R (2018) On sustainability, resilience, and the sustainable–resilient supply networks. Sustain Prod Consum 15:74–88. https://doi.org/10.1016/j.spc.2018.05.005
Ramezankhani MJ, Torabi SA, Vahidi F (2018) Supply chain performance measurement and evaluation: a mixed sustainability and resilience approach. Comput Ind Eng 126:531–548. https://doi.org/10.1016/j.cie.2018.09.054
Rossi E, Bertassini AC, Ferreira C Dos S et al (2020) Circular economy indicators for organizations considering sustainability and business models: plastic, textile and electro-electronic cases. J Clean Prod 247:119137. https://doi.org/10.1016/j.jclepro.2019.119137
Rowley J (2012) Conducting research interviews. Manag Res Rev 35:260–271. https://doi.org/10.1108/01409171211210154
Ruiz-Benitez R, López C, Real JC (2019) Achieving sustainability through the lean and resilient management of the supply chain. Int J Phys Distrib Logist Manag 49:122–155. https://doi.org/10.1108/IJPDLM-10-2017-0320
Saeed MA, Kersten W (2020) Sustainability performance assessment framework: a cross-industry multiple case study. Int J Sust Dev World 27:496–514. https://doi.org/10.1080/13504509.2020.1764407
Sahu AK, Datta S, Mahapatra SS (2017) Evaluation of performance index in resilient supply chain: a fuzzy-based approach. Benchmarking 24:118–142. https://doi.org/10.1108/BIJ-07-2015-0068
Saidani M, Yannou B, Leroy Y et al (2019) A taxonomy of circular economy indicators. J Clean Prod 207:542–559. https://doi.org/10.1016/j.jclepro.2018.10.014
Sarkis J, Cohen MJ, Dewick P, Schröder P (2020) A brave new world: lessons from the COVID-19 pandemic for transitioning to sustainable supply and production. Resour Conserv Recycl 159: 104894. https://doi.org/10.1016/j.resconrec.2020.104894
Sen DK, Datta S, Mahapatra SS (2018) On evaluation of supply chain's ecosilient (g-resilient) performance index: a fuzzy embedded decision support framework. Benchmarking 25:2370–2389. https://doi.org/10.1108/BIJ-03-2017-0038
Silva GM, Gomes PJ, Sarkis J (2019) The role of innovation in the implementation of green supply chain management practices. Bus Strateg Environ 28:819–832. https://doi.org/10.1002/bse.2283
Simão LE, Somensi K, Dávalos RV, Rodriguez CMT (2021) Measuring supply chain performance: the triple E model. Int J Product Perform Manag 71:2951. https://doi.org/10.1108/IJPPM-06-2020-0291
Stocker A, Gerold S, Seri FH et al (2015) The interaction of resource and labour. In: A scoping study. Sustainable Europe Research Institute (SERI) bio by Deloitte on behalf of European Commission, DG Environment. https://www.dosm.gov.my/v1/uploads/files/7_Publication/Technical_Paper/Occasional_Paper/Social_Demography/20-The-interaction-of-resource-and-labour-productivity.pdf
Subramanian N, Gunasekaran A (2015) Cleaner supply-chain management practices for twenty-first-century organizational competitiveness: practice-performance framework and research propositions. Int J Prod Econ 164:216–233. https://doi.org/10.1016/j.ijpe.2014.12.002
Suryawanshi P, Dutta P, Varun L et al (2021) Sustainable and resilience planning for the supply chain of online hyperlocal grocery services. Sustain Prod Consum 28:496–518. https://doi.org/10.1016/j.spc.2021.05.001
Swanson D, Goel L, Francisco K, Stock J (2018) An analysis of supply chain management research by topic. Supply Chain Manag 23:100–116. https://doi.org/10.1108/SCM-05-2017-0166
Tangen S (2005) Analysing the requirements of performance measurement systems. Meas Bus Excell 9:46–54. https://doi.org/10.1108/13683040510634835
Teece DJ (2014) Dynamic and ordinary capabilities in an (economic) theory of firms. Acad Manag Perspect 28:328–352

Tersine RJ, Hummingbird EA (1995) Lead-time reduction: the search for competitive advantage. Int J Oper Prod Manag 15:8–18. https://doi.org/10.1108/01443579510080382

Trianni A, Cagno E, Neri A (2017) Modelling barriers to the adoption of industrial sustainability measures. J Clean Prod 168:1482–1504. https://doi.org/10.1016/j.jclepro.2017.07.244

Um J, Han N (2020) Understanding the relationships between global supply chain risk and supply chain resilience: the role of mitigating strategies. Supply Chain Manag 26:240–255. https://doi.org/10.1108/SCM-06-2020-0248

Voss C, Tsikriktsis N, Frohlich M (2002) Case research in operations management. Int J Oper Prod Manag 22:195–219. https://doi.org/10.1108/01443570210414329

Wang J, Muddada RR, Wang H et al (2016) Toward a resilient holistic supply chain network system: concept, review and future direction. IEEE Syst J 10:410–421. https://doi.org/10.1109/JSYST.2014.2363161

Wicher P, Lenort R, Čech M (2015) Measurement of metallurgical supply chain resilience – case study. In: METAL 2015 – 24th international conference on metallurgy and materials, conference proceedings, pp 2125–2130

Witjes S, Vermeulen WJV, Cramer JM (2017) Exploring corporate sustainability integration into business activities. Experiences from 18 small and medium sized enterprises in The Netherlands. J Clean Prod 153:528–538. https://doi.org/10.1016/j.jclepro.2016.02.027

Woodside AG, Wilson EJ (2003) Case study research methods for theory building. J Bus Ind Mark 18:493–508. https://doi.org/10.1108/08858620310492374

Xu J, Zhuang J, Liu Z (2016) Modeling and mitigating the effects of supply chain disruption in a defender–attacker game. Ann Oper Res 236:255–270. https://doi.org/10.1007/s10479-015-1810-z

Xu M, Cui Y, Hu M et al (2019) Supply chain sustainability risk and assessment. J Clean Prod 225: 857–867. https://doi.org/10.1016/j.jclepro.2019.03.307

Yin RK (2003) Study design and methods, 3rd edn. Sage, Thousand Oak

Zainal Z (2007) Case study as a research method. Jurnal Kemanusiaan 5. https://doi.org/10.4135/9781473915480.n2

Zeng B, Yen BPC (2017) Rethinking the role of partnerships in global supply chains: a risk-based perspective. Int J Prod Econ 185:52–62. https://doi.org/10.1016/j.ijpe.2016.12.004

Chapter 4
An Assessment of Decision-Making in Resilient and Sustainable Projects Between Literature and Practice

Shahriar Tanvir Alam, Md Muzahid Khan, and Moddassir Khan Nayeem

Abstract Implementation of Resilient and Sustainable Green Supply Chain Management (RSGSCM) has been a quite challenging task in business as the global supply chain is facing an increased number of disruptions and multiple risks. Moreover, due to global warming, governments around the world are concerned with minimizing the carbon footprint and waste generation which eventually forces the firms to restructure their supply chain network from a social and ecological perspective. Over and above, the lack of proper assessment methodologies makes the implementation of RSGSCM even more challenging to confront the sudden disruption. Therefore, the significance of designing a supply chain network resilient to disruption and sustainable at the same time is of paramount importance to supply chain practitioners. Due to the paramount importance of making the supply chain (SC) resilient to disruptions and sustainable, this chapter identifies and assesses the risk factors and sub-factors. Also, risk prioritization is discussed to overcome the risk factors that will eventually help make the supply chain of any business resilient and increase the sustainability performance in disruptions. An integrated multi-criteria decision analysis approach has been used by combining the decision-making trial and evaluation laboratory (DEMATEL) method with intuitionistic fuzzy sets (IFS) to explore the key challenges of the RSSCM.

Keywords Resilient and sustainable supply chain · Multi-criteria decision-making · DEMATEL · Intuitionistic fuzzy sets

4.1 Introduction

A supply chain (SC) includes all exercises that transform raw materials into the final product and distribute them; the five SC stages involved are manufacturers, suppliers, retailers, warehouses, and customers (Zhang et al. 2021). Supply chain

S. T. Alam (✉) · M. M. Khan · M. K. Nayeem
Department of Industrial and Production Engineering, Military Institute of Science and Technology, Dhaka, Bangladesh
e-mail: tanvir.shahriar.tro@gmail.com; muzahidkhan.kuet@gmail.com; mknayeem.ie@gmail.com

© The Author(s), under exclusive license to Springer Nature Switzerland AG 2023
N. U. Ibne Hossain (ed.), *Data Analytics for Supply Chain Networks*, Greening of Industry Networks Studies 11, https://doi.org/10.1007/978-3-031-29823-3_4

management (SCM) professionals characterize SCM as arranging and dealing with all the exercises, including procurement, logistic activities, and sourcing (Kapoor 2018). Roughly, one in four international development projects (IDPs) financed by the World Bank (WB) doesn't make even moderate progress due to numerous potential risks (Independent Evaluation Group 2015; Rodríguez-Rivero et al. 2020). An investment of $143.22 billion in 2018 by the Organization for Economic Cooperation and Development (OECD) can reflect more about the significance of handling potential risk in a certain project (OECD 2020) as projects organized by the WB which can achieve satisfactory progress are only 3% (approximately 21 of 713 WB projects) (Rodríguez-Rivero et al. 2020).

In the most recent decades, natural issues have been expanding, which is a serious cause of environmental change and global warming. Green supply chain (GSC) alludes to incorporating sustainable natural processes into the conventional SC. This can incorporate cycles, for example, provider determination and buying material, product manufacturing, dissemination, and end-of-life executives (Abdul Rehman Khan 2019). GSC makes use of the key sustainable improvement strategy. Green supply chain management (GSCM) incorporates the following five aspects: "interior environmental administration," "green purchasing," "collaboration with clients," "investment recovery," and "eco-design" (Bhatia and Gangwani 2021; Zhu and Sarkis 2004).

Identifying the risk of any project comes in the planning phase that links to the overall project activities, and mitigating that risk is placed in the monitoring phase (Garwood and Poole 2018). Risk management can be defined as an uncertain group of events that will affect the accomplishment of the project goals (Settembre-Blundo et al. 2021; Stare 2014). Besides risk management being one of the most critical internal processes in an organization, it is considered as part of any organizational decision-making process (De Salvo et al. 2019). Consequently, it increases the resilience even during the prevention phase, and it is essential for ensuring process protection and security (Buganová and Šimíčková 2019). Moreover, the risk is divided into two main categories of positive and negative risk, in which positive risk consists of opportunities and negative risk means threat (Vasilieva 2021; Stare 2014). The projects are exposed to a variety of risks as a result of the internal and external environments. Each project comes with its own set of threats, which are present at all stages of the life cycle. Some risks are indirect, while others cause a direct threat to people and the environment (Buganová and Šimíčková 2019).

4.2 Literature Review

Risk management (RM) focuses on identifying the risk and mitigating them (Filippetto et al. 2021). The approach of risk management is about identifying, analyzing, evaluating, and then mitigating the risk (Oduoza 2020). According to Filippetto et al. (2021), RM is a standardized added percentage in the final cost of the product. Roy and Roy (2020) pointed out that to avoid losses, companies have to transfer the risk to insurance companies. Risk management follows an effective strategy to find out a project's SWOT by planning for unnatural scenarios (Komendantova et al. 2012). Moreover, to ensure the success of project, it's necessary to determine how to deal with any potential risk. Nevertheless, they classify the risk based on specific features by those methods without including their possible interactions (Marcelino-Sádaba et al. 2015).

Risk will cause a delay in investment decisions, which means that it will affect the project and cause severe failure (Kul et al. 2020). Risk can be classified as political, financial, social, and technical. Besides that, six key risks are impeding solar energy projects which can be divided into internal and external risks. Internal risks include technological risk, operational risk, and financial risk, while external risks incorporate political risk and regulatory risk, along with weather-related risk (Roy and Roy 2020). However, there are several key factors considered as risk mitigations: first is the improvement of energy efficiency, followed by investment in knowledge, depending on the technologies. In addition to that, it is necessary to specify the internal risk management function and collaboration strategies among stakeholders in the industry and ensure monitoring and maintenance (Kruger et al. 2019).

A study considering only environmental sustainability has been evaluated by applying the "interpretive ranking process (IRP)" modeling approach to illustrate interactions among the identified risks (Luthra et al. 2015). An assessment of challenges that affects green supply chain management (GSCM) has been studied based on the "structural equation modeling" approach (Jum'a et al. 2022). The result shows that environmental, supplier, cost, and customer factors mostly affect the overall GSCM. Furthermore, a systemic review has been conducted for a sustainable manufacturing industry focusing on GSCM. This study assimilates the similar arrangements of the USA, China, Japan, and the EU which uncover that the administration system of "multi-departmental cooperation" prompts the evasion and move of liability (Sheng et al. 2022). Moreover, research addressing "bi-level programming" that focused on resilience factors has been carried out for optimizing a sustainable supply chain.

Multiple-criteria decision-making (MCDM) has been generally adopted for settling on choices in complicated circumstances in any SCM. Among the various MCDM approaches, Aigbogun et al. (2014) proposed a model to increase the SC resilience of the pharmaceutical industry. In addition, Aigbogun et al. (2014) also proposed a fuzzy AHP model to identify the PSC's challenges. Meanwhile, Jaberidoost et al. (2015) developed AHP and simple additive weighting (SAW) methods to study PSC challenges' assessment. Later, Abbas (2018) proposed an interpretive structural modeling (ISM) to examine the challenges in reverse logistics practices in PSC. An essential technique that can compute the causal relations among different complicated challenges is expressed as the decision-making trial and evaluation laboratory (DEMATEL) method (Zhang and Deng 2019). A framework was developed for evaluating food SC risks in Bangladesh food preparing organizations utilizing the gray-DEMATEL method (Ali et al. 2019). This study combines the theory of gray system (TGS) with the DEMATEL method to evaluate the challenges in PSC. TGS deals with the uncertainty of individual choices with an absence of information (Memon et al. 2015; Scarlat and Delcea 2011). The advantage of TGS is its ability to produce consistent results with the assistance of data with a small set (Camelia 2015; Gong and Forrest 2014). DEMATEL recognizes the relations of cause and effect between challenges; however, it is unable to defeat the challenge's vagueness (Zhou et al. 2011; Shieh et al. 2010). The integrated TGS and DEMATEL, i.e., the gray-DEMATEL method, will eliminate the vagueness and uncertainty issues in the individual's choice and accordingly will work on the exactness of the outcome.

4.3 Methodology

A list of risks of RSGSCM is recognized based on previous literature and a questionnaire survey. Twenty risks are listed from the previous literature. To assess the RSGSCM risks, a total of 11 experts, including 2 academics, 3 advisors, and 6 practitioners (consisting of 2 buyers, 2 suppliers, and 2 distributors), are invited, as shown in Table 4.1. The academics are mainly researchers in supply chain management. The advisors and the practitioners are actively involved in the supply chain sectors of various organizations. Five risks are added and ten risks are eliminated by the experts. Table 4.2 lists these 15 risks and shows whether risks are gathered from literature or experts' suppositions. The listed 15 RSGSCM risks are provided to the experts for affirmation and approval. A second questionnaire survey is conducted to analyze the impacts of the identified risks. Figure 4.1 demonstrates the suggested methodology for this research.

Table 4.1 List of experts participating in this study

Expert no.	Role in PSC	Position	Type of organization	Years of experience
1	Advisor	Chairman	Public medical	30+
2	Advisor	Chief executive officer	Public medical	07
3	Advisor	Executive officer	Public medical	04
4	Academic	Assistant Professor	Business School	12
5	Academic	Professor	Public university	19
6	Practitioner	Chairman	Pharmaceutical company	25
7	Practitioner	General secretary	Pharmaceutical company	20
8	Practitioner	Managing director	Healthcare company	13
9	Practitioner	Executive, supply chain	Raw material supplier	10
10	Practitioner	Executive, production planning	Healthcare company	10
11	Practitioner	Senior merchandiser	Pharmaceutical company	07

Table 4.2 List of risks for RSGSCM

Main categories	Subcategories	Code	RSGSCM risks	References
Resilient	Reliable aspect	R_i1	Lack of control strategy	Gölgeci and Kuivalainen (2020)
		R_i2	Infrastructure-related risk	Expert opinion
		R_i3	Budget-related risk	Expert opinion
	Restorative aspect	R_i4	Backup resource risk	Hossain et al. (2019)
		R_i5	Limited green technology adaption	Karuppiah et al. (2021)
		R_i6	Resource restoration-related risk	Expert opinion
Sustainable	Economic aspect	R_i7	Capital investment risks	Negash et al. (2021)
		R_i8	Supply disruption risk	Kim and Chai (2017)
		R_i9	Production loss	Karuppiah et al. (2021)
	Environmental aspect	R_i10	Material consumption risk	Nicod et al. (2020)
		R_i11	Extensive greenhouse gas emission	Expert opinion
		R_i12	Waste management risk	Expert opinion
	Social aspect	R_i13	Negligence toward the safety of workers	Obiso et al. (2019)
		R_i14	Potential supplier failures	Kumar et al. (2015)
		R_i15	Lack of training and education of workers	Expert opinion

4.3.1 Theory of Intuitionistic Fuzzy Set (IFS)

The theory of fuzzy set (FS) was developed for managing vagueness along with uncertainty in evaluating data (Zadeh 1996), whereas IFS is an extension of the FS. In IFS, the data are presented through both membership function (MF) and nonmembership function (NMF), and it can also handle the vagueness of data (Gan and Luo 2017). The significant contrast between FS and IFS is that IFS can deal with the expert's vagueness (Govindan et al. 2015). Whenever decision-makers and specialists are not certainly confident about their perspectives, the IFS performs better compared to FS as IFS has the capacity to display uncertain and unknown data. The following definitions are the illustrations of IFS theory:

Definition 4.1 *A* which is a finite set may be presented as:

$$F = \{x, \mu_B(x), v_B(x) : x \in A\}, \text{ where } F \text{ is a Fuzzy Set (FS)}. \tag{4.1}$$

In Eq. 4.1, $v_B(x)$: $A \rightarrow [0,1]$ is NMF, and MF is $\mu_B(x)$: $A \rightarrow [0,1]$ which may be presented, considering $0 \leq \mu_B(x) + v_B(x) \leq 1$. Moreover, $B \subseteq A$ and $A = \{x, \mu_B(x): x \in A, \mu_B(x) \in [0,1]\}$, when $\mu_B(x)$ presents the degree of lack of information and v_B

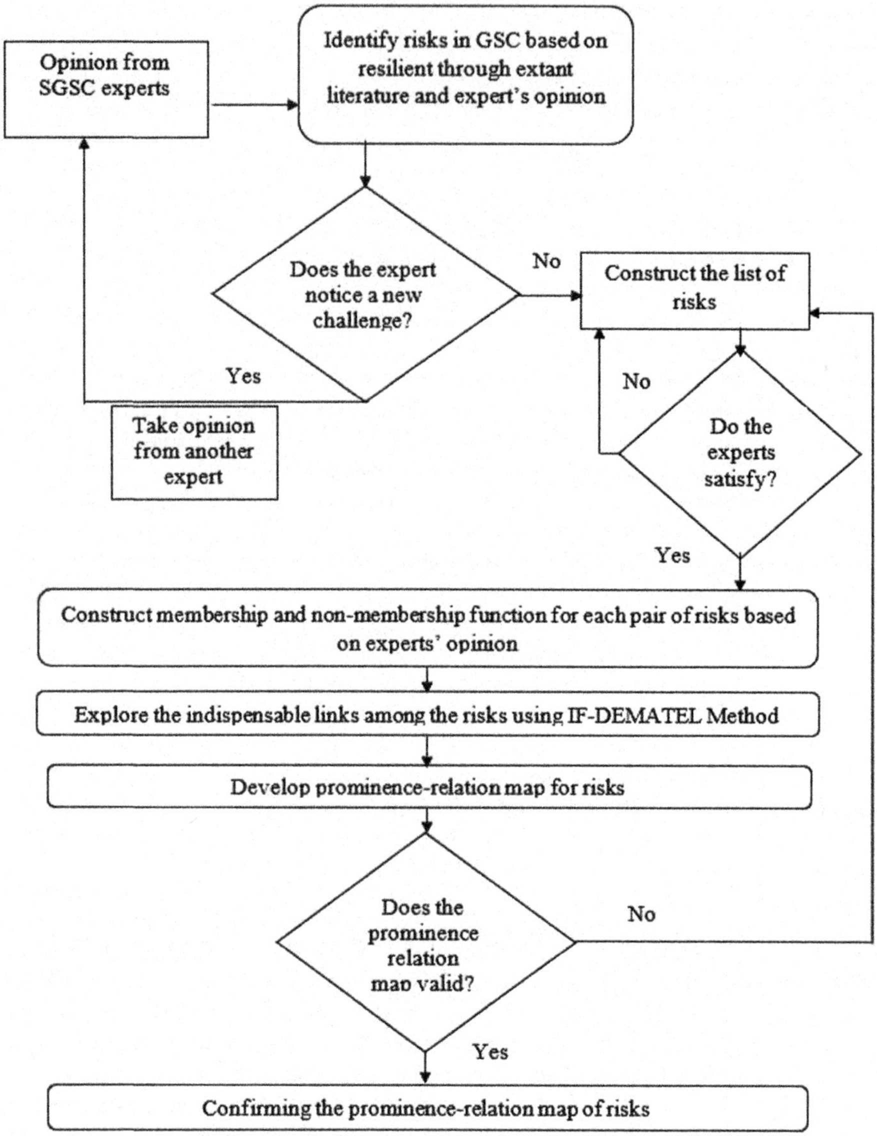

Fig. 4.1 The flow for evaluating RSGSCM risks in this study

(x) signifies the degree of element $x \in A$ to that fixed B (Atanassov 1999; Alam et al. 2021; Ocampo et al. 2020). The relation of μ_B (x) and ν_B (x) with π_B (x) can be presented by Eq. 4.2:

$$\pi_B(x) = 1 - \nu_B(x) - \mu_B(x), \text{ where } 0 \leq \mu_B(x) \leq 1 \tag{4.2}$$

Definition 4.2 A triangular fuzzy number (TFN) is expressed as a triplet $B = (l, m, u)$, $v_B(x)$ and $\mu_B(x)$ is presented as follows, where m, l, and u represent the most promising, the lowest, and the highest available data, respectively to explain a fuzzy event. (Balli and Korukoğlu 2009):

$$\mu B(x) = \begin{cases} 0 & x < 1 \\ \dfrac{(x-l)}{(m-l)} & l \leq x \leq m \\ \dfrac{(u-x)}{(u-m)} & m \leq x \leq u \\ 0 & x > u \end{cases} \tag{4.3}$$

Definition 4.3 The IFS can be presented as K (fixed universe) $\rightarrow$ [0,1] × [0,1], which is illustrated as $v_B(x)$ and $\mu_B(x)$, where $x \in K$ and MF and NMF are of x to B $\forall$ $\mu_B(x) + v_B(x) \leq 1$. The arrangement of [0,1] × [0,1] $\rightarrow M$ is presented by crispification operation (CO) when $K = M$. The CO can be explained as C_λ: [0,1] × [0,1] $\rightarrow M$ along with B is the IFS. The FS has assessed that set, utilizing a defuzzification technique, where FS is changed from B (Alam et al. 2021; Anzilli and Facchinetti 2016; Angelov 1995).

Definition 4.4 From Definition 4.3, the conversion of B into a standard FS can be expressed as:

$$C_\lambda(B) = \{x, \mu_B(x) + \lambda\pi_B(x), v_B(x) + (1-\lambda)\pi_B(x) \forall x \text{ in } A\} \tag{4.4}$$

where $\mu_\lambda(x)$ remains in $C_\lambda(B)$, and it is the addition of $\mu_B(x)$ along with $\lambda\mu_B(x)$, and λ may be assumed as a value from 0 to 1 (Ocampo et al. 2020). Assuming $\lambda = ½$, then the problem may be represented by $\min_{\lambda \in [0,1]} t(C_\lambda(B), B)$. Moreover, t is Euclidean intervals, where the set is illustrated through Eq. 4.5:

$$\mu_\lambda(x) = (\mu_B(x) + 1 - v_B(x))\, 0.5 \tag{4.5}$$

4.3.2 DEMATEL Method

The DEMATEL method can be noted as a diagram hypothetical method which presents causal connections between the number of listed risks (Alam et al. 2021). The basic phases of this method is explained below (Alam et al. 2021; Biswas and Gupta 2019):

Step 1: Development of a direct-relation matrix (DRM), which incorporates a pairwise correlation of causal connections among D number components. This correlation may be formulated from the presentation of a bunch of specialists including N individuals. The DRM $X^P = (x_{ij}^P)_{D \times D}$ for the *P*th specialist where x_{ij} presents the impact of the risks D_i on risks D_j.

The aggregate DRM (ADRM), *X*, $\forall X^P$ where $P = 1, 2, 3, \ldots$ N, $w_P \in R$ can be taken as the meaning of *P*th specialist. *X* is presented by Eq. 4.6:

$$X = (x_{ij})D \times D = \left(\frac{\sum_{P=1}^{N} W_P x_{ij}^P}{\sum_{P=1}^{N} W_P} \right) D \times D \tag{4.6}$$

Step 2: Normalization of ADRM, which can be calculated using Eqs. 4.7–4.8:

$E = f^{-1}X(4.7)$

$$f = \max\left(\max_{1 \le i \le D} \sum_{j=1}^{D} x_{ij}, \max_{1 \le j \le D} \sum_{i=1}^{D} x_{ij} \right) \tag{4.8}$$

Step 3: Evaluate the total relation matrix (TRM) R_T, that is determined from ($R_T = [r_{ij}]_{D \times D}$) by Eq. 4.9. R_T signifies the influential connections among the listed risks:

$$RT = E + E^2 + E^3 + E^4 + E^5 + \ldots = \sum_{i=1}^{\infty} E^i = E(I - E)^{-1} \tag{4.9}$$

where r_{ij} shows the element of R_T in *j*th column and *i*th row and *I* shows a $D \times D$ identity matrix (Alam et al. 2021).

Step 4: Prioritization of listed risks along with the cause. G_i presents the summation of rows when H_j presents a summation of columns. A graph based on cause and effect is accumulated through the arrangement of ($G_i + H_j$, $G_i - H_j$) value:

$$G_i = \left(\sum_{j=1}^{D} r_{ij} \right)_{D \times 1} = (r_i)_{D \times 1} \tag{4.10}$$

$$H_i = \left(\sum_{i=1}^{D} r_{ij} \right)_{1 \times D} = (r_j)_{1 \times D} \tag{4.11}$$

The "prominence vector" ($G_i + H_j$) signifies the relative importance of individual risks. Those risks in the "relation vector" ($G_i - H_j$) remain with net cause if $r_i - r_j > 0$, $i = j$. The risks remain with net effect when $r_i - r_j < 0$, $i = j$ (Alam et al. 2021).

Step 5: Development of the "prominence-relation map" (PRM). Figure 4.2 illustrates the ($G_i + H_j$, $G_i - H_j$) mapping of listed risks.

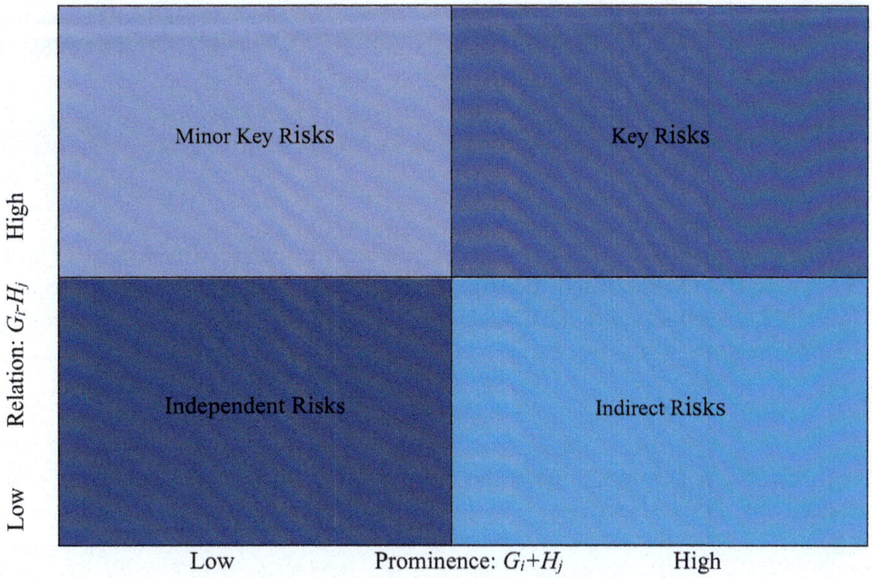

Fig. 4.2 The prominence-relation map (PRM)

4.3.3 Intuitionistic Fuzzy DEMATEL (IF-DEMATEL)

IF-DEMATEL can be explained as:

Phase 1: Investigate the RSGSCM risks.

The RSGSCM risks are investigated through semiorganized meetings/feedback based on the deductive and iterative procedures. The identified risks are listed in Table 4.2.

Phase 2: Develop the DRM.

DRM is formulated by a group of 11 experts. The specialists' group provides the value of x_{ij}. Specialist consultations and collaborations are directed to ensure that the values of these risks in the initial DRM (IDRM) aren't capricious. The specialists are asked to provide the value of membership function (μ_B (x)) along with nonmembership function (v_B (x)) on the causal impact of D_i on D_j. The numerical values of π_B (x) are calculated from Eq. 4.2. Table A.1 (shown in Appendix) presents IDRM in IFS, and all components may be indicated as a 2-tuple. The main idea of the 2-tuple is explained in Definition 4.3.

Phase 3: Calculate the corresponding membership function (CMF).

The CMF of the equivalent FS is calculated in this step. Calculation of the MF utilizes the defuzzification process (DP) of the IFS value (Alam et al. 2021; Ocampo et al. 2020). From Anzilli and Facchinetti (2016), a double-stage DP is applied in this calculation. First, IFS can be transformed into the FS utilizing Eq. 4.5. Table A.2 presents IDRM in FS. For example:

$$\mu(x_{24}) = \mu\ (0.5, 0) = (\mu_B\ (x) + 1 - \nu_B\ (x))\ 0.5 = 0.5\ (0.5 + 1 - 0) = 0.75$$

Phase 4: Evaluate the DP from FS.

Lastly, the defuzzification function k can be considered, which may arrange k: $\mu(x) \rightarrow$ R (L. Ocampo and Yamagishi 2020). The values of $\mu_B\ (x)$, present in Table A.2, are allocated to a TFN (l, m, u) = (0, 4, 4) (Alam et al. 2021). Equation 4.3 may be revised like Eq. 4.12:

$$\overline{x} = (m - 1)\mu(x) + 1 \tag{4.12}$$

In Eq. 12, $\mu\ (\overline{x})$ and $\overline{x}$ mean the membership function and the "crisp" value, respectively. For instance:

$$\overline{x}_{24} = l + \mu(\overline{x}_{24})(m - l) = 0 + 0.75 \times (4 - 0) = 3$$

Crisp values as IDRM are listed in Table A.3.

Phase 5: Develop the normalized DRM (NDRM).

The NDRM are calculated from Eqs. 4.7 to 4.8, where $f = 38.6$. The matrix is presented in Table A.4.

Phase 6: Construct the TRM.

TRM is constructed using Eq. 4.9 which is presented in Table A.5. The ($G_i - H_j$) and ($G_i + H_j$) vectors are listed in Table 4.2 along with Table 4.3 and can be calculated using Eqs. 4.10 and 4.11 (Kumar et al. 2020). The net cause (Y) and the net effect (Z) are categorized in Table 4.3.

Phase 7: Formulate the prominence-relation map (PRM).

Figure 4.4 illustrates the PRM, which is formulated based on ($G_i + H_j$, $G_i - H_j$) coordinates. Figure 4.3 shows all phases of the IF-DEMATEL.

4.4 Results and Analysis

Cause group risks along with effect group risks are categorized in this section. Moreover, risks are ranked in view of the "prominence vector." A PRM of risks to the RSGSCM is developed for uncovering the interactions among the risks (Alam et al. 2021).

Table 4.3 Relation vector ($G_i - H_j$)

Rank	Cause group	$G_i - H_j$	Rank	Effect group	$G_i - H_j$
1	R_i14	0.5242	1	R_i2	−0.0003
2	R_i7	0.3991	2	R_i5	−0.0172
3	R_i1	0.3869	3	R_i6	−0.0236
4	R_i11	0.3180	4	R_i8	−0.0909
5	R_i12	0.2713	5	R_i10	−0.1385
6	R_i13	0.1720	6	R_i4	−0.3079
			7	R_i15	−0.3081
			8	R_i3	−0.4729
			9	R_i9	−0.7119

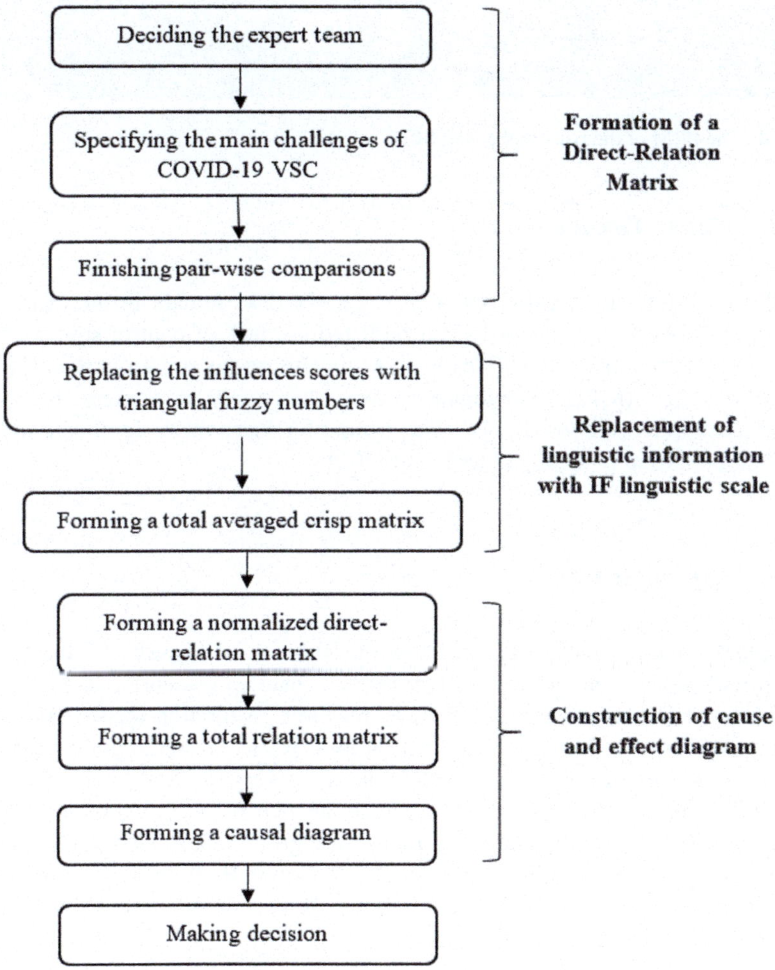

Fig. 4.3 Phases of IF-DEMATEL

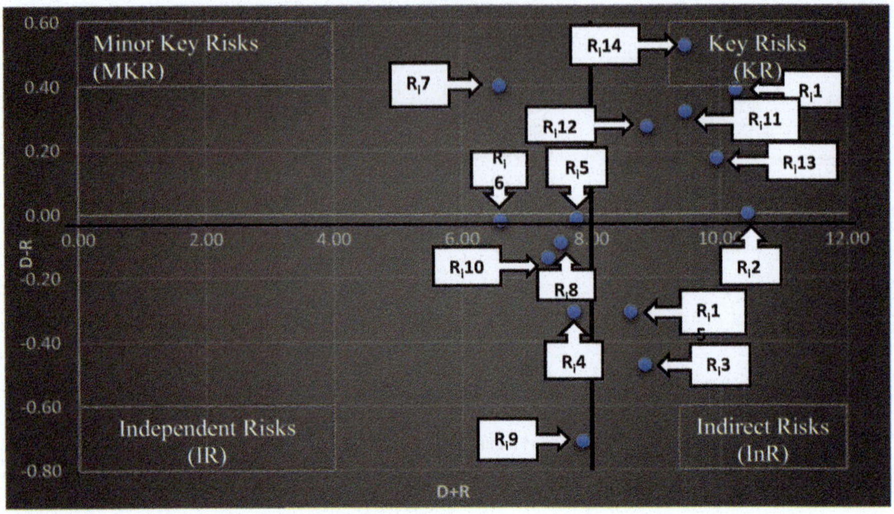

Fig. 4.4 The PRM of the risks to the RSGSCM

4.4.1 Cause Group

Based on the numerical value set ($G_i - H_j$), six risks remain in the cause group (shown in Table 4.3). These risks can be listed as "lack of control strategy (R;1)," "capital investment risks (R;7)," "extensive greenhouse gas emission (R;11)," "waste management risk (R;12)," "negligence toward the safety of workers (R;13)," and "potential supplier failures (R;14)." The group of Y has a more significant influence (G_i) than the group of Z (H_j) (Alam et al. 2021).

4.4.2 Effect Group

Nine risks remain in this group (as shown in Table 4.3). These are "infrastructure-related risk (R;2)," "budget-related risk (R;3)," "backup resource risk (R;4)," "limited green technology adaption (R;5)," "resource restoration-related risk (R;6)," "supply disruption risk (R;8)," "production loss (R;9)," "material consumption risk (R;10)," and "lack of training and education of workers (R;15)."

The Z group is relatively simple to impact as their (G_i - H_j) values are negative. These risks have to be considered after the cause group risk is handled. The ranking of the risks is (based on (G_i - H_j) values) as follows: **R; 14 > R; 7 > R; 1 > R; 11 > R; 12 > R; 13 > R; 2 > R; 5 > R; 6 > R; 8 > R; 10 > R; 4 > R; 15 > R; 3 > R; 9.**

Table 4.4 Prominence vector $(G_i + H_j)$

Rank	Risks	G_i	H_j	$(G_i + H_j)$
1	R_i2	5.2200	5.2203	10.4403
2	R_i1	5.3204	4.9335	10.2540
3	R_i13	5.0667	4.8948	9.9615
4	R_i11	4.8959	4.5779	9.4738
5	R_i14	4.9965	4.4723	9.4687
6	R_i12	4.5725	4.3012	8.8737
7	R_i3	4.1805	4.6534	8.8339
8	R_i15	4.1515	4.4595	8.6110
9	R_i9	3.5777	4.2896	7.8673
10	R_i5	3.8789	3.8961	7.7749
11	R_i4	3.7136	4.0215	7.7351
12	R_i8	3.7225	3.8134	7.5360
13	R_i10	3.6002	3.7387	7.3388
14	R_i6	3.2909	3.3145	6.6053
15	R_i7	3.4979	3.0988	6.5967

4.4.3 Prominence Vector

The $(G_i + H_j)$ values in Table 4.4 illustrate the "relative importance" of listed risks. The higher the value of $(G_i + H_j)$ for a particular factor, the larger the significance of that factor (Bai and Sarkis 2013). As shown in Table 4.4, "infrastructure-related risk (R2)" holds the most elevated $(G_i + H_j)$ value, signifying the most influential risk of the RSGSCM. The ranking of listed risks is (based on $(G_i + H_j)$ values) as follows: **R_i 2 > R_i 1 > R_i 13 > R_i 11 > R_i 14 > R_i 12 > R_i 3 > R_i 15 > R_i 9 > R_i 5 > R_i 4 > R_i 8 > R_i 10 > R_i 6 > R_i 7.**

4.4.4 Correlations Between the Challenges

Significant risks are identified through mapping them in the PRM (as shown in Fig. 4.4). In PRM, all listed risks can be sorted into four clusters: minor key risks (MKR) (HR, LP), key risks (KR) (HR, HP), independent risks (IR) (LR, LP), and indirect risks (InR) (LR, LP). HR signifies high relation and LR represents low relation; LP stands for low prominence and HP for high prominence (Alam et al. 2021). Risks arranged over x-axis remain in the cause group, and the risks which remain under x-axis are in the effect group. Effect group risks are influenced by the cause group risks.

It is evident from Fig. 4.4 the MKR cluster consists of one risk, and it is "capital investment risks (R_i7)." This risk has negligible influence on other risks, and their potential importance is also comparatively low. IR cluster contains six risks, and they are "backup resource risk (R_i4)," "limited green technology adaption (R_i5)," "resource restoration-related risk (R_i6)," "supply disruption risk (R_i8)," "production loss (R_i9)," and "material consumption risk (R_i10)." It means that these risks may not be influenced by other risks.

The InR cluster consists of three risks. These risks are (1) "infrastructure-related risk (R_i2)," (2) "budget-related risk (R_i3)," and (3) "lack of training and education of workers (R_i15)." InR has high influence but low relation. The KR cluster consists of five risks. These risks can be listed as (1) "lack of control strategy (R_i1)," (2) "extensive greenhouse gas emission (R_i11)," (3) "waste management risk (R_i12)," (4) "negligence toward the safety of workers (R_i13)," and (5) "potential supplier failures (R_i14)." Key risks have the most impact on other risks. All the risks of the cause group should be given the most priority.

4.5 Conclusion

The objective of this study is to provide a guideline for finding the main risks to RSGSCM. To make the research more concrete, a survey was conducted, and the investigations about RSGSCM were strategically corrected with the gathered information. This reflects that the most significant risk is "infrastructure-related risk" in RSGSCM. As RSGSCM is a unique and essential supply chain, this risk should be eliminated. The "infrastructure-related risk" is closely related to "lack of control strategy." "Lack of control strategy" along with "capital investment risks" are the main causes of the effect of "infrastructure-related risk." Moreover, "negligence toward the safety of workers" is also an influential risk, being the cause of "lack of training and education of workers" and "supply disruption risk."

This research can be a baseline for determining the risks that affect the RSGSCM in developing countries and underdeveloped countries. For further research in this area, it is highly recommended that a more detailed industrial survey needs to be conducted and a more in-depth investigation should be carried out. The restricted number of specialists utilized in this investigation opens doors for further investigation. Future examinations can incorporate more difficulties from various phases of the RSGSCM. Also, the recognized risks recorded simply mirror specialists' conclusions from developing countries; subsequently, the reasonableness of the outcomes might be restricted to lower economic nations. Consequently, a future study can be created, including specialists from developed nations. At last, applying distinctive MCDM to deal with the issues, difficulties, drivers, and hindrances identified with RSGSCM can be investigated in future examinations.

Appendix A

Table A.1 The IDRM in IFS

Risks	R_i 1	R_i 2	R_i 3	R_i 4	R_i 5	R_i 6	R_i 7	R_i 8	R_i 9	R_i 10	R_i 11	R_i 12	R_i 13	R_i 14	R_i 15
R_i 1	0 0	0.5 0	0.5 0	0.1 0.2	0.1 0.2	0.3 0.2	0.1 0.3	0.2 0.2	0.1 0.3	0.3 0.2	0.4 0.1	0.3 0.2	0.4 0	0.3 0	0.4 0.2
R_i 2	0.7 0.1	0 0	0.5 0	0.5 0	0.4 0.2	0.7 0.2	0.1 0.6	0.3 0.2	0.2 0.5	0.7 0	0.7 0.1	0.6 0.1	0.7 0.1	0.7 0.1	0.3 0.1
R_i 3	0.5 0	0.7 0	0 0	0.1 0	0.1 0	0.1 0.3	0.1 0.3	0.2 0.2	0.2 0	0.1 0.1	0 0.3	0.1 0	0.3 0.1	0.2 0	0.5 0.1
R_i 4	0.3 0.6	0.4 0.5	0.2 0.4	0 0	0.4 0.1	0.3 0	0.2 0	0 0.1	0.2 0.1	0.3 0	0.2 0.1	0.2 0.2	0.5 0	0.1 0.6	0.1 0.6
R_i 5	0.2 0	0.8 0	0.1 0	0.1 0.1	0 0	0.1 0	0.1 0	0 0.1	0.1 0	0.1 0	0.1 0	0.1 0	0.2 0.1	0.1 0.3	0.4 0.2
R_i 6	0.4 0.2	0.1 0.7	0.1 0.2	0.3 0	0 0	0 0	0.2 0	0 0	0.3 0.3	0.1 0.2	0 0.2	0 0.3	0 0	0 0.9	0 0.7
R_i 7	0.3 0	0.1 0.3	0.2 0.2	0.1 0.1	0.1 0	0.6 0.1	0 0	0.2 0	0.2 0.1	0.5 0.1	0.1 0.2	0 0.1	0 0	0.1 0.7	0.3 0.6
R_i 8	0.2 0.3	0.3 0.4	0 0	0.1 0	0 0.1	0 0.1	0.2 0	0 0	0.4 0.1	0 0.05	0.3 0.1	0.1 0.5	0.1 0	0.3 0.5	0.5 0.2
R_i 9	0.3 0.5	0.4 0.4	0.7 0	0.2 0	0 0	0.3 0.5	0.3 0.3	0.3 0.2	0 0	0 0.1	0 0.2	0.1 0	0.1 0.2	0.5 0.4	0.3 0.5
R_i 10	0.2 0.4	0.5 0	0 0	0.4 0	0.1 0	0.2 0.1	0 0.2	0 0.1	0.2 0.1	0 0	0 0.2	0.1 0.2	0.1 0	0.1 0.3	0.1 0.2
R_i 11	0.9 0	0.7 0	0.2 0.1	0.1 0.2	0 0	0.1 0.5	0.1 0	0.1 0.1	0.8 0.1	0.1 0.4	0 0	0.7 0	0 0.3	0.9 0	0.6 0.1
R_i 12	0.6 0.1	0.8 0	0.2 0	0.4 0.2	0.2 0.1	0.1 0.5	0.2 0.4	0 0.1	0.1 0	0 0.3	0.7 0.1	0 0	0.1 0	0.7 0.1	0.6 0
R_i 13	1 0	0.7 0	0.3 0.1	0.3 0	0.1 0.1	0.1 0	0 0.1	0.1 0	0.1 0	0.1 0	0.1 0	0.1 0	0 0	0.9 0	0.4 0.2
R_i 14	1 0	0.9 0	0.4 0.1	0 0.7	0.2 0	0 0.5	0.1 0.5	0.3 0.2	0.7 0	0.1 0.3	0.9 0.1	0.7 0.1	0.6 0.2	0 0	0.4 0.1
R_i 15	0.7 0.3	0.7 0	0.2 0	0.1 0.4	0.1 0.05	0.1 0.7	0.1 0.5	0.5 0.4	0.6 0.2	0.1 0.2	0.2 0	0.1 0	0.4 0.1	0.7 0.2	0 0

Table A.2 The IDRM in standard FS

Risks	R_i 1	R_i 2	R_i 3	R_i 4	R_i 5	R_i 6	R_i 7	R_i 8	R_i 9	R_i 10	R_i 11	R_i 12	R_i 13	R_i 14	R_i 15
R_i 1	0.000	0.750	0.750	0.450	0.450	0.550	0.400	0.500	0.400	0.550	0.650	0.550	0.700	0.650	0.600
R_i 2	0.800	0.000	0.750	**0.750**	0.600	0.750	0.250	0.550	0.350	0.850	0.800	0.750	0.800	0.800	0.600
R_i 3	0.750	0.850	0.000	0.550	0.550	0.400	0.400	0.500	0.600	0.500	0.350	0.550	0.600	0.600	0.700
R_i 4	0.350	0.450	0.400	0.000	0.650	0.650	0.600	0.450	0.550	0.650	0.550	0.500	0.750	0.250	0.250
R_i 5	0.600	0.900	0.550	0.500	0.000	0.550	0.550	0.450	0.550	0.550	0.550	0.550	0.550	0.400	0.600
R_i 6	0.600	0.200	0.450	0.650	0.500	0.000	0.600	0.500	0.500	0.450	0.400	0.350	0.500	0.050	0.150
R_i 7	0.650	0.400	0.500	0.500	0.550	0.750	0.000	0.600	0.550	0.700	0.450	0.450	0.500	0.200	0.350
R_i 8	0.450	0.450	0.500	0.550	0.450	0.450	0.600	0.000	0.650	0.475	0.600	0.300	0.550	0.400	0.650
R_i 9	0.400	0.500	0.850	0.600	0.500	0.400	0.500	0.550	0.000	0.450	0.400	0.550	0.450	0.550	0.400
R_i 10	0.400	0.750	0.500	0.700	0.550	0.550	0.400	0.450	0.550	0.000	0.400	0.450	0.550	0.400	0.450
R_i 11	0.950	0.850	0.550	0.450	0.500	0.300	0.550	0.500	0.850	0.350	0.000	0.850	0.350	0.950	0.750
R_i 12	0.750	0.900	0.600	0.600	0.550	0.300	0.400	0.450	0.550	0.350	0.800	0.000	0.550	0.800	0.800
R_i 13	1.000	0.850	0.600	0.650	0.500	0.550	0.450	0.550	0.550	0.550	0.550	0.550	0.000	0.950	0.600
R_i 14	1.000	0.950	0.650	0.150	0.600	0.250	0.300	0.550	0.850	0.400	0.900	0.800	0.700	0.000	0.650
R_i 15	0.700	0.850	0.600	0.350	0.525	0.200	0.300	0.550	0.700	0.450	0.600	0.550	0.650	0.750	0.000

Table A.3 The IDRM in crisp values

Risks	R_i 1	R_i 2	R_i 3	R_i 4	R_i 5	R_i 6	R_i 7	R_i 8	R_i 9	R_i 10	R_i 11	R_i 12	R_i 13	R_i 14	R_i 15
R_i 1	0	3	3	1.8	1.8	2.2	1.6	2	1.6	2.2	2.6	2.2	2.8	2.6	2.4
R_i 2	3.2	0	3	3	2.4	3	1	2.2	1.4	3.4	3.2	3	3.2	3.2	2.4
R_i 3	3	3.4	0	2.2	2.2	1.6	1.6	2	2.4	2	1.4	2.2	2.4	2.4	2.8
R_i 4	1.4	1.8	1.6	0	2.6	2.6	2.4	1.8	2.2	2.6	2.2	2	3	1	1
R_i 5	2.4	3.6	2.2	2	0	2.2	2.2	1.8	2.2	2.2	2.2	2.2	2.2	1.6	2.4
R_i 6	2.4	0.8	1.8	2.6	2	0	2.4	2	2	1.8	1.6	1.4	2	0.2	0.6
R_i 7	2.6	1.6	2	2	2.2	3	0	2.4	2.2	2.8	1.8	1.8	2	0.8	1.4
R_i 8	1.8	1.8	2	2.2	1.8	1.8	2.4	0	2.6	1.9	2.4	1.2	2.2	1.6	2.6
R_i 9	1.6	2	3.4	2.4	2	1.6	2	2.2	0	1.8	1.6	2.2	1.8	2.2	1.6
R_i 10	1.6	3	2	2.8	2.2	2.2	1.6	1.8	2.2	0	1.6	1.8	2.2	1.6	1.8
R_i 11	3.8	3.4	2.2	1.8	2	1.2	2.2	2	3.4	1.4	0	3.4	1.4	3.8	3
R_i 12	3	3.6	2.4	2.4	2.2	1.2	1.6	1.8	2.2	1.4	3.2	0	2.2	3.2	3.2
R_i 13	4	3.4	2.4	2.6	2	2.2	1.8	2.2	2.2	2.2	2.2	2.2	0	3.8	2.4
R_i 14	4	3.8	2.6	0.6	2.4	1	1.2	2.2	3.4	1.6	3.6	3.2	2.8	0	2.6
R_i 15	2.8	3.4	2.4	1.4	2.1	0.8	1.2	2.2	2.8	1.8	2.4	2.2	2.6	3	0

Table A.4 The NDRM

E	R_i 1	R_i 2	R_i 3	R_i 4	R_i 5	R_i 6	R_i 7	R_i 8	R_i 9	R_i 10	R_i 11	R_i 12	R_i 13	R_i 14	R_i 15
R_i 1	0.0000	0.0777	0.0777	0.0466	0.0466	0.0570	0.0415	0.0518	0.0415	0.0570	0.0674	0.0570	0.0725	0.0674	0.0622
R_i 2	0.0829	0.0000	0.0777	0.0777	0.0622	0.0777	0.0259	0.0570	0.0363	0.0881	0.0829	0.0777	0.0829	0.0829	0.0622
R_i 3	0.0777	0.0881	0.0000	0.0570	0.0570	0.0415	0.0415	0.0518	0.0622	0.0518	0.0363	0.0570	0.0622	0.0622	0.0725
R_i 4	0.0363	0.0466	0.0415	0.0000	0.0674	0.0674	0.0622	0.0466	0.0570	0.0674	0.0570	0.0518	0.0777	0.0259	0.0259
R_i 5	0.0622	0.0933	0.0570	0.0518	0.0000	0.0570	0.0570	0.0466	0.0570	0.0570	0.0570	0.0570	0.0570	0.0415	0.0622
R_i 6	0.0622	0.0207	0.0466	0.0674	0.0518	0.0000	0.0622	0.0518	0.0518	0.0466	0.0415	0.0363	0.0518	0.0052	0.0155
R_i 7	0.0674	0.0415	0.0518	0.0518	0.0570	0.0777	0.0000	0.0622	0.0570	0.0725	0.0466	0.0466	0.0518	0.0207	0.0363
R_i 8	0.0466	0.0466	0.0518	0.0570	0.0466	0.0466	0.0622	0.0000	0.0674	0.0492	0.0622	0.0311	0.0570	0.0415	0.0674
R_i 9	0.0415	0.0518	0.0881	0.0622	0.0518	0.0415	0.0518	0.0570	0.0000	0.0466	0.0415	0.0570	0.0466	0.0570	0.0415
R_i 10	0.0415	0.0777	0.0518	0.0725	0.0570	0.0570	0.0415	0.0466	0.0570	0.0000	0.0415	0.0466	0.0570	0.0415	0.0466
R_i 11	0.0984	0.0881	0.0570	0.0466	0.0518	0.0311	0.0570	0.0518	0.0881	0.0363	0.0000	0.0881	0.0363	0.0984	0.0777
R_i 12	0.0777	0.0933	0.0622	0.0622	0.0570	0.0311	0.0415	0.0466	0.0570	0.0363	0.0829	0.0000	0.0570	0.0829	0.0829
R_i 13	0.1036	0.0881	0.0622	0.0674	0.0518	0.0570	0.0466	0.0570	0.0570	0.0570	0.0570	0.0570	0.0000	0.0984	0.0622
R_i 14	0.1036	0.0984	0.0674	0.0155	0.0622	0.0259	0.0311	0.0570	0.0881	0.0415	0.0933	0.0829	0.0725	0.0000	0.0674
R_i 15	0.0725	0.0881	0.0622	0.0363	0.0544	0.0207	0.0311	0.0570	0.0725	0.0466	0.0622	0.0570	0.0674	0.0777	0.0000

Table A.5 The TRM

Risks	R_i 1	R_i 2	R_i 3	R_i 4	R_i 5	R_i 6	R_i 7	R_i 8	R_i 9	R_i 10	R_i 11	R_i 12	R_i 13	R_i 14	R_i 15
R_i 1	0.2976	0.3779	0.3348	0.2798	0.2803	0.2642	0.2367	0.2753	0.2959	0.2846	0.3180	0.3023	0.3277	0.3168	0.3014
R_i 2	0.4225	0.3563	0.3771	0.3463	0.3331	0.3170	0.2561	0.3165	0.3337	0.3495	0.3738	0.3613	0.3797	0.3713	0.3409
R_i 3	0.3658	0.3842	0.2613	0.2873	0.2878	0.2493	0.2347	0.2735	0.3111	0.2790	0.2887	0.2999	0.3174	0.3095	0.3081
R_i 4	0.2926	0.3086	0.2681	0.2075	0.2690	0.2493	0.2329	0.2418	0.2765	0.2653	0.2741	0.2641	0.2980	0.2438	0.2354
R_i 5	0.3487	0.3842	0.3117	0.2813	0.2313	0.2619	0.2474	0.2665	0.3037	0.2815	0.3035	0.2970	0.3090	0.2874	0.2955
R_i 6	0.2708	0.2400	0.2345	0.2362	0.2206	0.1559	0.2049	0.2136	0.2342	0.2134	0.2225	0.2132	0.2374	0.1870	0.1904
R_i 7	0.3195	0.3039	0.2782	0.2572	0.2600	0.2589	0.1748	0.2564	0.2767	0.2704	0.2651	0.2592	0.2759	0.2385	0.2451
R_i 8	0.3041	0.3113	0.2800	0.2610	0.2519	0.2303	0.2332	0.1992	0.2887	0.2502	0.2810	0.2481	0.2818	0.2608	0.2749
R_i 9	0.3019	0.3193	0.3142	0.2674	0.2587	0.2271	0.2248	0.2544	0.2270	0.2495	0.2650	0.2732	0.2752	0.2763	0.2547
R_i 10	0.3010	0.3407	0.2815	0.2778	0.2632	0.2422	0.2154	0.2448	0.2795	0.2057	0.2649	0.2636	0.2846	0.2619	0.2575
R_i 11	0.4173	0.4184	0.3449	0.3014	0.3085	0.2608	0.2693	0.2981	0.3630	0.2887	0.2826	0.3561	0.3219	0.3707	0.3406
R_i 12	0.3893	0.4124	0.3388	0.3071	0.3047	0.2536	0.2485	0.2851	0.3265	0.2808	0.3498	0.2659	0.3310	0.3487	0.3363
R_i 13	0.4247	0.4203	0.3509	0.3233	0.3111	0.2880	0.2629	0.3050	0.3376	0.3106	0.3382	0.3300	0.2898	0.3713	0.3276
R_i 14	0.4267	0.4326	0.3576	0.2778	0.3196	0.2584	0.2483	0.3051	0.3655	0.2956	0.3709	0.3547	0.3560	0.2866	0.3360
R_i 15	0.3624	0.3855	0.3202	0.2675	0.2847	0.2289	0.2246	0.2778	0.3210	0.2730	0.3121	0.3010	0.3208	0.3258	0.2419

References

Abbas H (2018) Barriers to reverse logistics practices in pharmaceutical supply chains: an ISM approach. Int J Bus Excell 16(1):47–60. https://doi.org/10.1504/IJBEX.2018.094574
Abdul Rehman Khan S (2019) Introductory chapter: Introduction of green supply chain management. In: Abdul Rehman Khan S (ed) Green practices and strategies in supply chain management. IntechOpen, New York. https://doi.org/10.5772/intechopen.81088
Aigbogun O, Ghazali Z, Razali R (2014) A framework to enhance supply chain resilience the case of Malaysian pharmaceutical industry. Glob Bus Manag Res 6(3):219–228
Alam ST, Ahmed S, Ali SM, Sarker S, Kabir G, ul-Islam, A. (2021) Challenges to COVID-19 vaccine supply chain: implications for sustainable development goals. Int J Prod Econ 239 (April):108193. https://doi.org/10.1016/j.ijpe.2021.108193
Ali SM, Moktadir MA, Kabir G, Chakma J, Rumi MJU, Islam MT (2019) Framework for evaluating risks in food supply chain: implications in food wastage reduction. J Clean Prod 228:786–800. https://doi.org/10.1016/j.jclepro.2019.04.322
Angelov P (1995) Crispification: defuzzification over intuitionistic fuzzy sets. Bull Stud Exch Fuzz Appl BUSEFAL 64:51–55
Anzilli L, Facchinetti G (2016) A new proposal of defuzzification of intuitionistic fuzzy quantities. In: Atanassov KT et al (eds) Novel developments in uncertainty representation and processing, Advances in intelligent systems and computing, vol 401. Springer, Cham, pp 185–195. https://doi.org/10.1007/978-3-319-26211-6_16
Atanassov, K. T. (1999). Intuitionistic fuzzy sets. In Atanassov, K. T. Intuitionistic fuzzy sets (pp. 1–137). Springer Heidelberg
Bai C, Sarkis J (2013) A grey-based DEMATEL model for evaluating business process management critical success factors. Int J Prod Econ 146(1):281–292. https://doi.org/10.1016/j.ijpe.2013.07.011
Balli S, Korukoğlu S (2009) Operating system selection using fuzzy AHP and topsis methods. Math Comp Appl 14(2):119–130. https://doi.org/10.3390/mca14020119
Bhatia MS, Gangwani KK (2021) Green supply chain management: Scientometric review and analysis of empirical research. J Clean Prod 284:124722. https://doi.org/10.1016/j.jclepro.2020.124722
Biswas B, Gupta R (2019) Analysis of barriers to implement blockchain in industry and service sectors. Comput Ind Eng 136:225–241. https://doi.org/10.1016/j.cie.2019.07.005
Buganová K, Šimíčková J (2019) Risk management in traditional and agile project management. Transp Res Procedia 40:986–993. https://doi.org/10.1016/j.trpro.2019.07.138
Camelia D (2015) Grey systems theory in economics – bibliometric analysis and applications' overview. Grey Syst Theor Appl 5(2):244–262. https://doi.org/10.1108/gs-03-2015-0005
De Salvo M, Capitello R, Gaudenzi B, Begalli D (2019) Risk management strategies and residual risk perception in the wine industry: A spatial analysis in Northeast Italy. Land Use Policy 83: 47–62. https://doi.org/10.1016/j.landusepol.2019.01.022
Filippetto AS, Lima R, Barbosa JLV (2021) A risk prediction model for software project management based on similarity analysis of context histories. Inf Softw Technol 131:106497. https://doi.org/10.1016/j.infsof.2020.106497
Gan J, Luo L (2017) Using DEMATEL and intuitionistic fuzzy sets to identify critical factors influencing the recycling rate of end-of-life vehicles in China. Sustainability (Switzerland) 9(10):1–22. https://doi.org/10.3390/su9101873
Garwood DA, Poole AH (2018) Project management as information management in interdisciplinary research: "Lots of different pieces working together". Int J Inf Manag 41:14–22. https://doi.org/10.1016/j.ijinfomgt.2018.03.002
Gölgeci I, Kuivalainen O (2020) Does social capital matter for supply chain resilience? The role of absorptive capacity and marketing-supply chain management alignment. Ind Mark Manag 84: 63–74. https://doi.org/10.1016/j.indmarman.2019.05.006

Gong Z, Forrest JYL (2014) Special issue on meteorological disaster risk analysis and assessment: on basis of grey systems theory. Nat Hazards 71(2):995–1000. https://doi.org/10.1007/s11069-013-0864-y

Govindan K, Khodaverdi R, Vafadarnikjoo A (2015) Intuitionistic fuzzy based DEMATEL method for developing green practices and performances in a green supply chain. Expert Syst Appl 42(20):7207–7220. https://doi.org/10.1016/j.eswa.2015.04.030

Hossain NUI, Jaradat R, Hosseini S, Marufuzzaman M, Buchanan RK (2019) A framework for modeling and assessing system resilience using a Bayesian network: A case study of an interdependent electrical infrastructure system. Int J Crit Infrastruct Prot 25:62–83. https://doi.org/10.1016/j.ijcip.2019.02.002

Independent Evaluation Group (2015) The World Bank group A to Z 2016. https://doi.org/10.1596/978-1-4648-0484-7

Jaberidoost M, Olfat L, Hosseini A, Kebriaeezadeh A, Abdollahi M, Alaeddini M, Dinarvand R (2015) Pharmaceutical supply chain risk assessment in Iran using analytic hierarchy process (AHP) and simple additive weighting (SAW) methods. J Pharm Policy Pract 8(1):1–10. https://doi.org/10.1186/s40545-015-0029-3

Jum'a L, Ikram M, Alkalha Z, Alaraj M (2022) Factors affecting managers' intention to adopt green supply chain management practices: evidence from manufacturing firms in Jordan. Environ Sci Pollut Res 29(4):5605–5621. https://doi.org/10.1007/s11356-021-16022-7

Kapoor D (2018) An overview on pharmaceutical supply chain: A next step towards good manufacturing practice. Drug Design Intell Prop Int J 1(2):49–54. https://doi.org/10.32474/ddipij.2018.01.000107

Karuppiah K, Sankaranarayanan B, Ali SM (2021) On sustainable predictive maintenance: exploration of key barriers using an integrated approach. Sustain Prod Consump 27:1537–1553. https://doi.org/10.1016/j.spc.2021.03.023

Kim M, Chai S (2017) Implementing environmental practices for accomplishing sustainable green supply chain management. Sustainability (Switzerland) 9(7). https://doi.org/10.3390/su9071192

Komendantova N, Patt A, Barras L, Battaglini A (2012) Perception of risks in renewable energy projects: the case of concentrated solar power in North Africa. Energy Policy 40(1):103–109. https://doi.org/10.1016/j.enpol.2009.12.008

Kruger W, Stritzke S, Trotter PA (2019) De-risking solar auctions in sub-Saharan Africa – A comparison of site selection strategies in South Africa and Zambia. Renew Sust Energ Rev 104:429–438. https://doi.org/10.1016/j.rser.2019.01.041

Kul C, Zhang L, Solangi YA (2020) Assessing the renewable energy investment risk factors for sustainable development in Turkey. J Clean Prod 276:124164. https://doi.org/10.1016/j.jclepro.2020.124164

Kumar S, Kumar P, Kumar M (2015) Risk analysis in green supply chain using fuzzy AHP approach: a case study. Resour Conserv Recycl 104:375–390

Kumar R, Singh RK, Dwivedi YK (2020) Application of industry 4.0 technologies in SMEs for ethical and sustainable operations: analysis of challenges. J Clean Prod 275:124063. https://doi.org/10.1016/j.jclepro.2020.124063

Luthra S, Garg D, Haleem A (2015) Critical success factors of green supply chain management for achieving sustainability in Indian automobile industry. Prod Plan Control 26(5):339–362. https://doi.org/10.1080/09537287.2014.904532

Marcelino-Sádaba S, González-Jaen LF, Pérez-Ezcurdia A (2015) Using project management as a way to sustainability. From a comprehensive review to a framework definition. J Clean Prod 99:1–16. https://doi.org/10.1016/j.jclepro.2015.03.020

Memon MS, Lee YH, Mari SI (2015) Group multi-criteria supplier selection using combined grey systems theory and uncertainty theory. Expert Syst Appl 42(21):7951–7959. https://doi.org/10.1016/j.eswa.2015.06.018

Negash YT, Sriplod T, Hassan AM (2021) A causal sustainable natural rubber development framework using a hierarchical structure with linguistic preferences in Thailand. J Clean Prod 305:127095. https://doi.org/10.1016/j.jclepro.2021.127095

Nicod T, Bathfield B, Bosc PM, Promkhambut A, Duangta K, Chambon B (2020) Households' livelihood strategies facing market uncertainties: how did Thai farmers adapt to a rubber price drop? Agric Syst 182:102846. https://doi.org/10.1016/j.agsy.2020.102846
Obiso JJA, Himang CM, Ocampo LA, Bongo MF, Caballes SAA, Abellana DPM, Deocaris CC, Padua R, Ancheta R (2019) Management of Industry 4.0-reviewing intrinsic and extrinsic adoption drivers and barriers. Int J Technol Manag 81(3–4):210–257. https://doi.org/10.1504/IJTM.2019.105310
Ocampo L, Yamagishi K (2020) Modeling the lockdown relaxation protocols of the Philippine government in response to the COVID-19 pandemic: an intuitionistic fuzzy DEMATEL analysis. Socio Econ Plan Sci 72:100911. https://doi.org/10.1016/j.seps.2020.100911
Oduoza CF (2020) Framework for sustainable risk management in the manufacturing sector. Proced Manufact 51:1290–1297. https://doi.org/10.1016/j.promfg.2020.10.180
OECD (2020) Official development assistance 2018. Int J Pest Manag B. https://doi.org/10.1080/05331846509432211
Rodríguez-Rivero R, Ortiz-Marcos I, Romero J, Ballesteros-Sánchez L (2020) Finding the links between risk management and project success: evidence from international development projects in Colombia. Sustainability (Switzerland) 12(21):1–19. https://doi.org/10.3390/su12219294
Roy NC, Roy NG (2020) Risk management in small hydropower (SHP) projects of Uttarakhand: an innovative approach: risk management in small hydropower projects. IIMB Manag Rev 32(3): 291–304. https://doi.org/10.1016/j.iimb.2019.10.012
Scarlat E, Delcea C (2011) Complete analysis of bankruptcy syndrome using grey systems theory. Grey Syst Theor Appl 1(1):19–32. https://doi.org/10.1108/20439371111106704
Settembre-Blundo D, González-Sánchez R, Medina-Salgado S, García-Muiña FE (2021) Flexibility and resilience in corporate decision making: A new sustainability-based risk management system in uncertain times. Glob J Flex Syst Manag 22(December):107–132. https://doi.org/10.1007/s40171-021-00277-7
Sheng X, Chen L, Yuan X, Tang Y, Yuan Q, Chen R, Wang Q, Ma Q, Zuo J, Liu H (2022) Green supply chain management for a more sustainable manufacturing industry in China: a critical review. Environment, Development and Sustainability:1–33. https://doi.org/10.1007/s10668-022-02109-9
Shieh JI, Wu HH, Huang KK (2010) A DEMATEL method in identifying key success factors of hospital service quality. Knowl-Based Syst 23(3):277–282. https://doi.org/10.1016/j.knosys.2010.01.013
Stare A (2014) Agile project management in product development projects. Procedia Soc Behav Sci 119:295–304
Vasilieva E (2021) Design thinking in the development of project management approaches and modeling of business processes of the organization. In: Zaramenskikh E, Fedorova A (eds) Digital transformation and new challenges. Springer, Cham, pp 3–16. https://doi.org/10.1007/978-3-030-71397-3_1
Zadeh LA (1996) Fuzzy sets. In: Klir GJ, Yuan B (eds) Fuzzy sets, fuzzy logic, and fuzzy systems: selected papers by Lotfi A Zadeh. World Scientific, Singapore, pp 394–432
Zhang W, Deng Y (2019) Combining conflicting evidence using the DEMATEL method. Soft Comput 23(17):8207–8216. https://doi.org/10.1007/s00500-018-3455-8
Zhang Y, Montenegro-Marin CE, Díaz VG (2021) Holistic cognitive conflict chain management framework in supply chain management. Environ Impact Assess Rev 88:106564. https://doi.org/10.1016/j.eiar.2021.106564
Zhou Q, Huang W, Zhang Y (2011) Identifying critical success factors in emergency management using a fuzzy DEMATEL method. Saf Sci 49(2):243–252. https://doi.org/10.1016/j.ssci.2010.08.005
Zhu Q, Sarkis J (2004) Relationships between operational practices and performance among early adopters of green supply chain management practices in Chinese manufacturing enterprises. J Oper Manag 22(3):265–289. https://doi.org/10.1016/j.jom.2004.01.005

Chapter 5
Barriers for Sustainable Supply Chain Management and Their Overcoming Strategies in Context of the Indian Automobile Industry

Sanjoy Kumar Paul, Koppiahraj Karuppiah, Bathrinath Sankaranarayanan, and Syed Mithun Ali

Abstract Sustaining in the present competitive business arena remains a challenge for global companies. The success of an organization and the final price of a product largely depend on the supply chain activities. Eliminating unproductive activities in SCM may bring down the product cost and improve efficiency. Recently, the integration of sustainability principles with SCM has gained significance and proven to be effective. However, several barriers remain as challenges in the integration. This study aims to collect and evaluate the barriers and strategies that will assist in integrating sustainability principles with SCM. For this, the study identified 15 barriers and 10 strategies related to sustainable SCM (SSCM) through literature review and experts' interview. Then, the fuzzy TOPSIS technique analyzes the strategies. The barriers and strategies considered in this study are evaluated using the inputs from the experts who are associated with the Indian automobile industry. Findings revealed robust supplier relationships, enhanced social fairness, ensuring economic opportunities, enhanced quality, and customer participation as the top five strategies that could assist in adopting SSCM practices. It is anticipated that the study's findings will equip the industries in SSCM adoption.

S. K. Paul
UTS Business School, University of Technology Sydney, Sydney, NSW, Australia
e-mail: Sanjoy.Paul@uts.edu.au

K. Karuppiah
Department of Mechanical Engineering, Saveetha School of Engineering, SIMATS, Chennai, Tamil Nadu, India

B. Sankaranarayanan (✉)
Department of Mechanical Engineering, Kalasalingam Academy of Research and Education, Krishnankoil, Tamil Nadu, India

S. M. Ali
Department of Industrial and Production Engineering, Bangladesh University of Engineering and Technology, Dhaka, Bangladesh

© The Author(s), under exclusive license to Springer Nature Switzerland AG 2023
N. U. Ibne Hossain (ed.), *Data Analytics for Supply Chain Networks*, Greening of Industry Networks Studies 11, https://doi.org/10.1007/978-3-031-29823-3_5

Keywords Sustainable supply chain management (SSCM) · Automobile industry · Fuzzy TOPSIS

5.1 Introduction

To stand in with the current competitive business environment, an organization needs to deliver the right product in the right quantity in the right place at the right time at a lower cost. In ensuring the above needs, the supply chain network plays a critical role (Autry 2021). The efficiency of a supply chain network largely depends on how the network is designed, i.e., how the manufacturing plants, warehouses, and distribution centers are connected, which controls 80% of the supply chain cost (Liu et al. 2020). In general, the cost of a product is determined by the cost involved in product transportation. Hence, there exists a direct relationship between the product cost and transportation cost. When the product movement is improved with effective supply chain management (SCM) approaches, it is possible to minimize the product cost. So, there is a substantial need to enhance the efficiency of the SCM. The significance of SCM is increasing with shortened life of products and globalization. Furthermore, with globalization and the emergence of new technologies, the supply chain network gets complicated owing to the interconnectedness among organizations (Li et al. 2020). For example, the US manufacturing industrial sector spends nearly 30% of the total product cost on transportation (Jasti and Kurra 2017; Marcucci et al. 2022). Generally, it is perceived that competition may be no longer between the organizations but between SCMs.

The manufacturing sector is facing challenges on three major fronts: adoption of the latest cutting-edge technologies, responding to volatile market demand, and abiding with the environmental norms (Kumar Singh and Modgil 2020). In combating these challenges, most industries are incorporating the sustainable concept in industrial activities and are witnessing improvement in performance and competitiveness. Initially, the Toyota company first developed the concept of lean strategy in the 1950s and incorporated the strategy in supply chain activities (Tseng et al. 2022). However, as the sustainability concept gained more attention, the industries are in a situation in which they need to incorporate the sustainability concept. As a result, the integration of sustainability concept in SCM was first introduced by Lamming (1996) to reduce the cost and delivery time as well as to improve effectiveness. Seuring and Müller (2008) stated sustainable supply chain management (SSCM) as "a set of activities directly linked by upstream and downstream flows of products, services, information and funds that work together to decrease cost and waste by efficiently pulling what is needed to meet the needs of individual customers." The main function of the SSCM is to improve the supply chain activities by considering the social, economic, and environmental aspects. Realizing the benefits of SSCM, most of the companies started adopting sustainable practices in SCM to stay ahead in the global competitive market environment (de Oliveira et al. 2022; Bota-Avram 2022).

With globalization paving the way for the venture of any company into any nation, worldwide, the automobile companies have restructured their SCM practices. As a result, the Indian automobile industry is faced with numerous challenges in terms of quality and delivery time. Realizing the situation, Indian automobile industries have started adopting SSCM practices. However, the adoption of SSCM demands more change in the organizational structure. The efficiency of SSCM not solely resides with the top management but also relies on the employees of the organization. Vanichchinchai (2019) argued that organizational characteristics, i.e., firm size, export size, and size of the SCM department, largely influence SSCM practice's efficiency. According to Moyano-Fuentes et al. (2019), the main challenge faced by the companies in implementing SSCM is the integration with key suppliers and customers. Another study by Singh and Kumar (2020) in analyzing the adoption level of SSCM by the Indian industries indicated that in terms of quick response and quality management, the Indian industries are lagging behind. Besides these challenges, the Indian industrial community faces challenges in terms of having supply chain managers with a modern skillset (Digalwar et al. 2020; Jell-Ojobor and Raha 2022).

From the above information, it is apparent that the industrial community needs to be updated with the latest technology either to thrive or to survive in the competitive business environment. One such technique is incorporating SM in SCM, i.e., SSCM. In developing countries, there are plenty of indicators for the need for SSCM. For instance, developing countries are regarded as the manufacturing hub by global countries. Hence, the industries in developing countries account for a huge proportion of global manufacturing. The need for SSCM is very critical for the automobile industry as it comprises multiple stages in the SCM activities, i.e., from procurement to transportation.

Given the importance of SSCM, this study acknowledges the critical gap in the literature. First, it identifies the barriers that are impediments in the incorporation of sustainable management in SCM. Second, the study evaluates strategies that could assist in overcoming the barriers in SSCM.

The main reason for carrying out this proposed study is to understand the significance of the critical barriers in SSCM adoption in the Indian automobile context as India ranks sixth in automobile production globally. Also, the automotive industry contributes almost 7% to the total GDP and 7% of the total employment in India. Further, the automobile industry accounts for approximately 40% of the total foreign investment in India. In light of the above discussion, this research raises the following questions:

RQ1. What are the barriers hindering SSCM in the Indian automobile sector?
RQ2. What strategies can be used to encounter the barriers to SSCM in the Indian automobile sector?

The study aims to analyze the barriers and strategies using the proposed multi-criteria decision-making (MCDM) technique framework model, and the fuzzy Technique for Order of Preference by Similarity to Ideal Solution (TOPSIS) has been used to evaluate the barriers and prioritize the strategies. The fuzzy TOPSIS method, introduced by Hwang and Yoon (1981), can assist industrial practitioners and policy-makers in the implementation of SSCM efficiently. In this study, fuzzy TOPSIS is used as a fuzzy concept that helps overcome vagueness and inconsistency in data. Many studies have been carried out using fuzzy TOPSIS like sustainability model selection (Samaie et al. 2020), data quality assessment challenges (Rafi et al. 2020), material selection for construction works (Saeli et al. 2020), and supply chain selection (Jellali et al. 2021).

5.2 Literature Review

5.2.1 *Sustainability and Supply Chain Management*

The sustainability concept is receiving growing interest among the industrial sector owing to the increased need for safeguarding the environment. However, considering the time required, in addition to the environmental aspects, the sustainability concept incorporated social and economic aspects. Adoption of the sustainable practice helps industries to gain a competitive advantage. Studies by Panwar et al. (2018) and Dave and Sohani (2019) indicated that incorporating sustainability principles in all industrial activities will enhance industrial performance. As a result, the sustainable approach has been incorporated in all industrial activities to the greatest extent possible. In supply chain management, adopting sustainability enhances supply chain activities with the latest technologies (Kumar Singh and Modgil 2020). Lamming (1996) was the first to stress the integration of sustainability concept with supply chain management. Depending on the operational capability and necessity, the supply chain model may be a single tier or multitier. Unnecessary activities in the supply chain may affect the efficiency of the process and also result in a time delay. In overcoming such delays and in improving supply chain resilience, the adoption of sustainability principles has helped the organizations witness significant results (Berger et al. 2018). Saudi et al.'s (2019) study recommended the adoption of SSCM as a solution to improve the performance of Malaysian electronics industries. A study by Ahmed and Huma (2021) indicated that market orientation and the need to maintain the quality of the product require SSCM adoption. Nath and Agrawal (2020) emphasized that adopting SSCM may act as a antecedent in achieving social sustainability.

5.2.2 *Indian Automobile Industry and Sustainable Supply Chain Management*

The automobile industry is a well-established industrial sector in India. After decades of sterile growth, the industry has taken giant strides since the inception of economic liberalization in the early 1990s. As the sixth largest automobile-producing country, India accounts for 2.37% of the total global production. As per the report of the Society of Indian Automobile Manufacturers (SIAM), 26,362,282 vehicles were produced in March–April 2020 (SIAM Report 2020). The contribution of the automobile industry to India's Gross Domestic Product (GDP) is 7.1% (Mishra et al. 2020; Chowdhury and Chatterjee 2020). Despite these credible achievements, accounting for 0.53% of global auto production at a rank of 26 among the world's automobile industries and parts exporting countries, India's automobile industries are still fairly low compared to other global players. This indicates that while the Indian automobile sector is in a healthy position, it has yet some way to go to before becoming a leader in the market. In this context, India's competitive edge can be honed through effective supply chain management. The automobile industry's supply chain network is intricate and involves multiple stakeholders such as raw material suppliers, component producers, assembly partners, distribution channels, and end customers (Tripathi and Talukder 2020). Such a complex network needs a robust and efficient management practice that must integrate all these links in a cost-efficient manner. Here, the incorporation of sustainable practices by the Indian automobile industries in supply chain management appears as a viable option.

Researchers such as Caiado et al. (2022) acknowledge that integration of sustainability in SCM activities faces a lot of obstacles. The existing structure of SSCM in the Indian automobile sector is different from that of the developed countries (Mukherjee et al. 2021). Thus, the entire structure of the SSCM should be redesigned to thrive in the global competition. Cultural differences and private variations are some of the common problems faced by the foreign automobile companies investing in India (Loaiza-Ramírez et al. 2022). According to Puche et al. (2019), stock buffers are mostly used to increase the effectiveness of the supply chain management. According to Butt (2021), organizations' top management should concentrate on important issues related to policies, and the implementation of SSCM also needs some improvement. According to Kazancoglu et al. (2021), the top management fails to perform according to the standards as the organization has poor foresight regarding its objective. This is also called unclear organization goals. As the Indian automobile sector has many barriers to implementing SSCM, it is important to investigate the barriers to SSCM and strategies for overcoming them in the Indian automobile sector. The list of barriers and strategies are given in Tables 5.1 and 5.2.

From earlier literature, it is understood that several earlier works have discussed the barriers in implementing SSCM. However, very few studies have discussed the strategies to mitigate the barriers to SSCM. Hence, it is important to identify both barriers and overcoming strategies and analyze them to determine the priority list of overcoming barriers. This study fulfills this gap.

5.3 Research Methodology

The research framework followed in the research is given in Fig. 5.1. By following literature review and experts' survey, the list of barriers and overcoming strategies are identified. The literature for review is searched in science databases like ScienceDirect, Google Scholar, EBSCO, Inderscience, Web of Science, and Taylor & Francis to identify the barriers and strategies. The following keywords and Boolean operators, problems faced by Indian automobile industries, problem OR barriers in adoption of SSCM, difficulties AND impediments in the adoption of SSCM, automobile industry AND SSCM, strategies OR techniques to overcome barriers in SSCM, and Indian automobile industry AND supply chain management, are used in literature collection. Initially, a total of 73 articles were collected. Among these 73 articles, 24 articles were rejected for replicating work, lack of work authenticity, and not being English. From the remaining 49 articles, 8 barriers and 10 strategies were identified. Then, these eight barriers were discussed with ten experts who have a profound knowledge of the sustainable practices. Initially, 25 experts were approached; however, only 10 of them reverted. The response rate

Table 5.1 Barriers in SSCM in the Indian automobile industry

Name of the barrier	Description	Source
Lack of commitment from top management (B1)	Commitment from the top management plays an important role in the success of SSCM	Expert survey
Poor training and education (B2)	Most workers engaged in SCM are not adequately trained on sustainability concept	Dai et al. (2021) + Expert survey
Employee demotivation (B3)	Most companies involved in supply chain management are not offering rewards to the employees	Mangla et al. (2022) + Expert survey
Organizational reluctance toward change (B4)	Non-willingness of organizations to incorporate sustainability practice	Seuring et al. (2022) + Expert survey
Cultural differences (B5)	Absence of integrity among the co-workers reduces the efficiency of supply chain management	Mangla et al. (2022) + Expert survey
Insufficient funds for implementing new techniques (B6)	There is a lack of financial resources to implement new techniques for SSCM	Seuring et al. (2022) + Expert survey
Ill-defined organizational goals (B7)	Most companies are perceived of short-term goals	Expert survey
Political instability (B8)	Changing political scenario impacts the efficiency of the SSCM	Mangla et al. (2022) + Expert survey
Lack of knowledge about SSCM (B9)	Organization is still following traditional SCM and is reluctant to learn about new techniques	Expert survey
Lack of awareness about the environmental issue (B10)	There is a lack of awareness for environmental issues due to the lack of understanding of the importance of the environment	Seuring et al. (2022) + Expert survey
Poor coordination in internal departments (B11)	Fragmentation of the internal departments results in miscommunication	Expert survey
Lack of vision about resources and capabilities (B12)	Being unaware of the sustainability concept, most industrial sectors are not concerned about the resource protection	Mangla et al. (2022) + Expert survey
Lack of employee involvement (B13)	There is a lack of employee involvement in decision-making and other activities	Mangla et al. (2022) + Expert survey
Poor organizational structure (B14)	Most of the supply chain organizations are functioning in an unstructured manner	Expert survey
Reluctant to share information with suppliers (B15)	Most of the organizations are feeling insecure about sharing information with the stakeholders.	Huang et al. (2022)

was 40%, which was satisfactory. Among the ten experts, seven are from industrial backgrounds, and three are from an academic background. The average work experience of the experts is 8 years. During the discussion, the experts further suggested seven barriers that need to be analyzed. Thus, 15 barriers and 10 strategies displayed in Tables 5.1 and 5.2 need to be analyzed. The barriers and strategies are then analyzed using a fuzzy TOPSIS method to determine the priority list

Table 5.2 List of overcoming strategies

Name of the strategy	Description	Source
Customer participation (S1)	Customers' demand for sustainable performance from the organization is crucial in SSCM	Seuring et al. (2022) + Expert survey
Constant improvement (S2)	Maintaining sustainability in the supply chain activities is a continuous process, and organizations must be prepared for it	Puche et al. (2019) + Expert survey
Proper supplier integration (S3)	Establish a rapport relationship with the stakeholders involved in the supply chain network	
Improved customer response (S4)	Responding to the customers' feedback improves the reliability of an organization	Najar (2022)
Robust supplier relationship (S5)	Information flow between the stakeholders is very crucial in supply chain activities	Seuring et al. (2022) + Expert survey
Ensure economic opportunity (S6)	Supply chain activities should be carried out in such a way that it should provide economic assistance	Tortorella et al. (2017)
Enhance social fairness (S7)	Well-established supply chain network ensures complete societal development	Expert survey
Enhance quality (S8)	Incorporation of sustainability increases the quality of the service	Seuring et al. (2022) + Expert survey
Improve workforce involvement (S9)	Periodical rewards and incentives may motivate the employees	Expert survey
Improve environmental performance (S10)	As the society is becoming more environment concerned, the supply chain organization must consider their environmental performance also	Seuring et al. (2022)

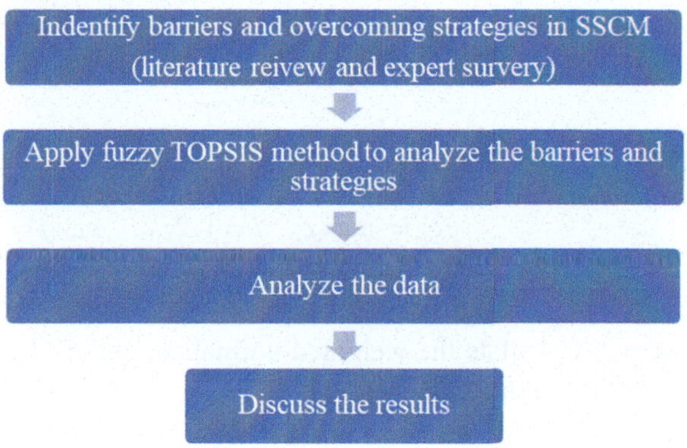

Fig. 5.1 Framework used in this study

of strategies. Fuzzy TOPSIS helps in overcoming the drawback of the TOPSIS method (Chowdhury and Paul 2020). The linguistic scale for the fuzzy TOPSIS method is presented in Table 5.3. The steps involved in fuzzy TOPSIS are as follows:

Table 5.3 Linguistic scale of the fuzzy TOPSIS method

Linguistic term	Fuzzy number
Very low	1,1,3
Low	1,3,5
Medium	3,5,7
High	5,7,9
Very high	7,7,9

Step 1: Establish a fuzzy decision matrix $\widetilde{D}$ for m strategies and n barriers: $\widetilde{x}_{ij} = (a_{ij}, b_{ij}, c_{ij})$ represents the triangular fuzzy numbers in linguistic terms. The fuzzy decision matrix $\widetilde{D}$ is represented as:

$$\widetilde{D} = \begin{bmatrix} \widetilde{x}_{11} & \cdots & \cdots & \widetilde{x}_{1n} \\ \vdots & \cdots & \cdots & \vdots \\ \vdots & \cdots & \cdots & \vdots \\ \widetilde{x}_{m1} & \cdots & \cdots & \widetilde{x}_{mn} \end{bmatrix} \tag{5.1}$$

Step 2: Calculate the normalized fuzzy decision matrix $\widetilde{R}$.

$$\widetilde{R} = \left[\widetilde{r}_{ij}\right]_{mxn} \tag{5.2}$$

The normalized $\widetilde{r}_{ij}$ is calculated as:

$$\widetilde{r}_{ij} = \left[\frac{a_{ij}}{c_j^*}, \frac{b_{ij}}{c_j^*}, \frac{c_{ij}}{c_j^*}\right] \tag{5.3}$$

where

$$c_j^* = \max c_{ij}. \ i = 1, 2, \ldots, m, \ j = 1, 2, \ldots, n. \tag{5.4}$$

Step 3: Calculate the weighted normalized matrix $\widetilde{V}$:
The weighted normalized value

$$\widetilde{v}_{ij} = w_i r_{ij}, \ i = 1, 2, \ldots, m, \ j = 1, 2, \ldots, n \tag{5.5}$$

$$\widetilde{V} = \left[\widetilde{v}_{ij}\right]_{\min}, \tag{5.6}$$

where w_i is the weight of the ith criterion and $\sum_{i=1}^{m} w_i = 1$.

Step 4: Calculate the fuzzy positive ideal solution A^* and fuzzy negative ideal solution A^-:

$$A^* = (\widetilde{v}_1^*,\ \widetilde{v}_2^*,\ \ldots,\ \widetilde{v}_n^*,) \tag{5.7}$$

$$A^- = (\widetilde{v}_1^-,\ \widetilde{v}_2^-,\ \ldots,\ \widetilde{v}_n^-,) \tag{5.8}$$

where $v_j^* = (1,\ 1,\ 1)$ and $v_j^- = (0,\ 0,\ 0)$, $j = 1, 2, \ldots, n$.

Step 5: Calculate the distance of each strategy using A^* and A^-:

$$D_j^* = \sum_{i=1}^{m} d(\widetilde{v}_{ij}, \widetilde{v}_i^*), j = 1, 2, \ldots, n \tag{5.9}$$

$$D_j^- = \sum_{i=1}^{m} d(\widetilde{v}_{ij}, \widetilde{v}_i^*), j = 1, 2, \ldots, n \tag{5.10}$$

where $d\left(\widetilde{A},\ \widetilde{B}\right) = \sqrt{\frac{1}{3}\left[(a_1 - b_1)^2 + (a_2 - b_2)^2 + (a_3 - b_3)^2\right]}$

Step 6: Calculate the relative closeness to the positive ideal solution:

$$RC_j^* = \frac{D_j^-}{D_j^* + D_j^-}, j = 1, 2, \ldots, n. \tag{5.11}$$

Step 7: Rank the strategies based on the preference order. The index values RC_j^* lie between 0 and 1. The larger the index value, the higher the rank of the strategies.

5.3.1 *Application of Fuzzy TOPSIS*

After finalizing the number of barriers and the strategies to be evaluated, the experts were requested to rate the barriers and the usefulness of strategies in overcoming the barriers. A response sheet (Appendix A) containing the barriers and strategies is given to each expert for getting the expert's opinion. The response of all the ten experts is given in Appendix A, Table A.1–A.10. The consolidated response of the experts is given in Appendix B, Table B.1. The normalized decision matrix is provided in Appendix C, Table C.1. The weight normalized decision matrix is given in Appendix D, Table D.1. The fuzzy positive and negative ideal solution is then calculated and given in Appendix E, Table E.1. Now, the distance of each strategy is calculated using A^* and A^- as given in Appendix F, Table F.1 and F.2. Finally, the relative closeness of the strategies is estimated and given in Table 5.4.

Table 5.4 Results from the fuzzy TOPSIS method and priority ranking of the strategies

Notation	Name of the strategy	d^+_i	d^-_i	Cci	Rank
S1	Customer participation	4.400692	5.140281	0.53876	5
S2	Constant improvement	5.093512	4.662722	0.47792	10
S3	Proper supplier integration	4.580312	5.087455	0.52623	7
S4	Improved customer response	4.543181	5.058389	0.52683	6
S5	Robust supplier relationship	3.412782	6.340047	0.65007	1
S6	Ensure economic opportunity	4.246732	5.523859	0.56536	3
S7	Enhance social fairness	3.682212	6.169301	0.62623	2
S8	Enhance quality	4.285222	5.514549	0.56272	4
S9	Improve workforce involvement	4.875691	4.75122	0.49354	9
S10	Improves environmental performance	4.605972	4.946448	0.51782	8

5.4 Results and Discussion

The strategies that could assist in overcoming the barriers in the implementation of SSCM are prioritized using fuzzy TOPSIS. For this, the ratings from the experts were used. The results obtained using fuzzy TOPSIS are provided in Table 5.4. The strategies are ranked based on the relative closeness index. According to Table 5.4, the strategy, robust supplier relationship (S5) with relative closeness 0.65007 secures the top rank. From this, it could be inferred that there is a need for industrial management to rapport a strong relationship with the stakeholders and multiple links involved in the network. This finding was endorsed by Mohapatra et al. (2021) in a study, which states that in maintaining a long-term relationship with the stakeholder and other critical links in the network, the top management's role is crucial. To maintain a robust supplier relationship, factors like mutual trust, open communication, long-time commitment, and ability to integrate information must be enhanced. Karuppiah et al. (2020) indicated that all the members involved in the supply chain network must adhere to the waste minimization or elimination principle. Such consensus in the working principle will drive the supply chain network in attaining the intended goal. To sustain and prosper in the rapidly changing market demand, establishing a robust supplier relationship is imperative for an organization (Sharma and Naude 2021; Huang et al. 2022). However, many organizations are reluctant to share information with suppliers (B15). Such kinds of barriers question the trust and understanding between the organization and various stakeholders. Adding to this, the barrier cultural difference (B5) between various links involved in the supply chain network remains an impediment in reaching a consensus goal.

Next, the strategy enhancing social equity (S7) could be adopted to adopt SSCM practices. Since sustainable practices intend to enhance the supply chain activities, there is a need to incorporate socially responsible practices. Hernandez-Martinez et al. (2020) stressed the need for social equity between the buyers and suppliers in the long run of SSCM adoption. In enhancing social equity, industrial organizations must strive to abolish child labor and ensure fair wage distribution for the workforce. Though sustainable practices intend to enhance the efficiency of the supply chain activities, it does not encourage workforce reduction. In most of the situations, the organization forcefully cut down the workforce as a practice in SSCM. The concept of social equity is restricted with labor welfare; it further extends to conclusive environmental benefits. In embracing environmental responsibility, the industrial community must move toward reverse supply chain management or circular economy. Such a transition may bolster the corporate social responsibility of the automobile industry. Instead, the industries are not willing to uplift the social status of the workforce. Barriers' poor education and training (B2) have been identified as the major reason for lacking social equity (Bhalaji et al. 2020).

Then, the strategy of ensure economic opportunity (S6) will provide great thrust in the adoption of SSCM. The adoption of any new strategy or technological upgradation will possess challenges to the industries. In such a situation, an organization that ensures economic opportunities will have a better chance of overcoming the new technology's initial hurdles. This was endorsed by Alazab et al. (2020) in a study that examined the ease of integrating blockchain technology in SSCM. The outcome of the study highlighted that the shift toward new technology needs surplus or sufficient financial assistance. However, many industries are running without sufficient funds for implementing new techniques (B6). This barrier results from organizational reluctance toward change (B4) and lack of commitment from the top management (B1). From this, it could be inferred that there is a strong relationship between the organization's commitment and every industrial progress. Hence, it could be perceived that the financial status of the industries to a large extent determines the capability of the industry in adopting new strategies and technologies.

Similarly, the strategy that improves quality (S8) is expected to contribute to SSCM practices in the automobile industry. Improvement in the quality of the product will increase customers' trust and preference for the product. So, to sustain the SSCM practice, improvisation of product quality is essential. Yadav et al. (2020) highlighted that the industries need to adopt the strategy of continuous process improvement to sustain in the competitive business arena. Improvisation of the industrial activity will offer many advantages for industries like market capturing and venturing into a new market. Single-minute exchange of die (SMED) has been suggested as a technique for improving product quality and time reduction (da Cunha et al. 2020). Another strategy that could prove to be feasible for the adoption of SSCM is customer participation (S1). Here, the strategy customer participation is prescribed as the customers are the end users and beneficiaries. To assess the level of impact of SSCM on industrial activities, the survey or feedback-based response collection from the customer will reveal the exact success of the adopted strategy (Celuch and Walz 2020). Since the ultimate aim of any organization is to earn the customer's trust and learn their preference, it is reasonable to receive the evaluation status of the industry's performance from the customer's perspective. When the industries become customer-centric, it focuses more on the opportunities and possibilities of maximizing the market size. Apart from the five discussed strategies, all other strategies mentioned in the study may have a sufficient impact on industrial performance.

5.5 Implications of the Study

5.5.1 Theoretical Implications

This research work offers some valuable insights into the literature on SSCM from an emerging economy context. In this study, fuzzy TOPSIS is used to prioritize the strategies that could help in overcoming the barriers to SSCM practice. It focuses on finding the best strategy by making a comparison between the barriers and the strategies. This comparison provides a better understanding of the barriers and also about the strategies.

The outcome of the study highlights that the success of SSCM mainly depends on the effectiveness of the relationship among the suppliers. It is obvious that only with the existence of consensus among the suppliers it is possible to meet the intended purpose of eliminating waste generation (Karuppiah et al. 2020). In a study, Takeda-Berger et al. (2021) highlighted that the relationship among the suppliers is very critical in the performance of SSCM practices. Though the outcome of the study stresses the need for effective relationships among the suppliers, most of the top management of the suppliers is not concerned about eliminating wastes. This reflects the poor understanding and unawareness of the sustainable concept by the top management of the suppliers.

Being unaware of the sustainable concept, there is a dire need to enhance the social equity among the suppliers. By familiarizing the suppliers with the sustainable concept, it is possible to steer the industrial activity toward eliminating or minimizing waste generation (Leksic et al. 2020). Hence, it is necessary to impart the knowledge of the sustainable concept among the Indian industrial sector. Such an initiative will also create environmental awareness among the industrial sector. When such kind of awareness is created, the industrial sector may show interest in the sustainable concept.

In this study, the fuzzy TOPSIS technique is used to evaluate the strategies that could prove to be feasible for SSCM by comparing the barriers. This work is a pioneering attempt to evaluate the strategies and barriers. This attempt reveals the importance of each barrier and also the ranks of strategies in overcoming the barriers to SSCM.

5.5.2 Managerial Implications

The outcomes of this study offer some implications for industrial practitioners and policy-makers.

Firstly, the knowledge of the sustainable concept has not been familiar among the industrial community. As a result, the industrial practitioners are not aware of the significance of incorporating the sustainable concept in industrial activities. From the government side, as an initiative in imparting the knowledge of the sustainable concept and also to mitigate adverse environmental impact caused by the industries, quality control of India (QCI) introduced a program called zero defect zero effect (ZED). This program aims at measuring the quality of the infrastructure available in an industry, the environmental performance, and certification and ratings. The introduction of the ZED program has driven the Indian industry's efforts in minimizing the waste generated and also in increasing the quality of the products developed (Huang et al. 2020). More kinds of such programs must be introduced by the government for raising the awareness of sustainable concept and minimizing waste generation.

Next, the consensus among the various stakeholders involved in the supply chain network must be attained. Only with a single converged goal it is possible to attain the success of sustainable supply chain management. For that, various stakeholders involved in the supply chain network must be imparted with the knowledge of the sustainable concept. Likewise, the stakeholders must think about the importance of conserving nature and also enhancing the quality of the production.

Finally, the study calls for more programs on imparting the knowledge of the sustainable concept among the industrial practitioners and also among the policymakers. Besides this, the various stakeholders involved in the network must be integrated and must work with a single intention of eliminating the wastes.

5.6 Conclusions

In answering the research questions mentioned in the introduction section, the researchers initially conducted an exhaustive literature review to access the principle of sustainable practice and its level of incorporation in SCM activities. The barriers and the strategies that could enable the adoption of SSCM were identified and scrutinized by keeping the Indian automobile industries in mind. The expert panel, comprising seven from the industrial background and three from the academic background, was formed, and the experts' inputs were obtained using the fuzzy TOPSIS approach. The fuzzy TOPSIS approach helped evaluate the strategies by making a pairwise comparison between the strategies and the identified barriers in SSCM practice. Such a comparison provides clarity in the ability of each strategy in handling each barrier.

The study's findings revealed robust supplier relationships, enhancing social equity, ensuring economic opportunity, improving quality, and customer participation are the top five strategies that could assist in overcoming the barriers faced by the automobile industry while incorporating SSCM practices. This ensures that the relationship between the organization and other links in the supply chain network is critical for seamless and efficient SSCM practice. In maintaining a relationship with the stakeholders, the role of top-level management is inevitable. Also, the involvement and commitment from each of the SSCM network members are important.

This study contributes immensely to the literature on sustainable management practices. One of the highlights of the chapter is that this study evaluates the effectiveness and capability of each strategy that is being followed in SSCM practices. Through this, the study ranks the strategies in the order of preference. Although the present study has some valuable contribution, it is confined to the automobile industry. So, when applying the findings of the study to other industries, caution must be practiced. Further, the study recommends multiple case studies as the study was carried out by considering only one automobile industry. Then, in this study, the strategies are prioritized using fuzzy TOPSIS. As a future study, a hybrid method of combining the best-worst method (BWM) or stepwise weight assessment ratio analysis (SWARA) with TOPSIS could be carried out as this combination estimates the weight importance of the barriers.

Appendixes

Appendix A: Data from Ten Experts for Fuzzy TOPSIS

Sample response sheet

	B1			B2			B3			B4			B5			B6			B7			B8			B9			B10			B11			B12			B13			B14			B15		
S1																																													
S2																																													
S3																																													
S4																																													
S5																																													
S6																																													
S7																																													
S8																																													
S9																																													
S10																																													

Table A.1 Response from expert 1

	B1			B2			B3			B4			B5			B6			B7			B8			B9			B10			B11			B12			B13			B14			B15		
S1	5	7	9	5	7	9	1	1	3	5	7	9	1	3	5	7	7	9	1	1	3	3	5	7	7	7	9	1	1	3	1	1	3	1	1	3	5	7	9	7	7	9	3	5	7
S2	1	3	5	1	1	3	1	1	3	7	7	9	5	7	9	3	5	7	7	7	9	1	3	5	1	3	5	7	7	9	5	7	9	7	7	9	5	7	9	1	1	3	7	7	9
S3	3	5	7	3	5	7	5	7	9	3	5	7	1	1	3	3	5	7	1	1	3	5	7	9	3	5	7	1	3	5	3	5	7	7	7	9	1	1	3	1	1	3	5	7	9
S4	1	1	3	3	5	7	1	1	3	3	5	7	1	1	3	3	5	7	1	1	3	1	1	3	1	3	5	3	5	7	1	1	3	1	1	3	3	5	7	3	5	7	1	1	3
S5	5	7	9	1	1	3	3	5	7	7	7	9	7	7	9	1	3	5	7	7	9	5	7	9	1	3	5	7	7	9	5	7	9	5	7	9	1	3	5	1	1	3	3	5	7
S6	7	7	9	1	3	5	1	3	5	5	7	9	5	7	9	1	1	3	1	1	3	7	7	9	3	5	7	1	3	5	5	7	9	7	7	9	1	3	5	3	5	7	3	5	7
S7	3	5	7	1	1	3	7	7	9	1	3	5	1	3	5	1	3	5	5	7	9	1	1	3	3	5	7	1	3	5	3	5	7	7	7	9	3	5	7	5	7	9	3	5	7
S8	1	3	5	3	5	7	5	7	9	3	5	7	1	1	3	1	3	5	1	3	5	5	7	9	3	5	7	1	1	3	7	7	9	1	1	3	1	1	3	5	7	9	5	7	9
S9	1	1	3	3	5	7	5	7	9	5	7	9	1	1	3	1	3	5	1	1	3	1	3	5	3	5	7	1	3	5	1	1	3	1	1	3	1	1	3	3	5	7	1	1	3
S10	1	1	3	5	7	9	7	7	9	7	7	9	7	7	9	1	1	3	1	1	3	7	7	9	5	7	9	1	1	3	1	3	5	7	7	9	5	7	9	7	7	9	5	7	9

Table A.2 Response from expert 2

	B1			B2			B3			B4			B5			B6			B7			B8			B9			B10			B11			B12			B13			B14			B15		
S1	1	3	5	1	1	3	1	3	5	1	1	3	1	3	5	5	7	9	3	5	7	5	7	9	7	7	9	7	7	9	3	5	7	1	3	5	1	3	5	7	7	9	3	5	7
S2	3	5	7	5	7	9	5	7	9	1	3	5	1	3	5	1	1	3	5	7	9	3	5	7	1	3	5	1	1	3	5	7	9	5	7	9	7	7	9	1	1	3	3	5	7
S3	5	7	9	3	5	7	1	3	5	3	5	7	3	5	7	3	5	7	3	5	7	7	7	9	3	5	7	3	5	7	7	7	9	7	7	9	1	1	3	1	1	3	1	3	5
S4	3	5	7	3	5	7	5	7	9	1	3	5	5	7	9	1	1	3	1	3	5	1	3	5	1	1	3	7	7	9	1	1	3	7	7	9	1	3	5	3	5	7	5	7	9
S5	3	5	7	7	7	9	5	7	9	3	5	7	1	3	5	3	5	7	7	7	9	1	3	5	5	7	9	5	7	9	7	7	9	7	7	9	5	7	9	3	5	7	7	7	9
S6	3	5	7	3	5	7	1	1	3	1	3	5	3	5	7	7	7	9	1	3	5	3	5	7	1	3	5	1	1	3	1	3	5	5	7	9	1	1	3	3	5	7	1	1	3
S7	7	7	9	1	1	3	7	7	9	1	1	3	1	3	5	1	1	3	5	7	9	7	7	9	1	1	3	1	1	3	5	7	9	5	7	9	3	5	7	7	7	9	7	7	9
S8	1	1	3	5	7	9	3	5	7	1	3	5	1	1	3	1	3	5	1	1	3	5	7	9	1	3	5	3	5	7	1	1	3	3	5	7	1	1	3	5	7	9	1	1	3
S9	3	5	7	5	7	9	3	5	7	1	3	5	1	3	5	3	5	7	1	3	5	1	3	5	3	5	7	1	3	5	3	5	7	3	5	7	5	7	9	1	3	5	3	5	7
S10	7	7	9	1	3	5	1	3	5	5	7	9	7	7	9	7	7	9	3	5	7	3	5	7	1	1	3	7	7	9	1	1	3	5	7	9	1	3	5	3	5	7	7	7	9

Table A.3 Response from expert 3

	B1			B2			B3			B4			B5		
S1	1	1	3	1	1	3	5	7	9	3	5	7	3	5	7
S2	7	7	9	5	7	9	7	7	9	1	1	3	1	3	5
S3	5	7	9	1	3	5	1	3	5	3	5	7	7	7	9
S4	3	5	7	1	1	3	3	5	7	7	7	9	1	3	5
S5	7	7	9	1	1	3	1	3	5	7	7	9	5	7	9
S6	3	5	7	3	5	7	1	1	3	1	1	3	1	1	3
S7	5	7	9	1	3	5	5	7	9	1	1	3	5	7	9
S8	3	5	7	5	7	9	3	5	7	7	7	9	3	5	7
S9	1	3	5	7	7	9	1	1	3	7	7	9	1	1	3
S10	1	3	5	1	1	3	3	5	7	5	7	9	5	7	9

	B6			B7			B8			B9			B10		
S1	1	3	5	1	3	5	1	1	3	7	7	9	3	5	7
S2	7	7	9	1	1	3	7	7	9	1	3	5	3	5	7
S3	5	7	9	7	7	9	5	7	9	5	7	9	1	3	5
S4	5	7	9	7	7	9	5	7	9	3	5	7	1	1	3
S5	1	3	5	7	7	9	3	5	7	1	1	3	1	3	5
S6	3	5	7	7	7	9	1	3	5	5	7	9	3	5	7
S7	3	5	7	5	7	9	7	7	9	1	3	5	7	7	9
S8	5	7	9	3	5	7	7	7	9	1	1	3	7	7	9
S9	1	3	5	7	7	9	1	3	5	1	1	3	3	5	7
S10	5	7	9	5	7	9	7	7	9	5	7	9	5	7	9

	B11			B12			B13			B14			B15		
S1	3	5	7	5	7	9	5	7	9	3	5	7	5	7	9
S2	1	1	3	1	3	5	1	3	5	3	5	7	1	3	5
S3	1	3	5	7	7	9	1	3	5	7	7	9	1	3	5
S4	5	7	9	1	1	3	5	7	9	3	5	7	5	7	9
S5	1	3	5	1	3	5	1	1	3	3	5	7	1	3	5
S6	1	1	3	5	7	9	3	5	7	5	7	9	1	1	3
S7	1	3	5	1	3	5	5	7	9	1	1	3	7	7	9
S8	7	7	9	7	7	9	7	7	9	5	7	9	7	7	9
S9	3	5	7	7	7	9	1	1	3	3	5	7	1	1	3
S10	5	7	9	5	7	9	1	3	5	5	7	9	1	3	5

Table A.4 Response from expert 4

	B1			B2			B3			B4			B5			B6			B7			B8			B9			B10			B11			B12			B13			B14			B15		
S1	1	3	5	1	1	3	5	7	9	7	7	9	3	5	7	3	5	7	5	7	9	7	7	9	5	7	9	3	5	7	1	3	5	1	3	5	1	3	5	3	5	7	7	7	9
S2	3	5	7	1	1	3	1	3	5	1	1	3	1	3	5	1	1	3	5	7	9	1	3	5	1	3	5	1	1	3	1	1	3	7	7	9	1	1	3	1	3	5	3	5	7
S3	7	7	9	1	3	5	1	1	3	7	7	9	7	7	9	5	7	9	3	5	7	5	7	9	1	3	5	1	3	5	5	7	9	7	7	9	3	5	7	5	7	9	5	7	9
S4	1	1	3	7	7	9	7	7	9	1	1	3	1	1	3	5	7	9	5	7	9	7	7	9	1	3	5	5	7	9	3	5	7	1	1	3	1	1	3	1	3	5	1	1	3
S5	5	7	9	5	7	9	1	3	5	3	5	7	1	3	5	7	7	9	1	1	3	3	5	7	3	5	7	3	5	7	1	3	5	1	1	3	1	1	3	1	1	3	5	7	9
S6	7	7	9	7	7	9	1	1	3	3	5	7	1	1	3	5	7	9	7	7	9	3	5	7	1	1	3	7	7	9	5	7	9	1	1	3	3	5	7	3	5	7	5	7	9
S7	7	7	9	7	7	9	1	3	5	1	3	5	1	3	5	7	7	9	5	7	9	7	7	9	1	1	3	7	7	9	1	3	5	1	3	5	1	1	3	3	5	7	5	7	9
S8	7	7	9	1	1	3	3	5	7	1	3	5	1	3	5	5	7	9	7	7	9	1	3	5	1	1	3	5	7	9	1	3	5	1	1	3	5	7	9	1	3	5	5	7	9
S9	5	7	9	1	3	5	1	3	5	7	7	9	5	7	9	5	7	9	1	1	3	1	1	3	1	1	3	3	5	7	1	3	5	5	7	9	1	3	5	7	7	9	3	5	7
S10	1	1	3	1	1	3	3	5	7	5	7	9	1	3	5	3	5	7	1	1	3	1	3	5	7	7	9	3	5	7	1	3	5	1	1	3	7	7	9	7	7	9	1	1	3

Table A.5 Response from expert 5

	B1			B2			B3			B4			B5			B6			B7			B8		
S1	1	1	3	1	3	5	5	7	9	7	7	9	5	7	9	7	7	9	1	1	3	7	7	9
S2	1	3	5	5	7	9	3	5	7	1	3	5	1	1	3	1	3	5	1	3	5	1	1	3
S3	5	7	9	1	1	3	3	5	7	3	5	7	5	7	9	1	1	3	3	5	7	1	1	3
S4	5	7	9	5	7	9	1	1	3	1	1	3	3	5	7	5	7	9	1	1	3	1	3	5
S5	1	1	3	5	7	9	1	3	5	7	7	9	1	1	3	5	7	9	5	7	9	5	7	9
S6	7	7	9	5	7	9	1	3	5	3	5	7	5	7	9	1	3	5	7	7	9	7	7	9
S7	1	3	5	7	7	9	1	1	3	7	7	9	3	5	7	5	7	9	7	7	9	1	3	5
S8	1	3	5	3	5	7	3	5	7	5	7	9	1	1	3	3	5	7	3	5	7	1	1	3
S9	5	7	9	1	1	3	7	7	9	7	7	9	5	7	9	5	7	9	3	5	7	3	5	7
S10	1	3	5	7	7	9	7	7	9	1	1	3	1	1	3	5	7	9	1	1	3	3	5	7

	B9			B10			B11			B12			B13			B14			B15		
S1	7	7	9	5	7	9	1	1	3	3	5	7	3	5	7	3	5	7	5	7	9
S2	1	3	5	1	3	5	3	5	7	5	7	9	1	1	3	5	7	9	5	7	9
S3	1	3	5	7	7	9	1	1	3	1	1	3	1	3	5	5	7	9	7	7	9
S4	7	7	9	3	5	7	1	3	5	7	7	9	1	3	5	1	1	3	1	3	5
S5	7	7	9	1	3	5	3	5	7	7	7	9	1	1	3	1	3	5	1	1	3
S6	5	7	9	1	3	5	5	7	9	5	7	9	7	7	9	5	7	9	7	7	9
S7	7	7	9	5	7	9	7	7	9	3	5	7	7	7	9	3	5	7	3	5	7
S8	5	7	9	7	7	9	5	7	9	1	1	3	1	1	3	5	7	9	1	3	5
S9	1	3	5	1	3	5	7	7	9	5	7	9	5	7	9	7	7	9	1	3	5
S10	5	7	9	1	3	5	1	3	5	1	3	5	1	1	3	5	7	9	5	7	9

Table A.6 Response from expert 6

	B1			B2			B3			B4			B5			B6			B7			B8			B9			B10			B11			B12			B13			B14			B15		
S1	5	7	9	3	5	7	5	7	9	5	7	9	7	7	9	5	7	9	5	7	9	1	1	3	1	1	3	5	7	9	5	7	9	3	5	7	3	5	7	7	7	9	5	7	9
S2	1	3	5	1	1	3	1	1	3	7	7	9	1	1	3	7	7	9	7	7	9	5	7	9	5	7	9	5	7	9	3	5	7	5	7	9	3	5	7	1	1	3	5	7	9
S3	3	5	7	1	1	3	3	5	7	7	7	9	1	3	5	1	1	3	1	1	3	3	5	7	7	7	9	1	3	5	1	3	5	5	7	9	3	5	7	1	3	5	7	7	9
S4	5	7	9	1	3	5	1	3	5	1	3	5	5	7	9	1	3	5	1	1	3	1	3	5	3	5	7	1	1	3	1	3	5	7	7	9	3	5	7	1	3	5	1	3	5
S5	7	7	9	7	7	9	3	5	7	3	5	7	7	7	9	1	1	3	3	5	7	7	7	9	7	7	9	7	7	9	3	5	7	7	7	9	1	1	3	1	3	5	7	7	9
S6	1	1	3	7	7	9	5	7	9	1	3	5	1	3	5	5	7	9	7	7	9	1	1	3	1	1	3	1	3	5	3	5	7	1	1	3	5	7	9	7	7	9	3	5	7
S7	3	5	7	1	1	3	3	5	7	7	7	9	7	7	9	7	7	9	3	5	7	1	3	5	3	5	7	1	1	3	1	3	5	7	7	9	3	5	7	3	5	7	3	5	7
S8	5	7	9	1	1	3	5	7	9	7	7	9	1	3	5	5	7	9	7	7	9	5	7	9	1	3	5	5	7	9	7	7	9	7	7	9	5	7	9	7	7	9	1	3	5
S9	3	5	7	7	7	9	7	7	9	1	3	5	5	7	9	5	7	9	1	1	3	3	5	7	1	3	5	3	5	7	5	7	9	1	3	5	1	1	3	1	3	5	1	1	3
S10	1	3	5	7	7	9	1	1	3	1	1	3	5	7	9	5	7	9	3	5	7	5	7	9	3	5	7	1	1	3	5	7	9	5	7	9	3	5	7	1	1	3	1	1	3

Table A.7 Response from expert 7

	B1			B2			B3			B4			B5			B6			B7			B8			B9			B10			B11			B12			B13			B14			B15		
S1	1	1	3	7	7	9	1	1	3	7	[illegible]	9	5	7	9	7	7	9	1	1	3	3	5	7	3	5	7	1	3	5	1	3	5	7	7	9	5	7	9	7	7	9	1	1	3
S2	5	7	9	1	1	3	1	1	3	3	[illegible]	7	5	7	9	3	5	7	1	1	3	1	1	3	7	7	9	7	7	9	1	1	3	5	7	9	5	7	9	7	7	9	1	3	5
S3	5	7	9	5	7	9	1	3	5	1	[illegible]	5	3	5	7	5	7	9	1	3	5	5	7	9	5	7	9	1	1	3	1	1	3	5	7	9	7	7	9	3	5	7	1	1	3
S4	1	1	3	3	5	7	7	7	9	7	[illegible]	9	5	7	9	1	3	5	5	7	9	5	7	9	7	7	9	5	7	9	1	3	5	7	7	9	3	5	7	3	5	7	5	7	9
S5	1	3	5	7	7	9	5	7	9	7	[illegible]	9	5	7	9	7	7	9	1	1	3	1	3	5	3	5	7	5	7	9	5	7	9	7	7	9	3	5	7	5	7	9	5	7	9
S6	1	3	5	7	7	9	7	7	9	1	[illegible]	5	5	7	9	7	7	9	1	1	3	5	7	9	1	3	5	1	1	3	1	1	3	1	3	5	7	7	9	5	7	9	1	1	3
S7	5	7	9	5	7	9	3	5	7	3	[illegible]	7	5	7	9	3	5	7	1	1	3	5	7	9	3	5	7	3	5	7	1	3	5	1	3	5	1	3	5	7	7	9	1	3	5
S8	3	5	7	1	1	3	1	3	5	7	[illegible]	9	3	5	7	1	1	3	5	7	9	1	1	3	1	1	3	1	3	5	3	5	7	3	5	7	5	7	9	7	7	9	3	5	7
S9	1	3	5	5	7	9	1	3	5	1	[illegible]	5	3	5	7	5	7	9	7	7	9	3	5	7	1	1	3	1	1	3	7	7	9	1	1	3	1	1	3	5	7	9	3	5	7
S10	5	7	9	7	7	9	1	1	3	1	[illegible]	3	5	7	9	5	7	9	1	3	5	1	1	3	1	1	3	1	3	5	1	3	5	1	1	3	5	7	9	3	5	7	3	5	7

Table A.8 Response from expert 8

	B1			B2			B3			B4			B5			B6			B7			B8			B9			B10			B11			B12			B13			B14			B15		
S1	1	1	3	3	5	7	1	1	3	1	1	3	7	7	9	5	7	9	1	1	3	1	3	5	7	7	9	1	3	5	3	5	7	5	7	9	1	3	5	5	7	9	3	5	7
S2	7	7	9	5	7	9	5	7	9	1	3	5	5	7	9	7	7	9	7	7	9	7	7	9	7	7	9	7	7	9	1	1	3	1	3	5	3	5	7	5	7	9	7	7	9
S3	1	3	5	1	3	5	1	3	5	1	3	5	7	7	9	7	7	9	5	7	9	3	5	7	3	5	7	1	1	3	3	5	7	1	1	3	5	7	9	7	7	9	1	1	3
S4	1	1	3	3	5	7	5	7	9	3	5	7	3	5	7	7	7	9	7	7	9	3	5	7	5	7	9	5	7	9	3	5	7	5	7	9	1	3	5	1	1	3	1	1	3
S5	5	7	9	5	7	9	5	7	9	5	7	9	5	7	9	1	1	3	1	1	3	3	5	7	3	5	7	5	7	9	1	3	5	1	3	5	1	3	5	5	7	9	1	3	5
S6	7	7	9	1	1	3	5	7	9	3	5	7	1	3	5	1	1	3	5	7	9	1	3	5	1	3	5	1	3	5	1	3	5	1	1	3	3	5	7	1	3	5	1	1	3
S7	3	5	7	7	7	9	1	3	5	1	1	3	3	5	7	1	3	5	7	7	9	3	5	7	3	5	7	7	7	9	7	7	9	5	7	9	1	1	3	1	1	3	3	5	7
S8	7	7	9	5	7	9	5	7	9	5	7	9	5	7	9	3	5	7	7	7	9	1	3	5	5	7	9	3	5	7	1	1	3	3	5	7	1	3	5	1	3	5	3	5	7
S9	3	5	7	5	7	9	1	1	3	5	7	9	1	1	3	3	5	7	5	7	9	1	3	5	3	5	7	3	5	7	5	7	9	7	7	9	1	3	5	3	5	7	5	7	9
S10	1	3	5	1	1	3	5	7	9	3	5	7	5	7	9	7	7	9	1	3	5	1	1	3	1	1	3	7	7	9	5	7	9	1	1	3	7	7	9	1	3	5	5	7	9

Table A.9 Response from expert 9

	B1			B2			B3			B4			B5			B6			B7			B8			B9			B10			B11			B12			B13			B14			B15		
S1	7	7	9	7	7	9	1	3	5	3	[illegible]	7	5	7	9	1	3	5	1	1	3	7	7	9	1	1	3	5	7	9	1	1	3	1	1	3	5	7	9	1	1	3	3	5	7
S2	3	5	7	7	7	9	1	3	5	5	[illegible]	9	1	1	3	7	7	9	7	7	9	5	7	9	1	1	3	3	5	7	1	1	3	1	3	5	7	7	9	5	7	9	1	3	5
S3	1	3	5	1	1	3	3	5	7	1	[illegible]	5	3	5	7	5	7	9	3	5	7	1	3	5	1	3	5	3	5	7	1	3	5	1	3	5	1	1	3	1	1	3	1	1	3
S4	5	7	9	1	3	5	5	7	9	5	[illegible]	9	7	7	9	5	7	9	3	5	7	5	7	9	1	1	3	3	5	7	1	3	5	5	7	9	1	3	5	1	3	5	1	1	3
S5	1	3	5	1	1	3	1	3	5	1	[illegible]	3	1	1	3	5	7	9	7	7	9	1	3	5	1	1	3	1	3	5	1	3	5	1	3	5	7	7	9	1	1	3	3	5	7
S6	1	1	3	3	5	7	5	7	9	5	[illegible]	9	3	5	7	1	3	5	3	5	7	7	7	9	3	5	7	1	1	3	1	3	5	3	5	7	1	1	3	5	7	9	3	5	7
S7	5	7	9	3	5	7	3	5	7	7	[illegible]	9	5	7	9	7	7	9	1	1	3	1	3	5	1	3	5	1	3	5	7	7	9	7	7	9	7	7	9	3	5	7	7	7	9
S8	1	3	5	5	7	9	1	1	3	1	[illegible]	5	5	7	9	1	1	3	3	5	7	1	1	3	1	3	5	7	7	9	1	3	5	1	3	5	5	7	9	5	7	9	3	5	7
S9	1	1	3	3	5	7	5	7	9	7	[illegible]	9	7	7	9	1	3	5	3	5	7	7	7	9	1	3	5	1	3	5	3	5	7	5	7	9	1	3	5	1	3	5	1	3	5
S10	7	7	9	1	3	5	5	7	9	1	[illegible]	3	1	1	3	5	7	9	7	7	9	1	3	5	5	7	9	7	7	9	3	5	7	1	3	5	5	7	9	7	7	9	1	1	3

Table A.10: Response from expert 10

	B1			B2			B3			B4			B5			B6			B7			B8			B9			B10			B11			B12			B13			B14			B15		
S1	1	3	5	3	5	7	3	5	7	3	5	7	7	7	9	1	3	5	3	5	7	1	1	3	3	5	7	1	1	3	3	5	7	1	3	5	1	3	5	1	3	5	3	5	7
S2	5	7	9	1	1	3	5	7	9	1	1	3	1	1	3	7	7	9	7	7	9	3	5	7	5	7	9	1	1	3	1	3	5	5	7	9	7	7	9	1	3	5	5	7	9
S3	1	3	5	1	3	5	5	7	9	1	3	5	1	3	5	3	5	7	1	1	3	1	3	5	1	1	3	3	5	7	3	5	7	3	5	7	5	7	9	1	3	5	5	7	9
S4	5	7	9	3	5	7	7	7	9	7	7	9	1	1	3	7	7	9	1	1	3	3	5	7	7	7	9	7	7	9	5	7	9	1	1	3	1	3	5	1	3	5	5	7	9
S5	1	1	3	1	1	3	3	5	7	5	7	9	1	1	3	7	7	9	7	7	9	1	3	5	1	3	5	1	1	3	1	3	5	5	7	9	5	7	9	1	1	3	1	1	3
S6	1	3	5	5	7	9	5	7	9	7	7	9	5	7	9	5	7	9	7	7	9	7	7	9	1	3	5	1	1	3	7	7	9	1	3	5	7	7	9	1	3	5	7	7	9
S7	5	7	9	7	7	9	1	1	3	5	7	9	1	3	5	1	3	5	1	1	3	1	3	5	3	5	7	1	3	5	1	3	5	1	1	3	1	3	5	5	7	9	7	7	9
S8	5	7	9	5	7	9	7	7	9	5	7	9	1	1	3	1	3	5	1	3	5	5	7	9	3	5	7	5	7	9	1	3	5	1	3	5	1	1	3	7	7	9	5	7	9
S9	1	3	5	5	7	9	3	5	7	3	5	7	1	1	3	7	7	9	1	3	5	3	5	7	3	5	7	7	7	9	7	7	9	3	5	7	5	7	9	5	7	9	1	1	3
S10	7	7	9	1	1	3	1	1	3	5	7	9	3	5	7	1	1	3	3	5	7	1	3	5	5	7	9	3	5	7	5	7	9	3	5	7	1	3	5	1	3	5	1	3	5

Appendix B

Table B.1 Consolidated response of the ten experts

Combined	B1			B2			B3			B4			B5		
S1	1	3.4	9	1	4.2	9	1	4.2	9	1	5.2	9	1	5.8	9
S2	1	5.2	9	1	4	9	1	4.2	9	1	3.8	9	1	3.4	9
S3	1	5.4	9	1	3.2	9	1	4.2	9	1	4.6	9	1	5	9
S4	1	4.2	9	1	4.6	9	1	5.2	9	1	4.6	9	1	4.4	9
S5	1	4.8	9	1	4.6	9	1	4.8	9	1	5.8	9	1	4.4	9
S6	1	4.6	9	1	5.4	9	1	4.4	9	1	4.6	9	1	4.6	9
S7	1	6	9	1	4.6	9	1	4.4	9	1	4.2	9	1	5	9
S8	1	4.8	9	1	4.8	9	1	5.2	9	1	5.6	9	1	3.4	9
S9	1	4	9	1	5.6	9	1	4.6	9	1	5.6	9	1	4	9
S10	1	4.2	9	1	3.8	9	1	4.4	9	1	4.4	9	1	5.2	9

Combined	B6			B7			B8			B9			B10		
S1	1	5.6	9	1	3.2	9	1	4.4	9	1	5.4	9	1	4.6	9
S2	1	5	9	1	5.4	9	1	4.6	9	1	4.4	9	1	4.4	9
S3	1	5.2	9	1	4	9	1	5.2	9	1	4.6	9	1	3.6	9
S4	1	5.4	9	1	4	9	1	4.8	9	1	4.6	9	1	5.2	9
S5	1	4.8	9	1	5	9	1	4.8	9	1	4.4	9	1	5	9
S6	1	4.8	9	1	5.2	9	1	5.2	9	1	3.8	9	1	2.8	9
S7	1	4.8	9	1	5	9	1	4.6	9	1	4	9	1	4.4	9
S8	1	4.2	9	1	5	9	1	4.4	9	1	3.6	9	1	5.6	9
S9	1	5.4	9	1	4	9	1	4	9	1	3.2	7	1	4	9
S10	1	5.6	9	1	3.8	9	1	4.2	9	1	5	9	1	4.6	9

Combined	B11			B12			B13			B14			B15		
S1	1	3.6	9	1	4.2	9	1	5	9	1	5.4	9	1	5.4	9
S2	1	3.2	9	1	5.8	9	1	5	9	1	4.2	9	1	5.4	9
S3	1	4	9	1	5.2	9	1	4	9	1	4.2	9	1	4.4	9
S4	1	3.8	9	1	4.6	9	1	3.8	9	1	3.4	7	1	3.8	9
S5	1	4.6	9	1	5.2	9	1	3.6	9	1	3.4	9	1	4.6	9
S6	1	4.4	9	1	4.2	9	1	4.8	9	1	5.6	9	1	4	9
S7	1	4.8	9	1	5	9	1	4.4	9	1	5	9	1	5.8	9
S8	1	4.4	9	1	3.8	9	1	4.2	9	1	6.2	9	1	5	9
S9	1	5.4	9	1	5	9	1	3.4	9	1	5.2	9	1	3.2	9
S10	1	4.6	9	1	4.2	9	1	5	9	1	5.2	9	1	4.2	9

Appendix C

Table C.1 Normalized decision matrix

	7	7	9	7	7	9	7	7	9	5	7	9
Combined	B1			B2			B3			B4		
S1	0.111111111	0.377777778	1	0.111111111	0.466666667	1	0.111111111	0.466666667	1	0.111111111	0.577777778	1
S2	0.111111111	0.577777778	1	0.111111111	0.444444444	1	0.111111111	0.466666667	1	0.111111111	0.422222222	1
S3	0.111111111	0.6	1	0.111111111	0.355555556	1	0.111111111	0.466666667	1	0.111111111	0.511111111	1
S4	0.111111111	0.466666667	1	0.111111111	0.511111111	1	0.111111111	0.577777778	1	0.111111111	0.511111111	1
S5	0.111111111	0.533333333	1	0.111111111	0.511111111	1	0.111111111	0.533333333	1	0.111111111	0.644444444	1
S6	0.111111111	0.511111111	1	0.111111111	0.6	1	0.111111111	0.488888889	1	0.111111111	0.511111111	1
S7	0.111111111	0.666666667	1	0.111111111	0.511111111	1	0.111111111	0.488888889	1	0.111111111	0.466666667	1
S8	0.111111111	0.533333333	1	0.111111111	0.533333333	1	0.111111111	0.577777778	1	0.111111111	0.622222222	1
S9	0.111111111	0.444444444	1	0.111111111	0.622222222	1	0.111111111	0.511111111	1	0.111111111	0.622222222	1
S10	0.111111111	0.466666667	1	0.111111111	0.422222222	1	0.111111111	0.488888889	1	0.111111111	0.488888889	1

	3	5	7	1	3	5	1	3	5
Combined	B9			B10			B11		
S1	0.111111111	0.6	1	0.111111111	0.511111111	1	0.111111111	0.4	1
S2	0.111111111	0.488888889	1	0.111111111	0.488888889	1	0.111111111	0.355555556	1
S3	0.111111111	0.511111111	1	0.111111111	0.4	1	0.111111111	0.444444444	1
S4	0.111111111	0.511111111	1	0.111111111	0.577777778	1	0.111111111	0.422222222	1
S5	0.111111111	0.488888889	1	0.111111111	0.555555556	1	0.111111111	0.511111111	1
S6	0.111111111	0.422222222	1	0.111111111	0.311111111	1	0.111111111	0.488888889	1
S7	0.111111111	0.444444444	1	0.111111111	0.488888889	1	0.111111111	0.533333333	1
S8	0.111111111	0.4	1	0.111111111	0.622222222	1	0.111111111	0.488888889	1
S9	0.111111111	0.355555556	0.777777778	0.111111111	0.444444444	1	0.111111111	0.6	1
S10	0.111111111	0.555555556	1	0.111111111	0.511111111	1	0.111111111	0.511111111	1

5	7	9	5	7	9	3	5	7	3	5	7
B5			B6			B7			B8		
0.111111111	0.644444444	1	0.111111111	0.622222222	1	0.111111111	0.355555556	1	0.111111111	0.488888889	1
0.111111111	0.377777778	1	0.111111111	0.555555556	1	0.111111111	0.6	1	0.111111111	0.511111111	1
0.111111111	0.555555556	1	0.111111111	0.577777778	1	0.111111111	0.444444444	1	0.111111111	0.577777778	1
0.111111111	0.488888889	1	0.111111111	0.6	1	0.111111111	0.444444444	1	0.111111111	0.533333333	1
0.111111111	0.488888889	1	0.111111111	0.533333333	1	0.111111111	0.555555556	1	0.111111111	0.533333333	1
0.111111111	0.511111111	1	0.111111111	0.533333333	1	0.111111111	0.577777778	1	0.111111111	0.577777778	1
0.111111111	0.555555556	1	0.111111111	0.533333333	1	0.111111111	0.555555556	1	0.111111111	0.511111111	1
0.111111111	0.377777778	1	0.111111111	0.466666667	1	0.111111111	0.555555556	1	0.111111111	0.488888889	1
0.111111111	0.444444444	1	0.111111111	0.6	1	0.111111111	0.444444444	1	0.111111111	0.444444444	1
0.111111111	0.577777778	1	0.111111111	0.622222222	1	0.111111111	0.422222222	1	0.111111111	0.466666667	1

1	3	5	1	1	3	1	1	3	1	1	3
B12			B13			B14			B15		
0.111111111	0.466666667	1	0.111111111	0.555555556	1	0.111111111	0.6	1	0.111111111	0.6	1
0.111111111	0.644444444	1	0.111111111	0.555555556	1	0.111111111	0.466666667	1	0.111111111	0.6	1
0.111111111	0.577777778	1	0.111111111	0.444444444	1	0.111111111	0.466666667	1	0.111111111	0.488888889	1
0.111111111	0.511111111	1	0.111111111	0.422222222	1	0.111111111	0.377777778	0.777777778	0.111111111	0.422222222	1
0.111111111	0.577777778	1	0.111111111	0.4	1	0.111111111	0.377777778	1	0.111111111	0.511111111	1
0.111111111	0.466666667	1	0.111111111	0.533333333	1	0.111111111	0.622222222	1	0.111111111	0.444444444	1
0.111111111	0.555555556	1	0.111111111	0.488888889	1	0.111111111	0.555555556	1	0.111111111	0.644444444	1
0.111111111	0.422222222	1	0.111111111	0.466666667	1	0.111111111	0.688888889	1	0.111111111	0.555555556	1
0.111111111	0.555555556	1	0.111111111	0.377777778	1	0.111111111	0.577777778	1	0.111111111	0.355555556	1
0.111111111	0.466666667	1	0.111111111	0.555555556	1	0.111111111	0.577777778	1	0.111111111	0.466666667	1

Appendix D

Table D.1 Weight normalized decision matrix

Combined	B1			B2			B3			B4		
S1	0.7	2.6	9	0.7	3.2	9	0.7	3.2	9	0.5	4.04	9
S2	0.7	4.044444444	9	0.777777778	3.111111111	9	0.777777778	3.266666667	9	0.555555556	2.955555556	9
S3	0.777777778	4.2	9	0.777777778	2.488888889	9	0.777777778	3.266666667	9	0.555555556	3.577777778	9
S4	0.777777778	3.266666667	9	0.777777778	3.577777778	9	0.777777778	4.044444444	9	0.555555556	3.577777778	9
S5	0.777777778	3.733333333	9	0.777777778	3.577777778	9	0.777777778	3.733333333	9	0.555555556	4.511111111	9
S6	0.777777778	3.577777778	9	0.777777778	4.2	9	0.777777778	3.422222222	9	0.555555556	3.577777778	9
S7	0.777777778	4.666666667	9	0.777777778	3.577777778	9	0.777777778	3.422222222	9	0.555555556	3.266666667	9
S8	0.777777778	3.733333333	9	0.777777778	3.733333333	9	0.777777778	4.044444444	9	0.555555556	4.355555556	9
S9	0.777777778	3.111111111	9	0.777777778	4.355555556	9	0.777777778	3.577777778	9	0.555555556	4.355555556	9
S10	0.777777778	3.266666667	9	0.777777778	2.955555556	9	0.777777778	3.422222222	9	0.555555556	3.422222222	9

Combined	B9			B10			B11		
S1	0.3	3	7	0.1	1.5	5	0.1	1.2	5
S2	0.333333333	2.444444444	7	0.111111111	1.466666667	5	0.111111111	1.066666667	5
S3	0.333333333	2.555555556	7	0.111111111	1.2	5	0.111111111	1.333333333	5
S4	0.333333333	2.555555556	7	0.111111111	1.733333333	5	0.111111111	1.266666667	5
S5	0.333333333	2.444444444	7	0.111111111	1.666666667	5	0.111111111	1.533333333	5
S6	0.333333333	2.111111111	7	0.111111111	0.933333333	5	0.111111111	1.466666667	5
S7	0.333333333	2.222222222	7	0.111111111	1.466666667	5	0.111111111	1.6	5
S8	0.333333333	2	7	0.111111111	1.866666667	5	0.111111111	1.466666667	5
S9	0.333333333	1.777777778	5.444444444	0.111111111	1.333333333	5	0.111111111	1.8	5
S10	0.333333333	2.777777778	7	0.111111111	1.533333333	5	0.111111111	1.533333333	5

B5			B6			B7			B8		
0.5	4.5	9	0.5	4.3	9	0.3	1.7	7	0.3	2.4	7
0.555555556	2.644444444	9	0.555555556	3.888888889	9	0.333333333	3	7	0.333333333	2.555555556	7
0.555555556	3.888888889	9	0.555555556	4.044444444	9	0.333333333	2.222222222	7	0.333333333	2.888888889	7
0.555555556	3.422222222	9	0.555555556	4.2	9	0.333333333	2.222222222	7	0.333333333	2.666666667	7
0.555555556	3.422222222	9	0.555555556	3.733333333	9	0.333333333	2.777777778	7	0.333333333	2.666666667	7
0.555555556	3.577777778	9	0.555555556	3.733333333	9	0.333333333	2.888888889	7	0.333333333	2.888888889	7
0.555555556	3.888888889	9	0.555555556	3.733333333	9	0.333333333	2.777777778	7	0.333333333	2.555555556	7
0.555555556	2.644444444	9	0.555555556	3.266666667	9	0.333333333	2.777777778	7	0.333333333	2.444444444	7
0.555555556	3.111111111	9	0.555555556	4.2	9	0.333333333	2.222222222	7	0.333333333	2.222222222	7
0.555555556	4.044444444	9	0.555555556	4.355555556	9	0.333333333	2.111111111	7	0.333333333	2.333333333	7

B12			B13			B14			B15		
0.1	1.4	5	0.1	0.5	3	0.1	0.6	3	0.1	0.6	3
0.111111111	1.933333333	5	0.111111111	0.555555556	3	0.111111111	0.466666667	3	0.111111111	0.6	3
0.111111111	1.733333333	5	0.111111111	0.444444444	3	0.111111111	0.466666667	3	0.111111111	0.488888889	3
0.111111111	1.533333333	5	0.111111111	0.422222222	3	0.111111111	0.377777778	2.333333333	0.111111111	0.422222222	3
0.111111111	1.733333333	5	0.111111111	0.4	3	0.111111111	0.377777778	3	0.111111111	0.511111111	3
0.111111111	1.4	5	0.111111111	0.533333333	3	0.111111111	0.622222222	3	0.111111111	0.444444444	3
0.111111111	1.666666667	5	0.111111111	0.488888889	3	0.111111111	0.555555556	3	0.111111111	0.644444444	3
0.111111111	1.266666667	5	0.111111111	0.466666667	3	0.111111111	0.688888889	3	0.111111111	0.555555556	3
0.111111111	1.666666667	5	0.111111111	0.377777778	3	0.111111111	0.577777778	3	0.111111111	0.355555556	3
0.111111111	1.4	5	0.111111111	0.555555556	3	0.111111111	0.577777778	3	0.111111111	0.466666667	3

Appendix E

Table E.1 Fuzzy positive ideal and fuzzy negative solution

Combined	B1			B2			B3			B4		
S1	0.777777778	2.644444444	9	0.777777778	3.266666667	9	0.777777778	3.266666667	9	0.555555556	4.044444444	9
S2	0.777777778	4.044444444	9	0.777777778	3.111111111	9	0.777777778	3.266666667	9	0.555555556	2.955555556	9
S3	0.777777778	4.2	9	0.777777778	2.488888889	9	0.777777778	3.266666667	9	0.555555556	3.577777778	9
S4	0.777777778	3.266666667	9	0.777777778	3.577777778	9	0.777777778	4.044444444	9	0.555555556	3.577777778	9
S5	0.777777778	3.733333333	9	0.777777778	3.577777778	9	0.777777778	3.733333333	9	0.555555556	4.511111111	9
S6	0.777777778	3.577777778	9	0.777777778	4.2	9	0.777777778	3.422222222	9	0.555555556	3.577777778	9
S7	0.777777778	4.666666667	9	0.777777778	3.577777778	9	0.777777778	3.422222222	9	0.555555556	3.266666667	9
S8	0.777777778	3.733333333	9	0.777777778	3.733333333	9	0.777777778	4.044444444	9	0.555555556	4.355555556	9
S9	0.777777778	3.111111111	9	0.777777778	4.355555556	9	0.777777778	3.577777778	9	0.555555556	4.355555556	9
S10	0.777777778	3.266666667	9	0.777777778	2.955555556	9	0.777777778	3.422222222	9	0.555555556	3.422222222	9
A*	0.7	4.6	9	0.78	4.3	9	0.7	4.0	9	0.5	4.5	9
A–	0.7	2.6	9	0.7	2.4	9	0.7	3.2	9	0.5	2.9	9

Combined	B9			B10			B11		
S1	0.333333333	3	7	0.111111111	1.533333333	5	0.111111111	1.2	5
S2	0.333333333	2.444444444	7	0.111111111	1.466666667	5	0.111111111	1.066666667	5
S3	0.333333333	2.555555556	7	0.111111111	1.2	5	0.111111111	1.333333333	5
S4	0.333333333	2.555555556	7	0.111111111	1.733333333	5	0.111111111	1.266666667	5
S5	0.333333333	2.444444444	7	0.111111111	1.666666667	5	0.111111111	1.533333333	5
S6	0.333333333	2.111111111	7	0.111111111	0.933333333	5	0.111111111	1.466666667	5
S7	0.333333333	2.222222222	7	0.111111111	1.466666667	5	0.111111111	1.6	5
S8	0.333333333	2	7	0.111111111	1.866666667	5	0.111111111	1.466666667	5
S9	0.333333333	1.777777778	5.444444444	0.111111111	1.333333333	5	0.111111111	1.8	5
S10	0.333333333	2.777777778	7	0.111111111	1.533333333	5	0.111111111	1.533333333	5
A*	0.3	3	7	0.1	1.8	5	0.1	1.8	5
A–	0.3	1.7	5.4	0.1	0.9	5	0.1	1.0	5

B5			B6			B7			B8		
0.555555556	4.511111111	9	0.555555556	4.355555556	9	0.333333333	1.777777778	7	0.333333333	2.444444444	7
0.555555556	2.644444444	9	0.555555556	3.888888889	9	0.333333333	3	7	0.333333333	2.555555556	7
0.555555556	3.888888889	9	0.555555556	4.044444444	9	0.333333333	2.222222222	7	0.333333333	2.888888889	7
0.555555556	3.422222222	9	0.555555556	4.2	9	0.333333333	2.222222222	7	0.333333333	2.666666667	7
0.555555556	3.422222222	9	0.555555556	3.733333333	9	0.333333333	2.777777778	7	0.333333333	2.666666667	7
0.555555556	3.577777778	9	0.555555556	3.733333333	9	0.333333333	2.888888889	7	0.333333333	2.888888889	7
0.555555556	3.888888889	9	0.555555556	3.733333333	9	0.333333333	2.777777778	7	0.333333333	2.555555556	7
0.555555556	2.644444444	9	0.555555556	3.266666667	9	0.333333333	2.777777778	7	0.333333333	2.444444444	7
0.555555556	3.111111111	9	0.555555556	4.2	9	0.333333333	2.222222222	7	0.333333333	2.222222222	7
0.555555556	4.044444444	9	0.555555556	4.355555556	9	0.333333333	2.111111111	7	0.333333333	2.333333333	7
0.5	4.5	9	0.5	4.3	9	0.333	3	7	0.3	2.8	7
0.5	2.6	9	0.5	3.2	9	0.3	1.7	7	0.3	2.2	7

B12			B13			B14			B15		
0.111111111	1.4	5	0.111111111	0.555555556	3	0.111111111	0.6	3	0.111111111	0.6	3
0.111111111	1.933333333	5	0.111111111	0.555555556	3	0.111111111	0.466666667	3	0.111111111	0.6	3
0.111111111	1.733333333	5	0.111111111	0.444444444	3	0.111111111	0.466666667	3	0.111111111	0.488888889	3
0.111111111	1.533333333	5	0.111111111	0.422222222	3	0.111111111	0.377777778	2.333333333	0.111111111	0.422222222	3
0.111111111	1.733333333	5	0.111111111	0.4	3	0.111111111	0.377777778	3	0.111111111	0.511111111	3
0.111111111	1.4	5	0.111111111	0.533333333	3	0.111111111	0.622222222	3	0.111111111	0.444444444	3
0.111111111	1.666666667	5	0.111111111	0.488888889	3	0.111111111	0.555555556	3	0.111111111	0.644444444	3
0.111111111	1.266666667	5	0.111111111	0.466666667	3	0.111111111	0.688888889	3	0.111111111	0.555555556	3
0.111111111	1.666666667	5	0.111111111	0.377777778	3	0.111111111	0.577777778	3	0.111111111	0.355555556	3
0.111111111	1.4	5	0.111111111	0.555555556	3	0.111111111	0.577777778	3	0.111111111	0.466666667	3
0.1	1.9	5	0.1	0.5	3	0.1	0.6	3	0.1	0.5	3
0.1	1.2	5	0.1	0.4	3	0.1	0.3	2.3	0.1	0.4	3

Appendix F

Table F.1 Distance between strategies and ideal positive reference point

Combined	B1	B2	B3	B4	B5	B6	B7	B8	B9	B10	B11	B12	B13	B14	B15	di*
S1	1.1675	0.6287	0.4491	0.2694	0.0000	0.0000	0.7057	0.2566	0.0000	0.1925	0.3464	0.3079	0.0000	0.0513	0.0257	4.3750
S2	0.3592	0.7185	0.4491	0.8981	1.0777	0.2694	0.0000	0.1925	0.3208	0.2309	0.4234	0.0000	0.0000	0.1283	0.0257	5.0679
S3	0.2694	1.0777	0.4491	0.5389	0.3592	0.1796	0.4491	0.0000	0.2566	0.3849	0.2694	0.1155	0.0642	0.1283	0.0385	4.5418
S4	0.8083	0.4491	0.0000	0.5389	0.6287	0.0898	0.4491	0.1283	0.2566	0.0770	0.3079	0.2309	0.0770	0.4247	0.0770	4.4662
S5	0.5389	0.4491	0.1796	0.0000	0.6287	0.3592	0.1283	0.1283	0.3208	0.1155	0.1540	0.1155	0.0898	0.1796	0.0257	3.3871
S6	0.6287	0.0898	0.3592	0.5389	0.5389	0.3592	0.0642	0.0000	0.5132	0.5389	0.1925	0.3079	0.0128	0.0385	0.0642	4.1826
S7	0.0000	0.4491	0.3592	0.7185	0.3592	0.3592	0.1283	0.1925	0.4491	0.2309	0.1155	0.1540	0.0385	0.0770	0.0513	3.6309
S8	0.5389	0.3592	0.0000	0.0898	1.0777	0.6287	0.1283	0.2566	0.5774	0.0000	0.1925	0.3849	0.0513	0.0000	0.0000	4.2852
S9	0.8981	0.0000	0.2694	0.0898	0.8083	0.0898	0.4491	0.3849	1.1422	0.3079	0.0000	0.1540	0.1026	0.0642	0.1155	4.7602
S10	0.8083	0.8083	0.3592	0.6287	0.2694	0.0000	0.5132	0.3208	0.1283	0.1925	0.1540	0.3079	0.0000	0.0642	0.0513	4.5547

Table F.2 Distance between strategies and ideal negative reference point

Combined	B1	B2	B3	B4	B5	B6	B7	B8	B9	B10	B11	B12	B13	B14	B15	di-
S1	0.0000	0.4491	0.0000	0.6287	1.0777	0.6287	0.0000	0.1283	1.1422	0.3464	0.0770	0.0770	0.0898	0.4057	0.0898	5.1403
S2	0.8083	0.3592	0.0000	0.0000	0.0000	0.3592	0.7057	0.1925	0.9771	0.3079	0.0000	0.3849	0.0898	0.3883	0.0898	4.6627
S3	0.8981	0.0000	0.0000	0.3592	0.7185	0.4491	0.2566	0.3849	1.0041	0.1540	0.1540	0.2694	0.0257	0.3883	0.0257	5.0875
S4	0.3592	0.6287	0.4491	0.3592	0.4491	0.5389	0.2566	0.2566	1.0041	0.4619	0.1155	0.1540	0.0128	0.0000	0.0128	5.0584
S5	0.6287	0.6287	0.2694	0.8981	0.4491	0.2694	0.5774	0.2566	0.9771	0.4234	0.2694	0.2694	0.0000	0.3849	0.0385	6.3400
S6	0.5389	0.9879	0.0898	0.3592	0.5389	0.2694	0.6415	0.3849	0.9185	0.0000	0.2309	0.0770	0.0770	0.4100	0.0000	5.5239
S7	1.1675	0.6287	0.0898	0.1796	0.7185	0.2694	0.5774	0.1925	0.9340	0.3079	0.3079	0.2309	0.0513	0.3984	0.1155	6.1693
S8	0.6287	0.7185	0.4491	0.8083	0.0000	0.0000	0.5774	0.1283	0.9072	0.5389	0.2309	0.0000	0.0385	0.4247	0.0642	5.5145
S9	0.2694	1.0777	0.1796	0.8083	0.2694	0.5389	0.2566	0.0000	0.0000	0.2309	0.4234	0.2309	0.0128	0.4018	0.0513	4.7512
S10	0.3592	0.2694	0.0898	0.2694	0.8083	0.6287	0.1925	0.0642	1.0677	0.3464	0.2694	0.0770	0.0898	0.4018	0.0128	4.9464

References

Ahmed W, Huma S (2021) Impact of lean and agile strategies on supply chain risk management. Total Qual Manag Bus Excell 32(1-2):33–56. https://doi.org/10.1080/14783363.2018.1529558

Alazab M, Alhyari S, Awajan A, Abdallah AB (2020) Blockchain technology in supply chain management: an empirical study of the factors affecting user adoption/acceptance. Clust Comput 1-19. https://doi.org/10.1007/s10586-020-03200-4

Autry CW (2021) Supply chain research: considering the discipline's evolving relationship with marketing, current issues, and future research directions. J Mark Theory Pract 1-13. https://doi.org/10.1080/10696679.2020.1860684

Berger SLT, Tortorella GL, Rodriguez CMT (2018) Lean supply chain management: a systematic literature review of practices, barriers and contextual factors inherent to its implementation. In: Davim J (ed) Progress in lean manufacturing. Management and industrial engineering. Springer, Cham. https://doi.org/10.1007/978-3-319-73648-8_2

Bhalaji RKA, Bathrinath S, Saravanasankar S (2020) A Fuzzy VIKOR method to analyze the risks in lean manufacturing implementation. Mater Today Proc. https://doi.org/10.1016/j.matpr.2020.05.123

Bota-Avram C (2022) Bibliometric analysis of sustainable business performance: where are we going? A science map of the field. Econ Res Ekonomska Istraživanja:1–40. https://doi.org/10.1080/1331677X.2022.2096094

Butt AS (2021) Strategies to mitigate the impact of COVID-19 on supply chain disruptions: a multiple case analysis of buyers and distributors. Int J Logist Manag. https://doi.org/10.1108/IJLM-11-2020-0455

Caiado RGG, Scavarda LF, Azevedo BD, Nascimento DLDM, Quelhas OLG (2022) Challenges and benefits of sustainable industry 4.0 for operations and supply chain management – a framework headed toward the 2030 agenda. Sustainability 14(2):830. https://doi.org/10.3390/su14020830

Celuch K, Walz AM (2020) The role of active identification in driving retail customer feedback. Serv Mark Q 41(2):163–181. https://doi.org/10.1080/15332969.2020.1742980

Chowdhury SG, Chatterjee S (2020) Determinants of Indian automobile industry growth. Eurasian J Bus Econ 13(26):65–91. https://doi.org/10.17015/ejbe.2020.026.04

Chowdhury P, Paul SK (2020) Applications of MCDM methods in research on corporate sustainability. Manag Environ Qual Int J 31(2):385–405. https://doi.org/10.1108/MEQ-12-2019-0284

da Cunha RF, Morais MM, Nogueira RL, Marzoque HJ, de Andrade Machado VE (2020) Smed to increase productivity: a case study in an automobile industry. Braz J Dev 6(6):42088–42108. https://doi.org/10.34117/bjdv6n6-664

Dai J, Xie L, Chu Z (2021) Developing sustainable supply chain management: the interplay of institutional pressures and sustainability capabilities. Sustain Prod Consump 28:254–268. https://doi.org/10.1016/j.spc.2021.04.017

Dave Y, Sohani N (2019) Improving productivity through lean practices in central India-based manufacturing industries. Int J Lean Six Sigma 10(2):601–621. https://doi.org/10.1108/IJLSS-10-2017-0115

de Oliveira UR, de Almeida Muniz M, Anaia LA, Rocha HM (2022) Medication supply chain risk management for a Brazilian home care provider: a business sustainability study. Clean Log Supp Chain 3:100018. https://doi.org/10.1016/j.clscn.2021.100018

Digalwar A, Raut RD, Yadav VS, Narkhede B, Gardas BB, Gotmare A (2020) Evaluation of critical constructs for measurement of sustainable supply chain practices in lean-agile firms of Indian origin: a hybrid ISM-ANP approach. Bus Strateg Environ 29(3):1575–1596. https://doi.org/10.1002/bse.2455

Hernandez-Martinez N, Mutlu N, Fransoo JC (2020) Social equity in supplier–buyer relationships in smallholder agri-food supply chains. Flex Serv Manuf J 1-63. https://doi.org/10.1007/s10696-020-09397-1

Huang J, Feng L, Liao J (2020, February) Analysis of the Indian Government's support for the development of medium, small, and micro enterprises. In International Academic Conference on Frontiers in Social Sciences and Management Innovation (IAFSM 2019) (pp 296–300). Atlantis Press. https://doi.org/10.2991/assehr.k.200207.046

Huang H, Chen K, Zhang H, Ren L (2022) Planning and coordinated response mechanism of economic and ecological services in urban expansion. Econ Res Ekonomska Istraživanja 1-21. https://doi.org/10.1080/1331677X.2022.2097112

Hwang C-L, Yoon K (1981) Multiple attribute decision making: methods and applications. In: Lecture notes in economics and mathematical systems. Springer, Now York

Jasti NVK, Kurra S (2017) An empirical investigation on lean supply chain management frameworks in Indian manufacturing industry. Int J Product Perform Manag 66(6):699–723. https://doi.org/10.1108/IJPPM-12-2015-0185

Jellali A, Hachicha W, Aljuaid AM (2021) Sustainable configuration of the Tunisian olive oil supply chain using a Fuzzy TOPSIS-based approach. Sustainability 13(2):722. https://doi.org/10.3390/su13020722

Jell-Ojobor M, Raha A (2022) Being good at being good – the mediating role of an environmental management system in value-creating green supply chain management practices. Bus Strateg Environ. https://doi.org/10.1002/bse.2993

Karuppiah K, Sankaranarayanan B, Ali SM, Chowdhury P, Paul SK (2020) An integrated approach to modeling the barriers in implementing green manufacturing practices in SMEs. J Clean Prod 265:121737. https://doi.org/10.1016/j.jclepro.2020.121737

Kazancoglu Y, Pala MO, Sezer MD, Luthra S, Kumar A (2021) Drivers of implementing big data analytics in food supply chains for transition to a circular economy and sustainable operations management. J Enterp Inf Manag. https://doi.org/10.1108/JEIM-12-2020-0521

Kumar Singh R, Modgil S (2020) Assessment of lean supply chain practices in Indian automotive industry. Glob Bus Rev 0972150919890234:10.1177%2F0972150919890234

Lamming R (1996) Squaring lean supply with supply chain management. Int J Oper Prod Manag 16(2):183–196. https://doi.org/10.1108/01443579610109910

Leksic I, Stefanic N, Veza I (2020) The impact of using different lean manufacturing tools on waste reduction. Adv Prod Eng Manag 15(1):10.14743/apem2020.1.351

Li Z, Guo H, Barenji AV, Wang WM, Guan Y, Huang GQ (2020) A sustainable production capability evaluation mechanism based on blockchain, LSTM, analytic hierarchy process for supply chain network. Int J Prod Res 58(24):7399–7419. https://doi.org/10.1080/00207543.2020.1740342

Liu S, Papageorgiou LG, Shah N (2020) Optimal design of low-cost supply chain networks on the benefits of new product formulations. Comput Ind Eng 139:106189. https://doi.org/10.1016/j.cie.2019.106189

Loaiza-Ramírez JP, Moreno-Mantilla CE, Reimer T (2022) Do consumers care about companies' efforts in greening supply chains? Analyzing the role of protected values and the Halo effect in product evaluation. Clean Log Supp Chain:100027. https://doi.org/10.1016/j.clscn.2021.100027

Mangla SK, Kazançoğlu Y, Yıldızbaşı A, Öztürk C, Çalık A (2022) A conceptual framework for blockchain-based sustainable supply chain and evaluating implementation barriers: a case of the tea supply chain. Bus Strateg Environ. https://doi.org/10.1002/bse.3027

Marcucci G, Mazzuto G, Bevilacqua M, Ciarapica FE, Urciuoli L (2022, January) Conceptual model for breaking ripple effect and cycles within supply chain resilience. Supp Chain Forum Int J: 1–20. https://doi.org/10.1080/16258312.2022.2031275

Mishra B, Mahanty B, Thakkar JJ (2020) A quantifiable quality enabled servitisation model: benchmarking Indian automobile manufacturers. Int J Prod Res 1-23. https://doi.org/10.1080/00207543.2020.1736721

Mohapatra P, Tripathy S, Dash A, Biswal A (2021) Supplier relationship management is a key to supply chain management. In: Pant P, Mishra SK, Mishra PC (eds) Advances in mechanical processing and design. Springer, Singapore, pp 661–670. https://doi.org/10.1007/978-981-15-7779-6_62

Moyano-Fuentes J, Bruque-Cámara S, Maqueira-Marín JM (2019) Development and validation of a lean supply chain management measurement instrument. Prod Plan Contr 30(1):20–32. https://doi.org/10.1080/09537287.2018.1519731

Mukherjee AA, Singh RK, Mishra R, Bag S (2021) Application of blockchain technology for sustainability development in agricultural supply chain: justification framework. Oper Manag Res:1–16. https://doi.org/10.1007/s12063-021-00180-5

Najar T (2022, January) Lean-Agile supply chain innovation performance; the mediating role of dynamic capability, innovation capacity, and relational embeddedness. Supp Chain Forum Int J:1–22. https://doi.org/10.1080/16258312.2022.2031276

Nath V, Agrawal R (2020) Agility and lean practices as antecedents of supply chain social sustainability. Int J Oper Prod Manag 40(10):1589–1611. https://doi.org/10.1108/IJOPM-09-2019-0642

Panwar A, Jain R, Rathore APS, Nepal B, Lyons AC (2018) The impact of lean practices on operational performance – an empirical investigation of Indian process industries. Prod Plan Contr 29(2):158–169. https://doi.org/10.1080/09537287.2017.1397788

Puche J, Costas J, Ponte B, Pino R, de la Fuente D (2019) The effect of supply chain noise on the financial performance of Kanban and Drum-Buffer-Rope: an agent-based perspective. Expert Syst Appl 120:87–102. https://doi.org/10.1016/j.eswa.2018.11.009

Rafi S, Yu W, Akbar MA, Alsanad A, Gumaei A (2020) Multicriteria based decision making of DevOps data quality assessment challenges using fuzzy TOPSIS. IEEE Access 8:46958–46980. https://doi.org/10.1109/ACCESS.2020.2976803

Saeli M, Micale R, Seabra MP, Labrincha JA, La Scalia G (2020) Selection of novel geopolymeric mortars for sustainable construction applications using fuzzy topsis approach. Sustainability 12(15):5987. https://doi.org/10.3390/su12155987

Samaie F, Meyar-Naimi H, Javadi S, Feshki-Farahani H (2020) Comparison of sustainability models in development of electric vehicles in Tehran using fuzzy TOPSIS method. Sustain Cities Soc 53:101912. https://doi.org/10.1016/j.scs.2019.101912

Saudi MHM, Juniati S, Kozicka K, Razimi MSA (2019) Influence of lean practices on supply chain performance. Pol J Manag Stud 19. https://doi.org/10.17512/pjms.2019.19.1.27

Seuring S, Müller M (2008) From a literature review to a conceptual framework for sustainable supply chain management. J Clean Prod 16(15):1699–1710. https://doi.org/10.1016/j.jclepro.2008.04.020

Seuring S, Aman S, Hettiarachchi BD, de Lima FA, Schilling L, Sudusinghe JI (2022) Reflecting on theory development in sustainable supply chain management. Clean Log Supp Chain 3:100016. https://doi.org/10.1016/j.clscn.2021.100016

Sharma MG, Naude MJ (2021) Interdependence analysis of supplier relationship challenges in the South African automotive industry. J Glob Oper Strat Sourc. https://doi.org/10.1108/JGOSS-01-2020-0005

SIAM (Society of Indian Automobile Manufacturers) Report (2020) Online. https://www.siam.in/statistics.aspx?mpgid=8&pgidtrail=9. Accessed on 2 Feb 2021

Singh RK, Kumar R (2020) Strategic issues in supply chain management of Indian SMEs due to globalization: an empirical study. BIJ 27(3):913–932. https://doi.org/10.1108/BIJ-09-2019-0429

Takeda-Berger SL, Tortorella GL, Rodriguez CMT, Frazzon EM, Yokoyama TT, Oliveira MAD (2021) Analysis of the relationship between barriers and practices in the lean supply chain management. Int J Lean Six Sigma. https://doi.org/10.1108/IJLSS-01-2019-0003

Tortorella GL, Miorando R, Marodin G (2017) Lean supply chain management: empirical research on practices, contexts and performance. Int J Prod Econ 193:98–112. https://doi.org/10.1016/j.ijpe.2017.07.006

Tripathi S, Talukder B (2020) Supply chain performance and profitability in Indian automobile industry: evidence of segmental difference. Glob Bus Rev:0972150919898302, 10.1177%2F0972150919898302

Tseng ML, Bui TD, Lim MK, Fujii M, Mishra U (2022) Assessing data-driven sustainable supply chain management indicators for the textile industry under industrial disruption and ambidexterity. Int J Prod Econ 245:108401. https://doi.org/10.1016/j.ijpe.2021.108401

Vanichchinchai A (2019) Exploring organizational contexts on lean manufacturing and supply chain relationship. J Manuf Technol Manag 31(2):236–259. https://doi.org/10.1108/JMTM-01-2019-0017

Yadav G, Luthra S, Huisingh D, Mangla SK, Narkhede BE, Liu Y (2020) Development of a lean manufacturing framework to enhance its adoption within manufacturing companies in developing economies. J Clean Prod 245:118726. https://doi.org/10.1016/j.jclepro.2019.118726

Chapter 6
Prioritizing Sustainability Criteria of Green Supply Chains Using the Best-Worst Method

Ertugrul Ayyildiz and Aslihan Yildiz

Abstract The planning and execution of the entire process in a way that protects the environment, beginning with the supply of the raw materials and ending with the waste or leftover product produced by the consumer, is known as the "green supply chain." The traditional supply chain has been replaced by a green one as a result of the rapid depletion of vital resources, the quick rise in environmental consciousness, and the growing significance of environmental preservation goals on a global and national scale. Supply chain sustainability is the control of the social, economic, and environmental effects of goods and/or services throughout their life cycles and the encouragement of ethical business conduct. The main objective of supply chain sustainability is to establish, preserve, and advance social, economic, and environmental values for all stakeholders in the supply chain. Along with these three primary objectives, risk concerns are also covered in this chapter, and key elements for the sustainability of green supply chains are examined. A hierarchical green supply chain sustainability framework is brought to the literature, with sub-criteria identified for each main criterion. The problem of weighing the importance of each criterion that makes up this framework is thereafter handled as a multi-criteria decision-making (MCDM) problem. The weights of the main and sub-criteria are determined using the best-worst method (BWM). The findings show that out of 36 sub-criteria, "cost," "environmental management system," and "pollution control" are the 3 that are the most crucial for sustainability.

Keywords Best-worst method · Green supply chain management · Risk · Sustainability

E. Ayyildiz (✉)
Department of Industrial Engineering, Karadeniz Technical University, Trabzon, Turkey
e-mail: ertugrulayyildiz@ktu.edu.tr

A. Yildiz
Department of Industrial Engineering, Yildiz Technical University, İstanbul, Turkey

© The Author(s), under exclusive license to Springer Nature Switzerland AG 2023

N. U. Ibne Hossain (ed.), *Data Analytics for Supply Chain Networks*, Greening of Industry Networks Studies 11, https://doi.org/10.1007/978-3-031-29823-3_6

6.1 Introduction

Numerous studies have been conducted in recent years to address supply chain management (SCM). Traditional SCM can be described as evaluating the coordinated management of the process from manufacturing to the delivery of products or services to the end consumer. But recently, researchers have asserted that the idea of sustainability is crucial to modern supply chain management (Shekarian et al. 2022). One of the key components for SCM success in businesses looking to gain a competitive edge is sustainability (Dalalah et al. 2022). The synchronization of economic, environmental, and social concerns is at the core of sustainability, which encourages the identification of environmental practices and efficient use of labor, knowledge, time, material, and financial resources (Sharma et al. 2021a, b).

The term sustainability, which was first introduced in 1987 in the United States, is a hot topic in the literature of business disciplines and operations in the current era (Sharma et al. 2021a, b). Today's sustainable supply chains must respond to changes in the needs of both urgent and current customers (Green et al. 2019). These days, customers are more concerned with a product or service's environmental impact or eco-friendliness than with delivering it at the right time and at the acceptable price. At this point, one of the important topics that scholars and practitioners commonly focus on, both in the literature and in business operations, is the idea of a green supply chain. With factors such as changing customer demands, technological conveniences, and increasing customer awareness, the biggest challenge faced by companies, regardless of power and size, is to develop and maintain a supply chain that is more responsible and sensitive to the environment, society, and the individual (Singh and Trivedi 2016). To comply with environmental standards, companies take steps to boost their performance throughout the supply chain. By taking into account the customer's environmental consciousness, these companies seek to lessen the environmental effect of their product or service operations, hence enhancing their ability to compete in the market (Rao and Holt 2005).

Sustainable green SCM is a popular organizational philosophy to increase ecological efficiency by considering the environmental concerns of companies, to respond correctly to customer needs, and to carry out production or service activities by giving priority to environmental responsibility and sensitivity (Cabral et al. 2012). It should not be forgotten that companies in the race to gain market shares and profitability can achieve their goals by considering environmental risks and effects on their supply chain activities.

Sustainable green SCM brings with it the vital requirements of effective green activities that can reduce waste and carbon emissions, minimize ecological damage and impact, and aim to ensure its sustainability, without compromising on profit, quality, customer satisfaction, effective production, and service activities (Srivastava 2007). Organizations must implement green SCM activities in response to ecological concerns for products or services produced using ecologically sustainable processes and in accordance with environmental standards (Murray 2000).

Due to the growing environmental damage in today's world, more and more people are becoming environmentally conscious, and there is a growing impression that environmentally friendly goods and services are superior. This has increased interest in green SC, which also examines the environmental impact of all processes throughout a product's life cycle rather than just the ones that the traditional supply chain focuses on (Banasik et al. 2018). The increasing interest of researchers and organizations in sustainable green SCM paves the way for studies in all areas of the supply chain, such as production, sales, logistics, information, and labor. Adopting the GSC in the organizational framework and ensuring its sustainability are of critical importance. Establishing a decision support system that can address multiple aspects of "green" concerns and sustainability in the implementation of a sustainable green SCM will facilitate the activities of organizations in this sense (Uygun and Dede 2016). It is frequently emphasized in the literature that there are multiple dimensions in providing a sustainable SCM where the "green" concerns are considered (Ayyildiz and Taskin Gumus 2021; Carter and Rogers 2008; Ecer and Pamucar 2020; Murray 2000; Naseem et al. 2017). While attempting to manage production, supply, logistics, sales, material, information, and cash flows effectively, a supply chain with sustainability at the forefront should try evaluating these processes in the economic, environmental, and social dimensions. The philosophy of "sustainability," which is based on three pillars, economic, environmental, and social, attracted increasing attention in this sense. Looking at this as a whole, it is an effort to balance the trade-offs between the goals that are equally desired to be achieved in these three dimensions (Ayyildiz 2022; Barbier 1987; Uygun and Dede 2016). These three dimensions of sustainability are typically, but not always, presented as intersecting circles, as summarized in Fig. 6.1. Sustainability can be expressed as the intersection of these three dimensions or as a whole in which all three dimensions are evaluated together.

Two significant contributions are provided by this chapter to the SCM literature. First, a comprehensive listing of in-depth literature evaluations for the assessment of sustainable green SCM is provided. Second and as the chapter's main contribution, the risk criterion has been added to the environmental, economic, and social criteria, which are the three pillars of sustainability, in the assessment of sustainable green SCM. In this context, the sustainable green SCM problem is addressed from four

Fig. 6.1 Three pillars in the sustainability framework

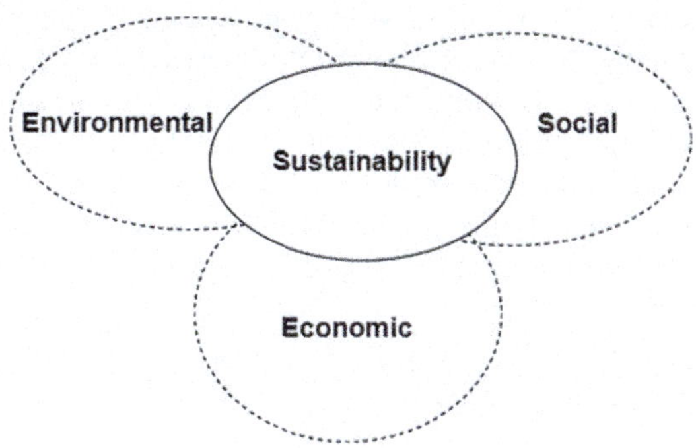

different dimensions. Sub-criteria are determined for each of these four main criteria, and the best-worst method (BWM) is integrated into the study for evaluating sustainable green SCM according to the criteria. Multi-criteria decision-making (MCDM) strategies can be used during the prioritization, assessment, and evaluation process for various complexities, since sustainable green SCM involves multidimensional considerations (Uygun and Dede 2016). When it comes to dealing with the examination of many factors in decision-making, MCDM approaches are useful and effective tools (Paul et al. 2021). The BWM is used by researchers frequently in the decision-making literature. A method is an effective tool that can be integrated into decision-making problems, thanks to its computational capability.

The remainder of this chapter is structured as follows: A thorough literature overview of the sustainability assessment of green SCM is presented in Sect. 6.2. Section 6.3 provides information about a sustainable green supply chain, its philosophy, benefits, terms, and related criteria. Section 6.4 presents an integrated BWM. In Sect. 6.5, the proposed method is illustrated with a case study in the assessment of a sustainable green SCM. Discussions and management implications are introduced in Sect. 6.6. Section 6.7 summarizes the conclusions and presents recommendations for future work.

6.2 Literature Review

Apart from SCM-related studies using the different MCDM approaches, it is necessary to systematically identify the studies that are used to highlight the differences and form the basis of this chapter. In this chapter, the PRISMA (Preferred Reporting Items for Systematic Reviews and Meta-Analyses) methodology is employed to perform a detailed literature review (Moher et al. 2009). The PRISMA methodology is composed of five basic steps: establishing criteria, locating sources, choosing the literature, gathering data, and choosing data elements (Santi and Putra 2018). This methodology is adopted to reduce the bias in literature research and to develop more systematic methods for achieving literature research (Satria et al. 2017). Studies focusing on sustainable supply chain management were investigated. Therefore, the use of MCDM approaches in these studies is analyzed, and studies that use BWM and its extensions are also analyzed in more details. Within the scope of this chapter, studies that adopt the MCDM approach in determining the sustainability of supply chains are examined in more details. The literature search was carried out between February 1, 2022, and August 1, 2022, with the keywords shown in Table 6.1.

With the keywords listed in Table 6.1, 243 papers in the Scopus database were identified. Some papers were repetitive because they were found in three or two searches. Therefore, repetitive studies were removed and 176 were left. After the Scopus search, the exclusion and inclusion criteria were established, and papers specifically relevant to this topic were identified. The criteria for exclusion and inclusion are presented in Table 6.2.

Table 6.1 Studies found in the literature review

Database	Details of the search	Number of studies
Scopus	(TITLE-ABS-KEY ("supply chain management" AND "sustainability") AND TITLE-ABS-KEY ("multi-criteria decision-making"))	115
	(TITLE-ABS-KEY ("supply chain management" AND "sustainability") AND TITLE-ABS-KEY ("MCDM"))	83
	(TITLE-ABS-KEY ("supply chain management" AND "sustainability") AND TITLE-ABS-KEY ("best worst method"))	45

Table 6.2 Criteria for inclusion and exclusion in the literature review

Inclusion	Exclusion
Sustainability assessment by MCDM analysis	Studies for which there are no full texts available
Supply chain resilience analysis by MCDM analysis	Studies that don't explicitly describe the methodology employed or the conclusions drawn
Green supply chain management by MCDM analysis	Studies written in languages other than English
Supply chain performance evaluation by the BWM	Studies published before 2020
	Conference papers, book chapters, reviews

Following the application of the criteria in Table 6.2, it is determined that 79 studies published in 2020 and later are pertinent to this chapter. Table 6.3 has been created to show and summarize the contributions of these studies to the literature. Thus, the originality and novelty of the proposed framework are emphasized.

When Table 6.3 is examined, Afrasiabi et al. (2022)'s study is found to be most similar to this study. To identify the most sustainable and resilient supplier, they created 16 sub-criteria under 4 main criteria: social, economic, environmental, and resilient. Fuzzy BWM was used to calculate the weights of the criteria, and the extended GRA-TOPSIS methodology was used to assess alternative providers. A limitation of this chapter is that they focused only on supplier selection not holistic SCM and risk was not evaluated comprehensively. In our proposed model, sustainable green SCM is analyzed under the three pillars of sustainability. We also extended the number of pillars with risk considering the post-pandemic era. For this purpose, a hierarchical structure of criteria is constructed with a specific purpose.

6.3 Green Supply Chain Sustainability Assessment Model

Nowadays, people are more aware of the environment than ever before, and there is a growing notion that eco-friendly goods and services are superior. This is due to the growing environmental destruction in the world. Due to this, there is now more interest in green supply chains, which additionally assess the environmental effects

Table 6.3 A summary of the remarkable studies

#	Author	Year	Problem	Adopted methodology	Fuzzy	Application area
1	Sharma and Joshi	2020	Digital supplier selection	SWARA, WASPAS	No	Manufacturing
2	Amiri et al.	2020	Sustainable supplier selection	BWM	Yes	Automotive
3	Khokhar et al.	2020	Social sustainability analysis	BWM	No	Manufacturing
4	Khan, Haleem, and Khan	2020	Risk analysis	BWM	Yes	Food
5	Yadav, Garg, and Luthra	2020a	Third-party logistics service provider selection	AHP	Yes	Generic
6	Ali et al.	2020	Green SCM adoption	TOPSIS	No	Construction
7	Baidya et al.	2020	E-waste management	AHP, QFD	No	Generic
8	Rouyendegh, Yildizbasi, and Üstünyer	2020	Green supplier selection	TOPSIS	Yes	Automotive
9	Garg and Sharma	2020	Supplier selection	BWM, VIKOR	No	Electronics
10	Abdel-Basset and Mohamed	2020	Risk analysis	TOPSIS, CRITIC	No	Telecommunications
11	Orji, Kusi-Sarpong, and Gupta	2020	Critical success factor analysis	BWM	No	Logistics
12	Mohammed	2020	Resiliency and greenness performance evaluation	DEMATEL, VIKOR	No	Chemical
13	Rajesh	2020	Sustainable SCM analysis	AHP, ANP	No	Generic
14	Yadav et al.	2020b	SCM analysis	BWM, ELECTRE	No	Automotive
15	Bajec, Tuljak-Suban, and Bajor	2020	Warehouse performance metrics framework	BWM	No	Generic
16	Hendiani, Mahmoudi, and Liao	2020	Sustainable supplier selection	BWM	Yes	Refinery
17	Ecer and Pamucar	2020	Sustainable supplier selection	BWM, CoCoSo	Yes	Home appliance manufacturers
18	H. Gupta, Kusi-Sarpong, and Rezaei	2020	Barrier analysis for sustainable supply chain innovation	BWM	No	Manufacturing

19	Öztürk and Yildizbaşi	2020	Barrier analysis for blockchain implementation	AHP, TOPSIS	Yes	Generic
20	Roy, Pamučar, and Kar	2020	Third-party logistics service provider selection	MABAC	Yes	Food
21	Sahebi, Masoomi, and Ghorbani	2020	Barrier analysis for blockchain implementation	BWM	No	Humanitarian
22	Vahabzadeh Najafi, Arshadi Khamseh, and Mirzazadeh	2020	Sustainable supplier selection	DEMATEL	No	Manufacturing
23	Ahmadi et al.	2020	Sustainable supplier selection	BWM, PROMETHEE	No	Manufacturing
24	Ghosh, Mandal, and Ray	2021a, b, c	Green supplier selection	COPRAS, GRA	No	Automotive
25	Fallahpour et al.	2021	Sustainable supplier selection	BWM, FIS	Yes	Textile
26	Sudipta Ghosh, Mandal, and Ray	2021a	Green supplier selection	COPRAS, TOPSIS, GRA	No	Generic
27	Mishra and Rani	2021	Sustainable partner assessment	ARAS	Yes	Recycling
28	Dursun	2021	Sustainable transportation service provider selection	BWM	Yes	Dye
29	Kazancoglu et al.	2021	Risk analysis	AHP, TODIM	Yes	Logistics
30	Tseng et al.	2021	SCM	BWM	Yes	Seafood
31	Sudipta Ghosh, Mandal, and Ray	2021b	Green supplier selection	TOPSIS	No	Automotive
32	Pınar, Erdebilli, and Özdemir	2021	Green supplier selection	TOPSIS	Yes	Automotive
33	Nabeeh, Abdel-Basset, and Soliman	2021	SCM	ANP, DEMATEL, TOPSIS	Yes	Manufacturing
34	Mahak Sharma et al.	2021a, b	Barrier analysis for industry 4.0 implementation	AHP, DEMATEL	No	Generic
35	Hoseini et al.	2021	Sustainable supplier selection	BWM, FIS	Yes	Construction
36	Moktadir et al.	2021	Risk analysis	BWM	No	Textile

(continued)

Table 6.3 (continued)

#	Author	Year	Problem	Adopted methodology	Fuzzy	Application area
37	Wang, Dang, and Nguyen	2021	Third-party logistics service provider selection	AHP, TOPSIS	Yes	E-commerce
38	Yıldızbaşı et al.	2021	Social sustainability analysis	AHP, TOPSIS	Yes	Automotive
39	Hendiani, Lev, and Gharehbaghi	2021	Social sustainability analysis	Novel methodology	Yes	Generic
40	Kumar et al.	2021a, b	Risk analysis	BWM	Yes	Food
41	Mahmud et al.	2021	Barrier analysis for supply chain collaboration	DEMATEL, BWM	Yes	Generic
42	Mohammadnazari and Ghannadpour	2021	Supply chain management	BWM, TOPSIS	No	Construction
43	Nguyen, Lin, and Dang	2021	Green supplier selection	AHP, VIKOR	Yes	Food
44	Elabed, Shamayleh, and Daghfous	2021	Sustainability-oriented innovation assessment	BWM	Yes	Healthcare
45	Cavalcante de Souza Feitosa, Ribeiro Carpinetti, and de Almeida-Filho	2021	Risk analysis	TOPSIS	Yes	Textile
46	Ali and Kaur	2021	Social sustainability analysis	BWM	No	Warehousing
47	Liaqait et al.	2021	Sustainable supplier selection	AHP, TOPSIS	Yes	Energy
48	Sarker, Moktadir, and Santibanez-Gonzalez	2021	Social sustainability analysis	BWM	No	Generic
49	Chand and Tarei	2021	Barrier analysis for multitier sustainable SCM	DEMATEL	No	Generic
50	Gunduz, Demir, and Paksoy	2021	Supply chain smartness and sustainability analysis	BWM, QFD	No	Automotive
51	P. Kumar et al.	2021a, b	Sustainable SCM challenge analysis	BWM	No	Automotive
52	Zarrinpoor	2021	Supply chain network design	BWM	No	Recycling
53	Perçin	2022	Supply chain capability analysis	DEMATEL, ANP, VIKOR	Yes	Food

54	Ahmed et al.	2022	Sustainable SCM challenge analysis	BWM	No	Textile
55	Agrawal et al.	2022	Success factor analysis for a sustainable SCM	AHP, TOPSIS, DEMATEL	No	Food
56	Han and Rani	2022	Barrier analysis for blockchain implementation	CRITIC, CoCoSo	Yes	Generic
57	Prashar	2022	Sustainable SCM driver analysis	DEMATEL	No	Food
58	Tolooie, Alvandi, and Arani	2022	Sustainable supplier selection	DEMATEL, ANP, VIKOR	Yes	Automotive
59	Mohseni, Baghizadeh, and Pahl	2022	Barrier and driver analysis for SCM	TOPSIS, AHP, COPRAS	No	Food
60	Bai, Zhu, and Sarkis	2022	Supplier selection	BWM	Yes	Manufacturing
61	Sudipta Ghosh, Mandal, and Ray	2022	Green SCM performance analysis	AHP, TOPSIS	Yes	Manufacturing
62	Pamucar, Torkayesh, and Biswas	2022b	Supplier selection	MACBETH	Yes	Healthcare
63	Yildizbasi and Arioz	2022	Green supplier selection	AHP, TOPSIS	Yes	Electronics
64	Kao et al.	2022	Supplier selection	AHP, WASPAS	Yes	Textile
65	Mondal and Roy	2022	Supplier selection	DEMATEL, MABAC	Yes	Automotive
66	Liaqait et al.	2022	Sustainable supplier selection	AHP, TOPSIS	Yes	Manufacturing
67	Chai and Zhou	2022	Sustainable supplier selection	AHP, TOPSIS	Yes	Energy
68	Zhang and Song	2022	Risk analysis for blockchain implementation	BWM, CoCoSo	No	Generic
69	Mastrocinque et al.	2022	Sustainable supply chain development analysis	FIS	Yes	Energy
70	Saraji and Streimikiene	2022	Circular supply chain evaluation	SWARA, COPRAS	Yes	Manufacturing
71	Caristi et al.	2022	Supplier selection	TOPSIS	Yes	Manufacturing
72	Yang et al.	2022	Sustainable SCM	CRITIC, VIKOR	Yes	Manufacturing
73	Dwivedi and Paul	2022	Digital supply chain adaptation analysis	BWM	Yes	Generic

(continued)

Table 6.3 (continued)

#	Author	Year	Problem	Adopted methodology	Fuzzy	Application area
74	Pamucar et al.	2022a, b	Sustainable SCM advantage prioritization to	WASPAS	Yes	Logistics
75	Salimian, Mousavi, and Antucheviciene	2022	Sustainable supplier selection	VIKOR, MARCOS	Yes	Healthcare
76	Hosseini, Flapper, and Pirayesh	2022	Sustainable supplier selection	BWM	No	Manufacturing
77	Erol, Ar, and Peker	2022	Blockchain implementation in a sustainable SCM	SWARA, COPRAS, EDAS	Yes	Generic
78	Anup Kumar et al.	2022	Sustainable supply chain indicator analysis	TOPSIS	Yes	Automotive
79	*Afrasiabi, Tavana, and Di Caprio*	2022	*Sustainable supplier selection*	*BWM, TOPSIS*	*Yes*	*Manufacturing*

of all operations throughout the life cycle rather than just those on which the traditional supply chain concentrates (Banasik et al. 2018). An SCM that provides environmental sustainability with legal regulations that encourage taking steps to improve the environment by considering ecological concerns and using technological innovations and conveniences is referred to as a green SCM.

The primary goal of green SCM can be summarized as reducing and eliminating the harmful and damaging consequences of the supply chain on the environment in operational and management domains such as production, supply, logistics, sales, and service (Andiç et al. 2012; Gupta et al. 2020; Sari 2017; Uygun and Dede 2016). Along with factors such as changing and developing technological opportunities, changing customer perceptions and expectations, increasing consumption, and easy and fast access to information, companies have evaluated the sustainability terminology in three pillars to realize sustainable development in terms of management and organization.

The philosophy of "sustainability," which is based on three pillars, economic, social, and environmental, has attracted increasing attention in this sense. Looking at this as a whole, it is an effort to balance the trade-offs between the goals that are equally desired to be achieved in these three dimensions (Ayyildiz 2022; Barbier 1987; Uygun and Dede 2016). It is essential to create a balance between economic, environmental, and social aspects of the supply chain and to properly run supply chain operations within this balance in order to create a sustainable supply chain. However, due to the complex nature of these dimensions and their intertwined relationships, it is a very difficult process for companies to establish and maintain this balance (Yildiz Çankaya and Sezen 2019).

Green supply chains, which are responsible for carrying out all operations in the supply chain focusing on environmental awareness, take into account the environmental dimension, which is one of the three pillars, at the point of ensuring sustainability. For a green supply chain to be sustainable, it is necessary to evaluate, implement, and maintain many interrelated performance criteria such as economic, environmental, and social dimensions. Our chapter's goal is to examine the green supply chain in the context of sustainability. At this point, the risk dimension has been introduced into the study as a fourth dimension for a sustainable green supply chain in addition to the economic, social, and environmental dimensions. The set of criteria, prepared for the problem based on the three dimensions, with the added risk dimension, is shown in Table 6.4. We assessed the sustainability of the green supply chain by specifying the sub-criteria for these four sustainability dimensions, which are called the main criteria within the framework of the problem and the applications discussed in the chapter. As a result of a literature research and experts' opinions, 36 sub-criteria and 4 main criteria were defined.

Table 6.4 Main and sub-criteria for the sustainability evaluation of green SCM

Environmental	
E1: Pollution production	Boutkhoum et al. (2016), Wang Chen et al. (2016), Yao et al. (2020) and Zhou et al. (2019)
E2: Pollution control	Deshmukh and Vasudevan (2018), Fallahpour et al. (2020) and Wang Chen et al. (2016)
E3: Resource consumption	Ayyildiz (2021) and Wang Chen et al. (2016)
E4: Environmental management system	Deshmukh and Vasudevan (2018), Tirkolaee et al. (2021), Wang Chen et al. (2016), Zhou and Chen (2020) and Zhou et al. (2019)
E5: Hazardous materials	Boutkhoum et al. (2016), Lu et al. (2007) and Zhou et al. (2019)
E6: Energy consumption	Fallahpour et al. (2020), Lu et al. (2007) and Yao et al. (2020)
E7: Reverse logistics (reuse and recycle)	Sari (2017), Sathiya Narayana et al. (2020) and Zhou et al. (2019)
E8: Green distribution	Deshmukh and Vasudevan (2018) and Zhou et al. (2019)
E9: Waste discharge	Ayyildiz (2021) and Yao et al. (2020)
Social	
S1: Culture	Wang Chen et al. (2016)
S2: Relationship	Sathiya Narayana et al. (2020) and Wang Chen et al. (2016)
S3: Human sources	Boutkhoum et al. (2016)
S4: Organizational structure	Ayyildiz (2021) and Boutkhoum et al. (2016)
S5: Innovation	Deshmukh and Vasudevan (2018)
S6: Learning	Deshmukh and Vasudevan (2018)
S7: Regulations	Ayyildiz (2021), Kumar et al. (2019), Malviya and Kant (2018) and Pourjavad et al. (2018)
S8: Competitiveness	Boutkhoum et al. (2016)
S9: Commercial operations	Pourjavad et al. (2018)
Economic	
C1: Cost	Ayyildiz (2021), Boutkhoum et al. (2016), Deshmukh and Vasudevan (2018), Tirkolaee et al. (2021) and Wang Chen et al. (2016)
C2: Quality	Ayyildiz (2021), Deshmukh and Vasudevan (2018), Sathiya Narayana et al. (2020), Wang Chen et al. (2016) and Zhou and Chen (2020)
C3: Technology	Ayyildiz (2021), Tirkolaee et al. (2021) and Wang Chen et al. (2016)
C4: Flexibility	Ayyildiz (2021), Tirkolaee et al. (2021) and Wang Chen et al. (2016)
C5: Productivity	Ayyildiz (2021) and Boutkhoum et al. (2016)
C6: Effectiveness	Kumar et al. (2019)
C7: Price	Tirkolaee et al. (2021) and Zhou and Chen (2020)
C8: Accuracy	Ayyildiz (2021)
C9: Responsiveness	Ayyildiz (2021)
Risk	
R1: Operational	Kumar et al. (2019) and Mangla et al. (2015)
R2: Supply	Kumar et al. (2019) and Mangla et al. (2015)
R3: Production	Kumar et al. (2019) and Mangla et al. (2015)

(continued)

Table 6.4 (continued)

R4: Financial	Kumar et al. (2019) and Mangla et al. (2015)
R5: Demand	Kumar et al. (2019) and Mangla et al. (2015)
R6: Technological	Boutkhoum et al. (2016)
R7: Delivery	Deshmukh and Vasudevan (2018) and Tirkolaee et al. (2021)
R8: Green	Deshmukh and Vasudevan (2018)
R9: Strategic barriers	Malviya and Kant (2018)

6.4 Best-Worst Method

In 2015, Rezaei introduced BWM in the MCDM literature (Rezaei 2015). The method is then employed as a successful tool to address decision-making issues (Ayyildiz and Taskin Gumus 2021). BWM relies on pairwise comparisons (Kannan et al. 2022). It requires fewer comparisons than the analytic hierarchy process (AHP), one of the most widely used pairwise comparison-based MCDM approaches, while permitting more reliable comparisons and outcomes (Wu et al. 2022). BWM requires $2n-3$ pairwise comparisons, while AHP requires $n(n-1)/2$ (Pamučar et al. 2017) and DEMATEL $n(n-1)$ (Xiao et al. 2022a, b). Moreover, BWM has a better performance than other pairwise comparison-based methodologies in maintaining the consistency (Xiao et al. 2022a, b). In summary, the method generates results by comparing the best and worst criteria with other criteria (Mohammadi and Rezaei 2020).

The best and worst criteria are first determined using this method, and then further factors are compared to these two criteria. The basic principle of BWM for calculating criterion weights is based on the differences and similarities of the best and worst criteria (Ijadi Maghsoodi et al. 2020). Furthermore, using only two vectors (comparisons with best and comparisons with worst), BWM reduces the complexity and computational time (Kannan et al. 2022). BWM has two main advantages over other MCDM methodologies: (i) a smaller number of pairwise comparisons must calculate criteria weights, and (ii) more consistent results are generated. Many academics use BWM to address various supply chain-related problems, such as resilience evaluation (Ayyildiz 2021), performance evaluation (Ayyildiz and Taskin Gumus 2021), external force affect analysis (Sadaghiani et al. 2015), supplier selection (Gupta and Barua 2017, 2018; Rezaei et al. 2016), social sustainability assessment (Badri Ahmadi et al. 2017), barrier analysis (Ghasemian Sahebi et al. 2017), green supplier selection (Lo et al. 2018; Tian et al. 2018), and supplier risk assessment (Er Kara and Firat 2018).

The application steps of the BWM methodology are presented as follows (Ayyildiz and Taskin Gumus 2021; Rezaei 2015):

Step 1. Define criteria $C = \{C_1, C_2, \ldots, C_n\}$,where C_i represents the criterion i.
Step 2. Select the best criterion.
Step 3. Select the worst criterion.

Table 6.5 The meaning of numbers 1–9

Number	Meaning
1	*Equal* importance
3	*Moderately* more important than
5	*Strongly* more important than
7	*Very strongly* important than
9	*Absolutely* more important than

Step 4. Construct best-to-others vector $A_B = \{a_{B1}, a_{B2}, \ldots, a_{Bn}\}$, where a_{Bi} is the comparison value between the best and criterion i. The best criterion is compared with other criteria using a scale between 1 and 9, which is given in Table 6.5.

Step 5. Construct others-to-worst vector $A_W = \{a_{1W}, a_{2W}, \ldots, a_{nW}\}^T$, where a_{iW} is the comparison value between criterion i and the worst. The worst is compared with other criteria using a scale between 1 and 9 which is given in Table 6.4.

Step 6. Finding the optimal weights of the criteria $\left(w_1^*, w_2^*, \ldots, w_n^*\right)$. $\frac{w_B}{w_j}$ and $\frac{w_j}{w_W}$, the values of $\frac{w_B}{w_j} = a_{Bj}$ and $\frac{w_j}{w_W} = a_{jW}$ are determined in previous steps for each pairwise comparison. The goal is determining the optimal criteria weights to find a solution with the maximum absolute value of $\left|\frac{w_B}{w_j} - a_{Bj}\right|$ and $\left|\frac{w_j}{w_W} - a_{jW}\right|$ for all j is minimized. Considering the nonnegativity constraints (Eq. 6.2) and the criterion weight summation (Eq. 6.3), the following model is solved (Yalcin Kavus et al. 2022) using:

$$\min \ \max_j \left\{ \left|\frac{w_B}{w_j} - a_{Bj}\right|, \left|\frac{w_j}{w_W} - a_{jW}\right| \right\} \tag{6.1}$$

Subject to:

$$w_j \geq 0, \text{for all } j \tag{6.2}$$

$$\sum_{j=1}^{n} w_j = 1 \tag{6.3}$$

This mathematical model can be represented as:

$$min \ \zeta \tag{6.4}$$

Subject to:

$$\left|\frac{w_B}{w_j} - a_{Bj}\right| \leq \zeta, \text{for all } j \tag{6.5}$$

Table 6.6 The consistency index values

Criteria number	1	2	3	4	5	6	7	8	9
Consistency index	0.00	0.44	1.00	1.63	2.3	3.00	3.73	4.47	5.23

Table 6.7 Description of the expert team

Expert	Experience	Education level	Title	Number of studies
E-1	5	MSc in Management	Researcher	2
E-2	9	PhD in Industrial Engineering	Associate Professor	12
E-3	14	PhD in Industrial Engineering	Associate Professor	18
E-4	11	MSc in Transportation Logistics	Researcher	4
E-5	12	PhD in Statistics	Manager	3
E-6	7	MSc in Industrial Engineering	Engineer	1

$$\left| \frac{w_j}{w_W} - a_{jW} \right| \leq \zeta, \text{ for all } j \tag{6.6}$$

The mathematical model is solved and criteria weights are calculated. The model's consistency is represented with ζ. Then, the consistency ratio (CR) is calculated via the consistency index (CI) values given in Table 6.6 by dividing ζ value by the CI. A smaller value of the consistency index means more reliable results.

6.5 Application and Results

In this chapter, an expert team is of paramount importance for the following: (i) specifying the criteria that influence supply chain sustainability, (ii) assessing the weight of criteria in a hierarchical structure, and lastly (iii) analyzing and validating results in the sustainable supply chain management context. In particular, a group of six qualified experts is constituted as an expert group. Experts are selected based on the following criteria: (i) experience in the related fields (at least a minimum of 5 years), (ii) broad education and knowledge in supply chain management-related areas, (iii) current academic position, and (iv) number of studies done on supply chain management. In this chapter, six experts (E-1, E-2, E-3, E-4, E-5, and E-6) who worked both as academics in different universities and as managers in different supply chains were required to join the team. Four academics and two professionals were consulted to gain their opinions. Information on the profile of the expert team is provided in Table 6.7.

In this section, the BWM technique is used to assess the importance of the criteria that influence the sustainability of the supply chain. The BWM stages outlined above are applied to calculate the weights of the criteria. First, the main criteria' weights are determined. The best (most important) and worst (least important) main criteria, as shown in Table 6.8, are determined by experts.

Four experts agree that the most crucial main criterion is environmental sustainability, and economic sustainability is determined as the most important by two experts. Additionally, except for Expert-3, all experts stated that the least important criterion is social sustainability. Expert-3 determined risk as the worst criterion. Then, the opinions taken from experts were used to construct best-to-others vectors and others-to-worst vectors as given in Table 6.9.

The matrix in Table 6.9 is the basis for a mathematical model, and it is through this model that the main criteria weights are determined, as shown in Table 6.10. In addition, the consistency ratio is calculated for each expert, and evaluations are determined as consistent.

Regarding the reputations of experts, the main criteria weights are combined. Their reputations are determined according to their expertise given in Table 6.7. The experts' weight is shown in Fig. 6.2.

Table 6.8 The best and worst criteria for each expert

	Best	Worst
E-1	Environmental	Social
E-2	Environmental	Social
E-3	Environmental	Risk
E-4	Economic	Social
E-5	Environmental	Social
E-6	Economic	Social

Table 6.9 Best-to-others and others-to-worst vectors

	Best-to-others	Others-to-worst
E-1	1, 6, 3, 4	6, 1, 4, 3
E-2	1, 4, 2, 3	4, 1, 3, 2
E-3	1, 4, 2, 7	6, 3, 5, 1
E-4	3, 9, 1, 2	6, 1, 9, 7
E-5	1, 7, 3, 3	7, 1, 4, 3
E-6	5, 7, 1, 3	3, 1, 7, 4

Table 6.10 Main criteria weights for each expert

	Environmental	Social	Economic	Risk
E-1	.550	.076	.214	.160
E-2	.466	.103	.259	.172
E-3	.500	.143	.286	.071
E-4	.188	.045	.484	.283
E-5	.533	.067	.200	.200
E-6	.132	.071	.578	.220

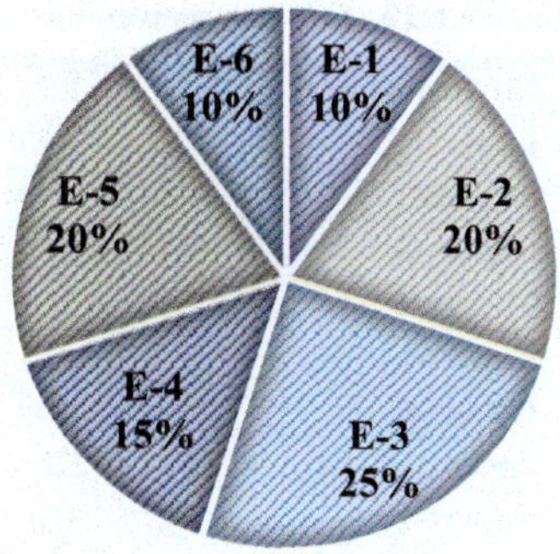

Fig. 6.2 The weight of the experts

Table 6.11 Best-to-others and others-to-worst vectors for sub-criteria

	Best-to-others	Others-to-worst	Best-to-others	Others-to-worst
	Sub-criteria of environmental		Sub-criteria of social	
E-1	1, 2, 2, 3, 1, 3, 4, 8, 5	8, 5, 3, 2, 4, 5, 2, 1, 2	8, 5, 3, 2, 4, 5, 2, 1, 2	1, 2, 5, 3, 8, 2, 4, 3, 2
E-2	3, 3, 2, 4, 5, 1, 3, 6, 6	4, 3, 6, 5, 3, 7, 4, 1, 2	4, 3, 6, 5, 3, 7, 4, 1, 2	3, 4, 2, 6, 4, 3, 3, 2, 1
E-3	6, 2, 5, 1, 3, 4, 7, 8, 9	3, 6, 4, 8, 5, 5, 2, 1, 1	3, 6, 4, 8, 5, 5, 2, 1, 1	4, 2, 1, 3, 8, 8, 6, 5, 2
E-4	4, 5, 6, 8, 1, 6, 5, 5, 3	3, 4, 3, 1, 8, 3, 4, 4, 6	3, 4, 3, 1, 8, 3, 4, 4, 6	1, 2, 6, 5, 6, 5, 8, 7, 7
E-5	3, 4, 4, 7, 5, 4, 1, 6, 5	4, 4, 3, 1, 6, 5, 7, 5, 4	4, 4, 3, 1, 6, 5, 7, 5, 4	5, 4, 6, 5, 5, 7, 8, 1, 4
E-6	3, 3, 4, 9, 1, 8, 8, 9, 5	8, 9, 6, 2, 9, 2, 3, 1, 2	8, 9, 6, 2, 9, 2, 3, 1, 2	3, 1, 3, 2, 7, 4, 9, 7, 8
	Sub-criteria of economic		Sub-criteria of risk	
E-1	2, 1, 2, 3, 2, 2, 3, 4, 8	4, 8, 3, 2, 3, 3, 3, 3, 1	3, 3, 1, 2, 2, 2, 4, 8, 2	4, 5, 8, 5, 3, 5, 3, 1, 2
E-2	1, 4, 3, 6, 4, 5, 2, 5, 4	6, 2, 3, 1, 3, 2, 5, 2, 3	1, 5, 2, 3, 4, 6, 6, 7, 8	8, 4, 6, 5, 4, 2, 3, 1, 1
E-3	3, 1, 4, 6, 5, 2, 7, 8, 9	5, 7, 4, 3, 3, 5, 2, 1, 1	1, 5, 3, 2, 4, 7, 6, 9, 8	8, 5, 6, 7, 5, 3, 4, 1, 3
E-4	1, 2, 6, 4, 5, 6, 2, 2, 4	6, 5, 1, 2, 2, 1, 2, 5, 3	5, 3, 4, 1, 2, 6, 5, 3, 7	3, 5, 4, 7, 6, 2, 3, 5, 1
E-5	1, 2, 4, 5, 5, 5, 2, 7, 6	7, 6, 6, 4, 4, 5, 2, 1, 5	4, 5, 5, 3, 5, 6, 5, 1, 4	4, 3, 3, 5, 5, 1, 4, 6, 4
E-6	1, 5, 7, 7, 2, 3, 2, 8, 9	9, 7, 6, 3, 7, 5, 8, 3, 1	6, 4, 2, 1, 2, 2, 7, 9, 8	3, 4, 8, 9, 8, 5, 2, 1, 3

The weights of sub-criteria are calculated using BWM again. The same experts constructed the best-to-others vectors and others-to-worst vectors for the sub-criteria of each of the four main criteria, as given in Table 6.11.

Table 6.12 Sub-criteria weights for each expert

Environmental	E1	E2	E3	E4	E5	E6	E7	E8	E9
Expert-1	.222	.137	.137	.092	.169	.092	.069	.029	.055
Expert-2	.111	.111	.167	.084	.067	.261	.111	.031	.056
Expert-3	.064	.191	.076	.318	.127	.095	.055	.032	.042
Expert-4	.103	.082	.068	.034	.342	.068	.082	.082	.137
Expert-5	.134	.101	.101	.030	.081	.101	.305	.067	.081
Expert-6	.144	.144	.108	.048	.336	.054	.054	.027	.086
Social	**S1**	**S2**	**S3**	**S4**	**S5**	**S6**	**S7**	**S8**	**S9**
Expert-1	.032	.070	.174	.174	.270	.044	.050	.070	.116
Expert-2	.088	.132	.066	.232	.132	.132	.088	.088	.040
Expert-3	.076	.054	.032	.063	.316	.190	.127	.095	.047
Expert-4	.026	.044	.103	.077	.103	.077	.258	.155	.155
Expert-5	.072	.072	.087	.072	.109	.145	.339	.030	.072
Expert-6	.057	.026	.057	.066	.100	.080	.316	.100	.199
Economic	**C1**	**C2**	**C3**	**C4**	**C5**	**C6**	**C7**	**C8**	**C9**
Expert-1	.136	.250	.114	.083	.114	.114	.091	.068	.030
Expert-2	.293	.084	.112	.042	.084	.067	.168	.067	.084
Expert-3	.126	.319	.095	.063	.076	.189	.054	.037	.042
Expert-4	.270	.154	.039	.077	.062	.051	.116	.154	.077
Expert-5	.281	.182	.091	.073	.073	.073	.139	.029	.061
Expert-6	.279	.072	.051	.051	.180	.120	.180	.045	.022
Risk	**R1**	**R2**	**R3**	**R4**	**R5**	**R6**	**R7**	**R8**	**R9**
Expert-1	.091	.091	.232	.137	.122	.137	.068	.027	.095
Expert-2	.311	.075	.188	.125	.094	.063	.063	.035	.047
Expert-3	.311	.077	.128	.192	.096	.055	.064	.030	.048
Expert-4	.068	.114	.085	.288	.171	.057	.068	.114	.034
Expert-5	.101	.081	.081	.134	.081	.035	.081	.307	.101
Expert-6	.053	.080	.160	.275	.160	.160	.046	.026	.040

Then, the experts' evaluations are analyzed for consistency and determined as consistent. The sub-criteria weights for each expert are determined by applying the BWM steps, as given in Table 6.12.

The weight of each sub-criteria is multiplied by the weight of the corresponding main criterion for each expert to determine the sub-criteria's score. After that, the weights are combined in relation to the reputations of the experts shown in Fig. 6.2. Final criteria weights and their ranks are presented in Table 6.13.

The most important sub-criteria is identified as "C1: Cost" when the outcomes of the criterion weights are analyzed. Cost for all stakeholders in the supply chain is a criterion that is ranked higher in terms of importance. The following criteria are found to be "E4: Environmental management system" and "E2: Pollution control." As can be observed, even in the post-pandemic era, environmental concerns have become more crucial to the sustainability of green supply chains. The group of criteria with the least weight is considered to be the "Social" sustainability criteria.

Table 6.13 Final criteria weights and their ranks

Environmental	Weight	[a]R	Social	Weight	[a]R
E1: Pollution production	.0496	8	S1: Culture	.0063	35
E2: Pollution control	.0567	3	S2: Relationship	.0067	34
E3: Resource consumption	.0467	9	S3: Human sources	.0061	36
E4: Environmental management sys.	.0573	2	S4: Organizational structure	.0104	30
E5: Hazardous materials	.0541	7	S5: Innovation	.0189	22
E6: Energy consumption	.0547	6	S6: Learning	.0129	27
E7: Reverse logistics	.0566	4	S7: Regulations	.0152	26
E8: Green distribution	.0183	23	S8: Competitiveness	.0079	32
E9: Waste discharge	.0271	14	S9: Commercial operations	.0068	33
Economic			**Risk**		
C1: Cost	.0740	1	R1: Operational	.0258	15
C2: Quality	.0551	5	R2: Supply	.0152	25
C3: Technology	.0244	16	R3: Production	.0228	17
C4: Flexibility	.0199	20	R4: Financial	.0336	11
C5: Productivity	.0299	13	R5: Demand	.0209	19
C6: Effectiveness	.0330	12	R6: Technological	.0127	28
C7: Price	.0388	10	R7: Delivery	.0115	29
C8: Accuracy	.0225	18	R8: Green	.0198	21
C9: Responsiveness	.0173	24	R9: Strategic barriers	.0103	31

[a]*R* ranking

6.6 Discussion

Today, companies coordinate their relations with their suppliers, who are as important as their customers among the stakeholders in the company's mission, with the perspective offered by the concept of SCM that emerged after the 1990s. The mass production-based and productivity-oriented system of the globalization and industrialization period in the world has changed. Thus, the manufacturer, which used to be the main actor determining the market, has been replaced by the supply chains closest to the customer. The level of competition has changed, the increase in competition brought about by globalization has begun to be effective, and competition has ceased to be a phenomenon experienced among individual organizations. In today's world, competition is experienced between global supply chains. In global supply chain competition, supply chains that want to maintain their power need to adapt quickly to changing ways of doing business. As a result of these changes, ideas like flexible supply chains, agile supply chains, resilient supply chains, sustainable supply chains, and others have evolved.

Along with the competitive advantage that SCM offers, the brutal nature of today's competition has forced businesses to take into account sustainability along with the supply chain in order to make a difference. Based on economic, environmental, and social considerations, a sustainable SCM satisfies the needs of stakeholders, including clients. With sustainable SCM, operations have become more adaptable, packaging protects the environment, lead times are reduced, and the company's reputation among customers and society as a whole is improved. Sustainable SCM is affected by the fact that the responsibilities of individual institutions now cover not only their own activities but also the activities of the supply chains in which they are located. Sustainable SCM manages the flow of materials, information, and capital along a supply chain and the collaboration of companies along the chain. Based on the expectations of stakeholders and customers, SCM simultaneously strives to fulfill economic, environmental, and social objectives related to the three pillars of sustainable development. Promotion of good governance standards and management of environmental, social, and economic repercussions over the course of a product or service's life cycle are the basics of supply chain sustainability.

The pandemic has caused great changes in people's lives. Although at first glance it seemed that the changes would be limited to health issues, there was a noticeable increase in people's awareness of their environment throughout this process. Today, many consumers are much more sensitive in their approach to environmental awareness than before the pandemic. For this reason, green, sustainable, carbon-free supply chains, environmentally friendly businesses, and similar values are now frequently encountered expectations in daily life. However, as customer awareness increases, they tend to be more selective about these values in their purchasing preferences. Therefore, these approaches and increasing environmental awareness are becoming more and more important for businesses and consumers.

Global environmental problems such as global warming and climate change force all stakeholders of the world to be more sensitive and take responsibility for environmental issues. Companies are under intense pressure to mitigate their negative environmental consequences from a variety of stakeholders, including the government and customers, as environmental consciousness grows throughout the world. Regardless of whether it is long or short, simple or complex, all companies engaged in supply chain management should integrate business practices in the service and manufacturing industry with sustainability and reduce end-to-end supply chain costs; in this way, it is necessary to reduce its environmental footprint. The green supply chain approach is extremely important in this context. It is possible to significantly reduce the global carbon footprint with green transformations in procurement processes.

People are now more concerned with the environmental impact of products' manufacturing, the size of the carbon footprint in the supply chain, and if a product is recyclable. This is one of the main indicators that businesses with a sustainable environmental approach will increase in the coming period and that the green supply chain approach will play an important role in this cycle. Green SCM combines the traditional supply chain strategy with environmentally friendly and sustainable practices. This chain covers all processes, from manufacturing and production to the procurement process and from material selection to waste management. Green SCM not only aims to lessen the adverse effects of the supply chain on the environment but also encourages value creation inside supply chain companies to lessen adverse effects on the environment. The main purpose of green supply chain management is generally to reduce CO_2 emissions and to create a balanced and sustainable ecosystem. But beyond that, there are different benefits it offers to businesses. Better integration with suppliers, lower prices, less waste, increased profitability, reuse of raw materials through recycling, positive added value created in customers' perception, and creation of a new eco-product market for companies are the main benefits of effective green SCM. The green supply chain approach can be applied to almost any industry, as there is always a greener method available for every product produced and consumed in any sector. The success here is how you incorporate sustainable eco-friendly methods into both supplier relationships and company culture. At this point, it is important to know the trends, methods, and how to apply them. The green supply chain strategy helps businesses finally achieve socioeconomic success in addition to protecting the environment. With more efficient procurement processes, waste can be reduced, and overall costs incurred throughout the product life cycle can be reduced. In this way, competitiveness in the market can be increased. In addition, responsible investors who consider the environmental performance of the institutions in their investments can be made to invest in the company. A wider customer portfolio can be created by creating a responsible brand image in the market.

6.7 Conclusion

The sustainability analysis of green supply chain operations has become an interesting research topic among the scholars and practitioners in the last two decades. The damage caused by companies to the environment has recently been a theme of a discussion. In green SCM, stakeholders make decisions taking environmental considerations into account along the entire value chain, from raw material and material supply needed for production activities to the waste produced by the end user of the product. Effective green supply chain management has numerous advantages for all supply chain stakeholders. It reduces the costs of the stakeholders, increases their productivity, and improves the quality of the products. A green supply chain involves less transportation, less energy, recycling, and more efficient use of space. Beyond these, businesses have a social duty to protect the environment by using green SCM practices in order to leave behind a world that is more livable for future generations. For these reasons, the sustainability of green supply chains is becoming more and more important every day. It is important to analyze and identify the obstacles and problems facing green SCM in terms of being a guide for both businesses and policy-makers.

This chapter examines the green supply chain's sustainability assessment in the post-COVID-19 era. The supply chain sustainability evaluation model is first built as a two-level hierarchical criterion framework for this objective. After that, the weights of each criterion in the hierarchical structure are determined using a BWM methodology. At this point, six experts are consulted to give their opinion on criteria. The results are discussed with the experts.

This chapter's main contributions can be summed up as follows: (i) three pillars of sustainability are extended with one more pillar, namely, risk, and proposed as sustainability evaluation model for green supply chains, especially post-pandemic era; (ii) the most noteworthy factors that contribute to the sustainability of the supply chain are identified and organized hierarchically; (iii) six professionals from the academic and business worlds have come together to form an expert panel that will advise on the significance of the criteria; (iv) the BWM assigns weights to the main and supporting criteria, which results in the determination of the relative importance of each criterion for the sustainability of the green supply chain; and (v) the goal is to help supply networks enhance their sustainability-related initiatives by using the proposed green supply chain sustainability model and weighting method as a guide.

Different decision-making approaches can be used in the weighting process as recommendations for future studies, and the results can be compared with the findings of this study. The specified criteria are the only ones considered in the suggested decision-making framework for the sustainability assessment of green supply chains; however, the number of criteria can be expanded or lowered depending on the needs or circumstances of the problem.

References

Abdel-Basset M, Mohamed R (2020) A novel plithogenic TOPSIS-CRITIC model for sustainable supply chain risk management. J Clean Prod 247. https://doi.org/10.1016/j.jclepro.2019.119586

Afrasiabi A, Tavana M, Di Caprio D (2022) An extended hybrid fuzzy multi-criteria decision model for sustainable and resilient supplier selection. Environ Sci Pollut Res 29(25):37291–37314. https://doi.org/10.1007/S11356-021-17851-2

Agrawal V, Mohanty RP, Agarwal S, Dixit JK, Agrawal AM (2022) Analyzing critical success factors for sustainable green supply chain management. Environ Dev Sustain. https://doi.org/10.1007/s10668-022-02396-2

Ahmadi HB, Lo H-W, Gupta H, Kusi-Sarpong S, Liou JJH (2020) An integrated model for selecting suppliers on the basis of sustainability innovation. J Clean Prod 277. https://doi.org/10.1016/j.jclepro.2020.123261

Ahmed T, Karmaker CL, Nasir SB, Moktadir MA (2022) Identifying and analysis of key flexible sustainable supply chain management strategies toward overcoming the post-COVID-19 impacts. Int J Emerg Mark. https://doi.org/10.1108/IJOEM-12-2021-1830

Ali SS, Kaur R (2021) Effectiveness of corporate social responsibility (CSR) in implementation of social sustainability in warehousing of developing countries: a hybrid approach. J Clean Prod 324. https://doi.org/10.1016/j.jclepro.2021.129154

Ali Y, Saad TB, Sabir M, Muhammad N, Salman A, Zeb K (2020) Integration of green supply chain management practices in construction supply chain of CPEC. Manag Environ Qual Int J 31(1): 185–200. https://doi.org/10.1108/MEQ-12-2018-0211

Amiri M, Hashemi-Tabatabaei M, Ghahremanloo M, Keshavarz-Ghorabaee M, Zavadskas EK, Banaitis A (2020) A new fuzzy BWM approach for evaluating and selecting a sustainable supplier in supply chain management. Int J Sust Dev World 1–18. https://doi.org/10.1080/13504509.2020.1793424

Andiç E, Yurt Ö, Baltacioğlu T (2012) Green supply chains: efforts and potential applications for the Turkish market. Resour Conserv Recycl 58:50–68. https://doi.org/10.1016/J.RESCONREC.2011.10.008

Ayyildiz E (2021) Interval valued intuitionistic fuzzy analytic hierarchy process-based green supply chain resilience evaluation methodology in post COVID-19 era. Environ Sci Pollut Res 0123456789. https://doi.org/10.1007/s11356-021-16972-y

Ayyildiz E (2022) Fermatean fuzzy step-wise weight assessment ratio analysis (SWARA) and its application to prioritizing indicators to achieve sustainable development goal-7. Renew Energy 193:136–148. https://doi.org/10.1016/j.renene.2022.05.021

Ayyildiz E, Taskin Gumus A (2021) Interval-valued Pythagorean fuzzy AHP method-based supply chain performance evaluation by a new extension of SCOR model: SCOR 4.0. Complex Intell Syst 7(1):559–576. https://doi.org/10.1007/s40747-020-00221-9

Badri Ahmadi H, Kusi-Sarpong S, Rezaei J (2017) Assessing the social sustainability of supply chains using Best Worst Method. Resour Conserv Recycl 126:99–106. https://doi.org/10.1016/J.RESCONREC.2017.07.020

Bai C, Zhu Q, Sarkis J (2022) Circular economy and circularity supplier selection: a fuzzy group decision approach. Int J Prod Res. https://doi.org/10.1080/00207543.2022.2037779

Baidya R, Debnath B, Ghosh SK, Rhee S-W (2020) Supply chain analysis of e-waste processing plants in developing countries. Waste Manag Res 38(2):173–183. https://doi.org/10.1177/0734242X19886633

Bajec P, Tuljak-Suban D, Bajor I (2020) A warehouse social and environmental performance metrics framework | Socialni in okoljski kriteriji za merjenje uspešnosti v skladišču. Promet Traffic Traffico 32(4):513–526. https://doi.org/10.7307/ptt.v32i4.3390

Banasik A, Bloemhof-Ruwaard JM, Kanellopoulos A, Claassen GDH, van der Vorst JGAJ (2018) Multi-criteria decision making approaches for green supply chains: a review. Flex Serv Manuf J 30(3):366–396. https://doi.org/10.1007/S10696-016-9263-5

Barbier EB (1987) The concept of sustainable economic development. Environ Conserv 14(2): 101–110. https://doi.org/10.1017/S0376892900011449

Boutkhoum O, Hanine M, Boukhriss H, Agouti T, Tikniouine A (2016) Multi-criteria decision support framework for sustainable implementation of effective green supply chain management practices. Springerplus 5(1). https://doi.org/10.1186/s40064-016-2233-2

Cabral I, Grilo A, Cruz-Machado V (2012) A decision-making model for Lean, Agile, Resilient and Green supply chain management. Int J Prod Res 50(17):4830–4845. https://doi.org/10.1080/00207543.2012.657970

Caristi G, Boffardi R, Ciliberto C, Arbolino R, Ioppolo G (2022) Multicriteria approach for supplier selection: evidence from a case study in the fashion industry. Sustainability (Switzerland) 14(13). https://doi.org/10.3390/su14138038

Carter CR, Rogers DS (2008) A framework of sustainable supply chain management: moving toward new theory. Int J Phys Distrib Logist Manag 38(5):360–387. https://doi.org/10.1108/09600030810882816

Cavalcante de Souza Feitosa IS, Ribeiro Carpinetti LC, de Almeida-Filho AT (2021) A supply chain risk management maturity model and a multi-criteria classification approach. Benchmarking 28(9):2636–2655. https://doi.org/10.1108/BIJ-09-2020-0487

Chai N, Zhou W (2022) A novel hybrid MCDM approach for selecting sustainable alternative aviation fuels in supply chain management. Fuel 327. https://doi.org/10.1016/j.fuel.2022.125180

Chand P, Tarei PK (2021) Do the barriers of multi-tier sustainable supply chain interact? A multi-sector examination using resource-based theory and resource-dependence theory. J Purch Supply Manag 27(5). https://doi.org/10.1016/j.pursup.2021.100722

Dalalah D, Khan SA, Al-Ashram Y, Albeetar S, Ali YA, Alkhouli E (2022) An integrated framework for the assessment of environmental sustainability in wood supply chains. Environ Technol Innov 27:102429. https://doi.org/10.1016/J.ETI.2022.102429

Deshmukh AJ, Vasudevan H (2018) A combined approach for supplier selection using AHP and fuzzy AHP in Indian gear manufacturing MSMEs. IOP Conf Ser: Mater Sci Eng 376(1). https://doi.org/10.1088/1757-899X/376/1/012122

Dursun M (2021) BWM integrated intuitionistic fuzzy approach for sustainable transportation service provider selection. J Mult Valued Log Soft Comput 37(3–4):277–294

Dwivedi A, Paul SK (2022) A framework for digital supply chains in the era of circular economy: implications on environmental sustainability. Bus Strateg Environ 31(4):1249–1274. https://doi.org/10.1002/bse.2953

Ecer F, Pamucar D (2020) Sustainable supplier selection: a novel integrated fuzzy best worst method (F-BWM) and fuzzy CoCoSo with Bonferroni (CoCoSo'B) multi-criteria model. J Clean Prod 266. https://doi.org/10.1016/j.jclepro.2020.121981

Elabed S, Shamayleh A, Daghfous A (2021) Sustainability-oriented innovation in the health care supply chain. Comput Ind Eng 160. https://doi.org/10.1016/j.cie.2021.107564

Er Kara M, Firat SO (2018) Supplier risk assessment based on best-worst method and k-means clustering: a case study. Sustainability (Switzerland) 10(4). https://doi.org/10.3390/SU10041066

Erol I, Ar IM, Peker I (2022) Scrutinizing blockchain applicability in sustainable supply chains through an integrated fuzzy multi-criteria decision making framework. Appl Soft Comput 116. https://doi.org/10.1016/j.asoc.2021.108331

Fallahpour A, Wong KY, Rajoo S, Mardani A (2020) An integrated fuzzy carbon management-based model for suppliers' performance evaluation and selection in green supply chain management. Int J Fuzzy Syst 22(2):712–723. https://doi.org/10.1007/s40815-019-00759-y

Fallahpour A, Wong KY, Rajoo S, Fathollahi-Fard AM, Antucheviciene J, Nayeri S (2021) An integrated approach for a sustainable supplier selection based on industry 4.0 concept. Environ Sci Pollut Res. https://doi.org/10.1007/s11356-021-17445-y

Garg CP, Sharma A (2020) Sustainable outsourcing partner selection and evaluation using an integrated BWM–VIKOR framework. Environ Dev Sustain 22(2):1529–1557. https://doi.org/10.1007/s10668-018-0261-5

Ghasemian Sahebi I, Arab A, Sadeghi Moghadam MR (2017) Analyzing the barriers to humanitarian supply chain management: a case study of the Tehran Red Crescent Societies. Int J Disaster Risk Red 24:232–241. https://doi.org/10.1016/J.IJDRR.2017.05.017

Ghosh S, Mandal MC, Ray A (2021a) Green supply chain management framework for supplier selection: an integrated multi-criteria decision-making approach. Int J Manag Sci Eng Manag. https://doi.org/10.1080/17509653.2021.1997661

Ghosh S, Mandal MC, Ray A (2021b) Selection of environmental-conscious sourcing: an empirical investigation. Benchmarking 28(6):2130–2155. https://doi.org/10.1108/BIJ-08-2020-0416

Ghosh S, Mandal MC, Ray A (2021c) Strategic sourcing model for green supply chain management: an insight into automobile manufacturing units in India. Benchmarking. https://doi.org/10.1108/BIJ-06-2021-0333

Ghosh S, Mandal MC, Ray A (2022) A PDCA based approach to evaluate green supply chain management performance under fuzzy environment. Int J Manag Sci Eng Manag. https://doi.org/10.1080/17509653.2022.2027292

Green KW, Inman RA, Sower VE, Zelbst PJ (2019) Impact of JIT, TQM and green supply chain practices on environmental sustainability. J Manuf Technol Manag 30(1):26–47. https://doi.org/10.1108/JMTM-01-2018-0015

Gunduz MA, Demir S, Paksoy T (2021) Matching functions of supply chain management with smart and sustainable tools: a novel hybrid BWM-QFD based method. Comput Ind Eng 162. https://doi.org/10.1016/j.cie.2021.107676

Gupta H, Barua MK (2017) Supplier selection among SMEs on the basis of their green innovation ability using BWM and fuzzy TOPSIS. J Clean Prod 152:242–258. https://doi.org/10.1016/J.JCLEPRO.2017.03.125

Gupta H, Barua MK (2018) A novel hybrid multi-criteria method for supplier selection among SMEs on the basis of innovation ability. Int J Log Res Appl 21(3):201–223. https://doi.org/10.1080/13675567.2017.1382457

Gupta H, Kusi-Sarpong S, Rezaei J (2020) Barriers and overcoming strategies to supply chain sustainability innovation. Resour Conserv Recycl 161. https://doi.org/10.1016/j.resconrec.2020.104819

Han X, Rani P (2022) Evaluate the barriers of blockchain technology adoption in sustainable supply chain management in the manufacturing sector using a novel Pythagorean fuzzy-CRITIC-CoCoSo approach. Oper Manag Res. https://doi.org/10.1007/s12063-021-00245-5

Hendiani S, Mahmoudi A, Liao H (2020) A multi-stage multi-criteria hierarchical decision-making approach for sustainable supplier selection. Appl Soft Comput J 94. https://doi.org/10.1016/j.asoc.2020.106456

Hendiani S, Lev B, Gharehbaghi A (2021) Diagnosing social failures in sustainable supply chains using a modified Pythagorean fuzzy distance to ideal solution. Comput Ind Eng 154. https://doi.org/10.1016/j.cie.2021.107156

Hoseini SA, Fallahpour A, Wong KY, Mahdiyar A, Saberi M, Durdyev S (2021) Sustainable supplier selection in construction industry through hybrid fuzzy-based approaches. Sustainability (Switzerland) 13(3):1–19. https://doi.org/10.3390/su13031413

Hosseini ZS, Flapper SD, Pirayesh M (2022) Sustainable supplier selection and order allocation under demand, supplier availability and supplier grading uncertainties. Comput Ind Eng 165. https://doi.org/10.1016/j.cie.2021.107811

Ijadi Maghsoodi A, Soudian S, Martínez L, Herrera-Viedma E, Zavadskas EK (2020) A phase change material selection using the interval-valued target-based BWM-CoCoMULTIMOORA approach: a case-study on interior building applications. Appl Soft Comput 95:106508. https://doi.org/10.1016/J.ASOC.2020.106508

Kannan D, Solanki R, Kaul A, Jha PC (2022) Barrier analysis for carbon regulatory environmental policies implementation in manufacturing supply chains to achieve zero carbon. J Clean Prod 358:131910. https://doi.org/10.1016/J.JCLEPRO.2022.131910

Kao J-C, Wang C-N, Nguyen VT, Husain ST (2022) A fuzzy mcdm model of supplier selection in supply chain management. Intell Autom Soft Comput 31(3):1451–1466. https://doi.org/10.32604/IASC.2022.021778

Kazancoglu Y, Ozkan-Ozen YD, Sagnak M, Kazancoglu I, Dora M (2021) Framework for a sustainable supply chain to overcome risks in transition to a circular economy through Industry 4.0. Prod Plan Control. https://doi.org/10.1080/09537287.2021.1980910

Khan S, Haleem A, Khan MI (2020) Assessment of risk in the management of Halal supply chain using fuzzy BWM method. Supply Chain Forum:1–17. https://doi.org/10.1080/16258312.2020.1788905

Khokhar M, Hou Y, Rafique MA, Iqbal W (2020) Evaluating the social sustainability criteria of supply chain management in manufacturing industries: a role of BWM in MCDM | Ocena kryteriów zrównoważoności społecznej w zarządzaniu łańcuchem dostaw w przemyśle wytwórczym: Rola BWM w MCDM. Problemy Ekorozwoju 15(2):185–194

Kumar A, Zavadskas EK, Mangla SK, Agrawal V, Sharma K, Gupta D (2019) When risks need attention: adoption of green supply chain initiatives in the pharmaceutical industry. Int J Prod Res 57(11):3554–3576. https://doi.org/10.1080/00207543.2018.1543969

Kumar A, Mangla SK, Kumar P, Song M (2021a) Mitigate risks in perishable food supply chains: learning from COVID-19. Technol Forecast Soc Chang 166. https://doi.org/10.1016/j.techfore.2021.120643

Kumar P, Singh RK, Paul J, Sinha O (2021b) Analyzing challenges for sustainable supply chain of electric vehicle batteries using a hybrid approach of Delphi and Best-Worst Method. Resour Conserv Recycl 175. https://doi.org/10.1016/j.resconrec.2021.105879

Kumar A, Shrivastav S, Adlakha A, Vishwakarma NK (2022) Appropriation of sustainability priorities to gain strategic advantage in a supply chain. Int J Product Perform Manag 71(1):125–155. https://doi.org/10.1108/IJPPM-06-2020-0298

Liaqait RA, Warsi SS, Zahid T, Ghafoor U, Ahmad MS, Selvaraj J (2021) A decision framework for solar PV panels supply chain in context of sustainable supplier selection and order allocation. Sustainability (Switzerland) 13(23). https://doi.org/10.3390/su132313216

Liaqait RA, Warsi SS, Agha MH, Zahid T, Becker T (2022) A multi-criteria decision framework for sustainable supplier selection and order allocation using multi-objective optimization and fuzzy approach. Eng Optim 54(6):928–948. https://doi.org/10.1080/0305215X.2021.1901898

Lo HW, Liou JJH, Wang HS, Tsai YS (2018) An integrated model for solving problems in green supplier selection and order allocation. J Clean Prod 190:339–352. https://doi.org/10.1016/J.JCLEPRO.2018.04.105

Lu LYY, Wu CH, Kuo TC (2007) Environmental principles applicable to green supplier evaluation by using multi-objective decision analysis. Int J Prod Res 45(18–19):4317–4331. https://doi.org/10.1080/00207540701472694

Mahmud P, Paul SK, Azeem A, Chowdhury P (2021) Evaluating supply chain collaboration barriers in smalland medium-sized enterprises. Sustainability (Switzerland) 13(13). https://doi.org/10.3390/su13137449

Malviya RK, Kant R (2018) Prioritising the solutions to overcome the barriers of green supply chain management implementation: a hybrid fuzzy AHP-VIKOR framework approach. J Decis Syst 27(4):275–320. https://doi.org/10.1080/12460125.2019.1603597

Mangla SK, Kumar P, Barua MK (2015) Risk analysis in green supply chain using fuzzy AHP approach: a case study. Resour Conserv Recycl 104:375–390. https://doi.org/10.1016/j.resconrec.2015.01.001

Mastrocinque E, Ramírez FJ, Honrubia-Escribano A, Pham DT (2022) Industry 4.0 enabling sustainable supply chain development in the renewable energy sector: a multi-criteria intelligent approach. Technol Forecast Soc Chang 182. https://doi.org/10.1016/j.techfore.2022.121813

Mishra AR, Rani P (2021) A q-rung orthopair fuzzy ARAS method based on entropy and discrimination measures: an application of sustainable recycling partner selection. J Ambient Intell Humaniz Comput. https://doi.org/10.1007/s12652-021-03549-3

Mohammadi M, Rezaei J (2020) Bayesian best-worst method: a probabilistic group decision making model. Omega 96:102075. https://doi.org/10.1016/J.OMEGA.2019.06.001

Mohammadnazari Z, Ghannadpour SF (2021) Sustainable construction supply chain management with the spotlight of inventory optimization under uncertainty. Environ Dev Sustain 23(7): 10937–10972. https://doi.org/10.1007/s10668-020-01095-0

Mohammed A (2020) Towards 'gresilient' supply chain management: a quantitative study. Resour Conserv Recycl 155. https://doi.org/10.1016/j.resconrec.2019.104641

Moher D, Liberati A, Tetzlaff J, Altman DG, Group TP (2009) Preferred reporting items for systematic reviews and meta-analyses: the PRISMA statement. PLoS Med 6(7):e1000097. https://doi.org/10.1371/JOURNAL.PMED.1000097

Mohseni S, Baghizadeh K, Pahl J (2022) Evaluating barriers and drivers to sustainable food supply chains. Math Probl Eng 2022. https://doi.org/10.1155/2022/4486132

Moktadir MA, Dwivedi A, Khan NS, Paul SK, Khan SA, Ahmed S, Sultana R (2021) Analysis of risk factors in sustainable supply chain management in an emerging economy of leather industry. J Clean Prod 283. https://doi.org/10.1016/j.jclepro.2020.124641

Mondal A, Roy SK (2022) Application of Choquet integral in interval type-2 Pythagorean fuzzy sustainable supply chain management under risk. Int J Intell Syst 37(1):217–263. https://doi.org/10.1002/int.22623

Murray JG (2000) Effects of a green purchasing strategy: the case of Belfast City Council. Supply Chain Manag 5(1):37–44. https://doi.org/10.1108/13598540010312954

Nabeeh NA, Abdel-Basset M, Soliman G (2021) A model for evaluating green credit rating and its impact on sustainability performance. J Clean Prod 280. https://doi.org/10.1016/j.jclepro.2020.124299

Naseem A, Shah STH, Khan SA, Malik AW (2017) Decision support system for optimum decision making process in threat evaluation and weapon assignment: current status, challenges and future directions. Annu Rev Control 43:169–187. https://doi.org/10.1016/J.ARCONTROL.2017.03.003

Nguyen NBT, Lin G-H, Dang T-T (2021) A two phase integrated fuzzy decision-making framework for green supplier selection in the coffee bean supply chain. Mathematics 9(16). https://doi.org/10.3390/math9161923

Orji IJ, Kusi-Sarpong S, Gupta H (2020) The critical success factors of using social media for supply chain social sustainability in the freight logistics industry. Int J Prod Res 58(5): 1522–1539. https://doi.org/10.1080/00207543.2019.1660829

Öztürk C, Yildizbaşi A (2020) Barriers to implementation of blockchain into supply chain management using an integrated multi-criteria decision-making method: a numerical example. Soft Comput 24(19):14771–14789. https://doi.org/10.1007/s00500-020-04831-w

Pamučar D, Mihajlović M, Obradović R, Atanasković P (2017) Novel approach to group multi-criteria decision making based on interval rough numbers: hybrid DEMATEL-ANP-MAIRCA model. Expert Syst Appl 88:58–80. https://doi.org/10.1016/j.eswa.2017.06.037

Pamucar D, Deveci M, Gokasar I, Popovic M (2022a) Fuzzy Hamacher WASPAS decision-making model for advantage prioritization of sustainable supply chain of electric ferry implementation in public transportation. Environ Dev Sustain 24(5):7138–7177. https://doi.org/10.1007/s10668-021-01742-0

Pamucar D, Torkayesh AE, Biswas S (2022b) Supplier selection in healthcare supply chain management during the COVID-19 pandemic: a novel fuzzy rough decision-making approach. Ann Oper Res. https://doi.org/10.1007/s10479-022-04529-2

Paul A, Shukla N, Paul SK, Trianni A (2021) Sustainable supply chain management and multi-criteria decision-making methods: a systematic review. Sustainability (Switzerland) 13(13). https://doi.org/10.3390/su13137104

Perçin S (2022) Evaluating the circular economy–based big data analytics capabilities of circular agri-food supply chains: the context of Turkey. Environ Sci Pollut Res. https://doi.org/10.1007/s11356-022-21680-2

Pınar A, Erdebilli BDRB, Özdemir YS (2021) Q-rung orthopair fuzzy topsis method for green supplier selection problem. Sustainability (Switzerland) 13(2):1–14. https://doi.org/10.3390/su13020985

Pourjavad E, Torkabadi AM, Mayorga RV (2018) Analysis of green supply chain management initiatives: a study of critical implementation barriers. WSEAS Trans Syst Control 13:153–160

Prashar A (2022) Supply chain sustainability drivers for fast-moving consumer goods (FMCG) sector: an Indian perspective. Int J Product Perform Manag. https://doi.org/10.1108/IJPPM-10-2021-0582

Rajesh R (2020) Sustainable supply chains in the Indian context: an integrative decision-making model. Technol Soc 61. https://doi.org/10.1016/j.techsoc.2020.101230

Rao P, Holt D (2005) Do green supply chains lead to competitiveness and economic performance? Int J Oper Prod Manag 25(9):898–916. https://doi.org/10.1108/01443570510613956

Rezaei J (2015) Best-worst multi-criteria decision-making method. Omega 53:49–57

Rezaei J, Nispeling T, Sarkis J, Tavasszy L (2016) A supplier selection life cycle approach integrating traditional and environmental criteria using the best worst method. J Clean Prod 135:577–588. https://doi.org/10.1016/J.JCLEPRO.2016.06.125

Rouyendegh BDBD, Yildizbasi A, Üstünyer P (2020) Intuitionistic fuzzy TOPSIS method for green supplier selection problem. Soft Comput 24(3):2215–2228. https://doi.org/10.1007/s00500-019-04054-8

Roy J, Pamučar D, Kar S (2020) Evaluation and selection of third party logistics provider under sustainability perspectives: an interval valued fuzzy-rough approach. Ann Oper Res 293(2): 669–714. https://doi.org/10.1007/s10479-019-03501-x

Sadaghiani S, Ahmad KW, Rezaei J, Tavasszy L (2015) Evaluation of external forces affecting supply chain sustainability in oil and gas industry using Best Worst Method. In: 2015 international Mediterranean gas and oil conference (MedGO), pp 1–4. https://doi.org/10.1109/MedGO.2015.7330322

Sahebi IG, Masoomi B, Ghorbani S (2020) Expert oriented approach for analyzing the blockchain adoption barriers in humanitarian supply chain. Technol Soc 63. https://doi.org/10.1016/j.techsoc.2020.101427

Salimian S, Mousavi SM, Antucheviciene J (2022) An interval-valued intuitionistic fuzzy model based on extended VIKOR and MARCOS for sustainable supplier selection in organ transplantation networks for healthcare devices. Sustainability 14(7):3795. https://doi.org/10.3390/SU14073795

Santi RP, Putra H (2018) A systematic literature review of business intelligence technology, contribution and application for higher education. In: 2018 international conference on information technology systems and innovation, ICITSI 2018 – proceedings, pp 404–409. https://doi.org/10.1109/ICITSI.2018.8696019

Saraji MK, Streimikiene D (2022) Evaluating the circular supply chain adoption in manufacturing sectors: a picture fuzzy approach. Technol Soc 70. https://doi.org/10.1016/j.techsoc.2022.102050

Sari K (2017) A novel multi-criteria decision framework for evaluating green supply chain management practices. Comput Ind Eng 105:338–347. https://doi.org/10.1016/j.cie.2017.01.016

Sarker MR, Moktadir MA, Santibanez-Gonzalez EDR (2021) Social sustainability challenges towards flexible supply chain management: post-COVID-19 perspective. Glob J Flex Syst Manag 22:199–218. https://doi.org/10.1007/s40171-021-00289-3

Sathiya Narayana SN, Parthiban P, Anbuchezhian, Dhanalakshmi R (2020) Assessment and implementation of lean and green supply chain in medium scale automobile industries using AHP and fuzzy TOPSIS. J Sci Ind Res 79(8):720–726

Satria D, Sensuse DI, Noprisson H (2017) A systematic literature review of the improved agile software development. In: 2017 international conference on information technology systems and innovation, ICITSI 2017 – proceedings, 2018, January, pp 94–99. https://doi.org/10.1109/ICITSI.2017.8267925

Sharma M, Joshi S (2020) Digital supplier selection reinforcing supply chain quality management systems to enhance firm's performance. TQM J. https://doi.org/10.1108/TQM-07-2020-0160

Sharma M, Kamble S, Mani V, Sehrawat R, Belhadi A, Sharma V (2021a) Industry 4.0 adoption for sustainability in multi-tier manufacturing supply chain in emerging economies. J Clean Prod 281. https://doi.org/10.1016/j.jclepro.2020.125013

Sharma VK, Sachdeva A, Singh LP (2021b) A meta-analysis of sustainable supply chain management from different aspects. Int J Oper Prod Manag 8(3):289–313. https://doi.org/10.22034/IJSOM.2021.3.4

Shekarian E, Ijadi B, Zare A, Majava J (2022) Sustainable supply chain management: a comprehensive systematic review of industrial practices. Sustainability (Switzerland) 14(13). https://doi.org/10.3390/su14137892

Singh A, Trivedi A (2016) Sustainable green supply chain management: trends and current practices. Compet Rev 26(3):265–288. https://doi.org/10.1108/CR-05-2015-0034

Srivastava SK (2007) Green supply-chain management: a state-of-the-art literature review. Int J Manag Rev 9(1):53–80. https://doi.org/10.1111/j.1468-2370.2007.00202.x

Tian ZP, Zhang HY, Wang JQ, Wang TL (2018) Green supplier selection using improved TOPSIS and best-worst method under intuitionistic fuzzy environment. Informatica (Netherlands) 29(4):773–780. https://doi.org/10.15388/INFORMATICA.2018.192

Tirkolaee EB, Dashtian Z, Weber GW, Tomaskova H, Soltani M, Mousavi NS (2021) An integrated decision-making approach for green supplier selection in an agri-food supply chain: threshold of robustness worthiness. Mathematics 9(11):1–30. https://doi.org/10.3390/math9111304

Tolooie M, Alvandi M, Arani MS (2022) Sustainable supplier evaluation and selection in developing countries: an integrated fuzzy framework. Int J Integr Supply Manag 15(2):151–183. https://doi.org/10.1504/IJISM.2022.121952

Tseng M-L, Tran TPT, Fujii M, Lim MK, Negash YT (2021) Modelling hierarchical circular supply chain management enablers in the seafood processing industry in Vietnam under uncertainties. Int J Log Res Appl. https://doi.org/10.1080/13675567.2021.1965105

Uygun Ö, Dede A (2016) Performance evaluation of green supply chain management using integrated fuzzy multi-criteria decision making techniques. Comput Ind Eng 102:502–511. https://doi.org/10.1016/J.CIE.2016.02.020

Vahabzadeh Najafi N, Arshadi Khamseh A, Mirzazadeh A (2020) An integrated sustainable and flexible supplier evaluation model under uncertainty by game theory and subjective/objective data: Iranian casting industry. Glob J Flex Syst Manag 21(4):309–322. https://doi.org/10.1007/s40171-020-00250-w

Wang Chen HM, Chou SY, Luu QD, Yu THK (2016) A fuzzy MCDM approach for green supplier selection from the economic and environmental aspects. Math Probl Eng 2016. https://doi.org/10.1155/2016/8097386

Wang C-N, Dang T-T, Nguyen N-A-T (2021) Outsourcing reverse logistics for e-commerce retailers: a two-stage fuzzy optimization approach. Axioms 10(1). https://doi.org/10.3390/axioms10010034

Wu J, Liu C, Wu Y, Cao M, Liu Y (2022) A novel hotel selection decision support model based on the online reviews from opinion leaders by best worst method. Int J Comput Intell Syst 15(1): 1–20. https://doi.org/10.1007/S44196-022-00073-W/FIGURES/6

Xiao L, Huang G, Pedrycz W, Pamucar D, Martínez L, Zhang G (2022a) A q-rung orthopair fuzzy decision-making model with new score function and best-worst method for manufacturer selection. Inf Sci 608:153–177. https://doi.org/10.1016/J.INS.2022.06.061

Xiao L, Huang G, Zhang G (2022b) Toward an action-granularity-oriented modularization strategy for complex mechanical products using a hybrid GGA-CGA method. Neural Comput & Applic 34(8):6453–6487. https://doi.org/10.1007/S00521-021-06796-9/FIGURES/18

Yadav S, Garg D, Luthra S (2020a) Selection of third-party logistics services for internet of things-based agriculture supply chain management. Int J Logist Syst Manag 35(2):204–230. https://doi.org/10.1504/IJLSM.2020.104780

Yadav G, Luthra S, Jakhar SK, Mangla SK, Rai DP (2020b) A framework to overcome sustainable supply chain challenges through solution measures of industry 4.0 and circular economy: an automotive case. J Clean Prod 254. https://doi.org/10.1016/j.jclepro.2020.120112

Yalcin Kavus B, Gulum Tas P, Ayyildiz E, Taskin A (2022) A three-level framework to evaluate airline service quality based on interval valued neutrosophic AHP considering the new dimensions. J Air Transp Manag 99:102179. https://doi.org/10.1016/J.JAIRTRAMAN.2021.102179

Yang K, Duan T, Feng J, Mishra AR (2022) Internet of things challenges of sustainable supply chain management in the manufacturing sector using an integrated q-Rung Orthopair Fuzzy-CRITIC-VIKOR method. J Enterp Inf Manag 35(4–5):1011–1039. https://doi.org/10.1108/JEIM-06-2021-0261

Yao J, Shi H, Liu C (2020) Optimising the configuration of green supply chains under mass personalisation. Int J Prod Res 0(0):1–19. https://doi.org/10.1080/00207543.2020.1723814

Yildiz Çankaya S, Sezen B (2019) Effects of green supply chain management practices on sustainability performance. J Manuf Technol Manag 30(1):98–121. https://doi.org/10.1108/JMTM-03-2018-0099

Yildizbasi A, Arioz Y (2022) Green supplier selection in new era for sustainability: a novel method for integrating big data analytics and a hybrid fuzzy multi-criteria decision making. Soft Comput 26(1):253–270. https://doi.org/10.1007/s00500-021-06477-8

Yıldızbaşı A, Öztürk C, Efendioğlu D, Bulkan S (2021) Assessing the social sustainable supply chain indicators using an integrated fuzzy multi-criteria decision-making methods: a case study of Turkey. Environ Dev Sustain 23(3):4285–4320. https://doi.org/10.1007/s10668-020-00774-2

Zarrinpoor N (2021) Designing a sustainable supply chain network for producing high-value products from waste glass. Waste Manag Res 39(12):1489–1500. https://doi.org/10.1177/0734242X21994669

Zhang F, Song W (2022) Sustainability risk assessment of blockchain adoption in sustainable supply chain: an integrated method. Comput Ind Eng 171. https://doi.org/10.1016/j.cie.2022.108378

Zhou F, Chen TY (2020) An integrated multicriteria group decision-making approach for green supplier selection under pythagorean fuzzy scenarios. IEEE Access 8:165216–165231. https://doi.org/10.1109/ACCESS.2020.3022377

Zhou Y, Xu L, Shaikh GM (2019) Evaluating and prioritizing the green supply chain management practices in Pakistan: based on Delphi and fuzzy AHP approach. Symmetry 11(11). https://doi.org/10.3390/sym11111346

Chapter 7
Economic Performance Analysis of a Resilient and Sustainable Supply Chain: Adoption of Electric Vehicles as a Sustainable Logistics Option

Md Muzahid Khan, Md Doulotuzzaman Xames, Priom Mahmud, Shahriar Tanvir Alam, and Niamat Ullah Ibne Hossain

Abstract Electric vehicles (EVs) can be seen as a significant technology that can help to create a more resilient and sustainable supply chain by reducing greenhouse gas (GHG) emissions and air pollution. With the recent focus on green sustainability, EVs have grown in popularity in many counties such as the United States, Germany, and China because of their zero emission. Driven by the incentives of electrified power trains, many companies are shifting from internal combustion engine vehicles (ICEVs) to EVs. Using the economic performance analysis, this research seeks to investigate the feasibility of adopting EVs into company logistics along with the identification of the key elements that influence its feasibility. A comparative study based on the economic aspects has been conducted between the ICEV, hybrid electric vehicle (HEV), and EV. This chapter implies that both EV and HEV outperform the traditional ICEV in terms of economic performance and they also have enormous potential as sustainable and resilient mobility solutions in a supply chain.

Keywords Electric vehicle · Sustainability · Transportation · Supply chain

M. M. Khan (✉) · S. T. Alam
Department of Industrial and Production Engineering, Military Institute of Science and Technology, Dhaka, Bangladesh
e-mail: tanvir.shahriar.tro@gmail.com

M. D. Xames
Department of Industrial and Systems Engineering, Virginia Tech, Blacksburg, VA, USA
e-mail: xames@vt.edu

P. Mahmud
Department of Systems and Industrial Engineering, University of Arizona, Tucson, AZ, USA
e-mail: priom1996@arizona.edu

N. U. Ibne Hossain
Engineering Management, College of Engineering and Computer Science, Arkansas State University, Jonesboro, AR, USA
e-mail: nibnehossain@astate.edu

© The Author(s), under exclusive license to Springer Nature Switzerland AG 2023

N. U. Ibne Hossain (ed.), *Data Analytics for Supply Chain Networks*, Greening of Industry Networks Studies 11, https://doi.org/10.1007/978-3-031-29823-3_7

7.1 Introduction

The logistics industries are substantial contributors to social and economic advancement in today's societies and act as a crucial sector in global economies. On the other hand, the logistics and transportation industries account for over a quarter of global GHG emissions, and these sectors are considered one of the most significant sources of air pollution in cities (Onat et al. 2019). Global CO_2 emissions from transportation are 9000 billion tons, with 18% coming from man-made sources, and these are anticipated to increase to 60% by 2050 (Liimatainen et al. 2019). The transportation sector is the second largest source of GHG emissions, responsible for more than 60% of total GHG emissions (Zhao et al. 2016). As residential and industrial energy or fuel needs are growing over time, the consequent increase in petroleum fuel usage has exacerbated worries about global warming and energy reliance. As a consequence, while fossil fuels remain the most common source of energy, today concepts like renewable energy and green mobility have gotten a lot of attention both in academia and in industry (Noori et al. 2015). CO_2 emissions from transportation systems will play a large role in decreasing total GHG emissions. These difficulties necessitate transportation system development since it will reduce environmental, economic, and social impacts, resulting in a significant improvement in achieving sustainable and efficient mobility (Onat et al. 2019). The European Union (EU) has set a goal for lowering GHG emissions by 80% within the year 2050 and set several climate change policies to reduce carbon emissions by lowering the use of conventionally fueled-based cars and achieving CO_2-free city logistics (Fiori and Marzano 2018; Osieczko et al. 2021).

Electric vehicles (EVs) are seen as a significant technology in the automotive industry that can help to achieve sustainable development by reducing greenhouse gas emissions, reducing air pollution for residents, and improving firm profitability. Multiple aspects are encouraging the adoption of electric vehicles, including incentives for companies to lower their carbon footprint; greater variability of oil-based goods and long-term risk involved with reliance on oil-based sources of energy; government allowances to lower acquisition costs; and advancements in alternative energy technologies (Juan et al. 2016). So far, no comprehensive study of the impact of electrified power trains on the automobile industry's supply chain and logistical sustainability has been conducted in an integrated way (Günther et al. 2015). This research aims to show the economic performance evaluations of EVs and seeks to forecast the most appropriate combination of ICEV, HEV, and EV that could be used for global logistics, taking into account economic expenses and environmental damage costs, using the most likely range of values.

7.2 Literature Review

EVs, including HEVs, BEVs, and the recently introduced electric range-extended vehicles (EREVs), have been extensively pushed by the governments as a widely recognized green and sustainable transportation method (Günther et al. 2015; Hu et al. 2013; Zhao et al. 2016). Kennedy and Philbin (2019) conducted a techno-economic analysis of the wide acceptance of electric-powered vehicles including comparisons to hybrid, gasoline, and diesel vehicles in their research work. The findings showed that EVs have a strong place in the near future because of the intricacy of electricity supply, facility standards, and overall life cycle cost considerations.

Another group of researchers (Yusof et al. 2021) also performed a techno-economic analysis, sensitivity analysis, and environmental impact analysis of EVs, HEVs, and ICEVs. The researcher conducted a life cycle cost (LCC) analysis to examine the feasibility of adopting electric vehicles (EVs) in Brunei. As the outcome of the study, the researcher found that EVs can enter and compete in the Bruneian market with the established ICEV and HEV. Günther et al. (2015) studied the potential development prospects for the automobile industry supply chain in Europe and China over the next 20 years, with a focus on EVs. Based on real-world data, a model-based quantitative analysis was undertaken by the researchers utilizing a thorough linear optimization. The researchers also evaluated the use of electric vehicles in an end-to-end supply chain to explore the industry's potential sustainability performance based on quantitative analysis. Onat et al. (2019) analyzed the sustainability impacts of EVs in Qatar. They developed a unique life cycle sustainability assessment (LCSA) model to identify and quantify the impacts on the aspects of sustainability. The researcher found that EVs can save up to 28% in LCC, 71% in GHG emissions, 51% in photochemical oxidant production, and 63% in the effects on human health. Apart from that, Juan et al. (2016) explored major challenges to the adoption of EVs in logistics. They briefly reviewed the environmental issues, and rising strategic- and operational-related challenges, and also examined how the adoption of EVs leads to a new dimension of vehicle routing problems. They also proposed that metaheuristics and simheuristics models can be used to solve complex logistics network optimization problems.

Moll et al. (2020) explored the use of heavy-duty battery electric trucks for sustainable urban logistics. They developed a system dynamics (SD) model to find out whether 24-h delivery with battery electric trucks is beneficial for logistics companies or not. The researchers discovered that deploying battery-powered electric trucks for 24-h delivery is highly profitable and it's a promising solution to creating sustainable urban logistics. In another research, Osieczko et al. (2021) pointed out possible ways for growing low-emission cargo transportation in the European Union (EU). They suggested few factors to quantify the scale of development of low-carbon emission logistics with the use of electric delivery vehicles. The suggested factors include the installation of charging stations, introducing new power transmission system with the incorporation of new technologies, and financial incentives for the organization that will be active in facilitating, assisting, and enabling road transport electrification. The researchers concluded that the expanding adoption of electric vehicles for delivery led to the growth of low-emission logistics in the EU.

7.3 Methodology

7.3.1 *Vehicle Selection and Data Collection*

We have chosen the Toyota Proace model, Ford Transit PHEV, and Nissan E NV2000 as the IC, HEV, and EV options, respectively. We have collected all the relevant data from the manufacturers of our selected vehicles. A summary of the data is represented in Table 7.1.

7.3.2 *Economic Analysis*

To evaluate the economic impact of EV, the life cycle cost (LCC) analysis is considered (Qiao et al. 2020). LCC includes all sorts of costs throughout a product's lifetime containing the time value of money (Abas and Mahlia 2019). LCC considers all costs associated with vehicle acquisition to the disposal phase within the life cycle of an EV (Ayodele and Mustapa 2020; Maciel Fuentes and González 2021). These costs include acquisition costs, operating costs, maintenance costs, etc. The present values of these costs are considered to incorporate the time value of money as follows (Ascher et al. 2020):

Table 7.1 Data required for economic analysis

	Toyota Proace (compact)	Ford Transit PHEV	Nissan E NV200
Vehicle types	ICEV	HEV	EV
Price (€)	31,340	39,990	34,105
Kerb weight (kg), $M_{vehicle}$	1660	2235	1515
Battery capacity (kWh)	50.0	13.6	40.0
Battery weight (kg), $M_{battery}$	–	53	239
Battery size (kWh)	50.0	13.6	40.0
Engine (cc)	1499	1952	–
Vehicle efficiency (VE)	0.063	0.042	0.210
Annual distance traveled (km), TD	12,000	12,000	12,000
Fuel cost (€/kWh or (€/L), EC or GC	1.88	0.32	0.32
Charging efficiency (%), CE	–	–	80%
Maintenance rate (€/km), MR	0.039	0.0285	0.026
Tire repairing cost (€), TRC	258	268	268
Tire average lifetime (km)	40,000	40,000	40,000
Battery lifetime (years)	8	8	8
Battery repairing cost (€), BRC	285.55	285.55	285.55
Scrap value for the vehicle (€), SC	21.55	21.55	21.55
Annual vehicle license fee (€), LC	29.47	29.47	29.47
Annual insurance cover (€), IC	69.05	69.05	69.05

$$\text{PW(present worth)} = \frac{\text{FW(future worth)}}{(1+q)^i} \tag{7.1}$$

Here, q is the rate of interest and i is the time period in years.

If we have taken the cumulative present value by inserting lifetime (n), Eq. (7.1) can be rewritten as:

$$\text{CPW} = \sum\nolimits_{i=1}^{n} \frac{\text{FW}_i}{(1+q)^i} \tag{7.2}$$

The acquisition cost (AQC) is calculated without considering the additional tax, and the related transportation-related cost is considered inside the acquisition cost (Moon and Lee 2019). Tax cost depends upon engine capacity, but this cost does not apply to EVs. The acquisition costs of vehicles are shown in Table 7.1.

7.3.2.1 Operating Cost

Operating costs for electric vehicles include fuel cost (FC), yearly license cost (LC), and yearly insurance cost (IC) (Parker et al. 2021). EVs do not require any fuel, and so the fuel cost of EVs includes electricity cost (EC) instead of gasoline cost (GC). On the other hand, HEVs and ICEVs simply consider gasoline costs for evaluating FC (Compostella et al. 2020). Fuel cost also considers travel distances (TD) and vehicle efficiency (VE) (Moon et al. 2018). FCs of EV, HEV, and ICEV can be represented as seen in Eqs. 7.3 and 7.4:

$$\text{FC(EV)} = \text{VE} \times \text{TD} \times \frac{\text{EC}}{\text{CE}} \tag{7.3}$$

$$\text{FC(HEV or ICEV)} = \text{VE} \times \text{TD} \times \text{GC} \tag{7.4}$$

Here, CE means charging efficiency of EV. Overall operating cost (OPC) can be calculated as seen below:

$$\text{OPC} = \sum\nolimits_{i-1}^{n} \frac{\text{FC}_i + \text{LC}_i + \text{IC}_i}{(1+q)^i} \tag{7.5}$$

7.3.2.2 Maintenance Cost

The periodic maintenance cost (PMC) is an important part of total maintenance cost (MC). It is calculated with a maintenance rate (MR) per travel distance as shown below (Yoon et al. 2019):

$$\text{PMC} = \text{MR} \times \text{TD} \tag{7.6}$$

Battery and tire replacement is dependent on maintenance, so this replacement cost is included in MC (Sun et al. 2020). Equation 7.7 is used to calculate the MC:

$$\mathrm{MC} = \sum\nolimits_{i=1}^{n} \frac{\mathrm{PMC}_i + \mathrm{BRC}_i + \mathrm{TRC}_i}{(1+q)^i} \tag{7.7}$$

Here, BRC and TRC indicate battery and tire replacement costs, respectively.

7.3.2.3 Salvage Value

The salvage value (SV) of a vehicle stands for the scrap value (SC) for that vehicle during its lifetime (Figliozzi et al. 2011). It can be represented as seen in Eq. 7.8:

$$\mathrm{SV} = \sum\nolimits_{i=1}^{n} \frac{\mathrm{SC}_i}{(1+q)^i} \tag{7.8}$$

7.3.2.4 Life Cycle Cost

The life cycle cost (LCC) includes the present values of all types of costs discussed above. By summing up AQC, OPC, and MC and by subtracting SV from that summation, we can get LCC (Zhao et al. 2019). It is calculated using Eq. 7.9:

$$\mathrm{LCC} = \mathrm{AQC} + \mathrm{OPC} + \mathrm{MC} - \mathrm{SV} \tag{7.9}$$

7.3.3 *Environmental Impact Analysis*

To analyze the environmental effect of electric vehicles, the greenhouse gas (GHG) emission of vehicles is considered. GHG emissions during the production and the operational stages of EVs, HEV, and ICEV have been calculated and compared. The environmental impact of the end-of-life stage of these vehicles has been included within the GHG emission of the production stage.

GHG emission for the operational stage contains an annual fuel consumption rate (FCR). FCR of EV considers the electricity required for the vehicle. The electricity needed for EVs (E_{EV}) is computed as follows:

$$\mathrm{E_{EV}} = \frac{\mathrm{VE} \times \mathrm{TD}}{\mathrm{CE} \times \mathrm{TE}} \tag{7.10}$$

Here, TE is the efficiency of the transmission system which has been taken to be 28.4% (Valladolid et al. 2022). FCR is then calculated using the value of E_{EV}. This also considers the heating value (HV) of the fuel used in the vehicle and the efficiency of power plant. The related equation for FCR is given below:

$$FCR_{EV} = \frac{E_{EV}}{HV \times PE(93.586\%)} \tag{7.11}$$

However, FCR is calculated differently for ICEV and HEV, that is:

$$FCR_{ICEV/HEV} = VE \times TD \tag{7.12}$$

GHG does not include only CO_2 gas. It also has some non-CO_2 gases such as methane (CH_4) and nitrous oxide (N_2O). Those non-CO_2 emissions are converted into equivalent CO_2 gas emissions for calculating fuel consumption (Helmers et al. 2019). The following equation is used for determining the total equivalent CO_2 (GHG) emission for the operational phase (GHG_{op}) of vehicles:

$$GHG_{op} = FCR \times (M_{CO_2} + 21 \times M_{CH_4} + 310 \times M_{N_2O}) \tag{7.13}$$

The necessary data for operational GHG emissions are given in Table 7.2. Natural gas is used for EVs and gasoline is used in the case of ICEV and HEV.

GHG emissions for the production stage including end-of-life emissions can be calculated by considering the impact on the environment from both the battery and the rest of the body part of an EV (Nanaki and Koroneos 2013). The same procedure goes for HEV. However, ICEV considers only the environmental effect of the main body part of the vehicle. Nanaki and Koroneos (2013) showed that the total GHG emission during manufacturing and at the end of life of a typical vehicle body is 3.172 kg per kg of the curb weight of that vehicle. On the other hand, the GHG emission from battery production and end-of-life stage is 254 gm/TD and 226 gm/TD for EV and HEV, respectively, throughout the lifetime of a certain vehicle (Ambrose and Kendall 2016). The required equations to calculate GHG emission at the production stage (GHG_{pr}) are shown below:

$$GHG_{pr}(EV \text{ or } HEV) = (M_{vehicle} - M_{battery})ENV_{body} + M_{battery}ENV_{battery} \tag{7.14}$$

$$GHG_{pr}(ICEV) = (M_{vehicle})ENV_{body} \tag{7.15}$$

Here, M and ENV stand for mass and GHG emission, respectively.

Table 7.2 Data for GHG emission calculation for the operational stage

	Fuel type	
Parameters	Natural gas (scf)	Vehicle gasoline (liter)
Heating value (kWh per fuel unit), *HV*	0.300922	9.685114224
Mass of CO_2 emission (kg/fuel unit), M_{CO_2}	0.05444	2.31943016
Mass of CH_4 emission (g/fuel unit), M_{CH_4}	0.00103	0.10038536
Mass of N_2O emission (g/fuel unit), M_{N_2O}	0.0001	0.02113376

7.4 Results and Discussion

The results shown in Table 7.3 indicate that life cycle cost is the lowest for HEV which is €962 less compared to EV. The acquisition costs for all three models are the most dominant among the four cost components. For HEV, the acquisition cost is the highest (80.52% of LCC), followed by EV (67.36%) and ICEV (59.64%). The breakdown of the LCC costs is illustrated in Fig. 7.1. The operating cost is also observed to be the lowest for HEV. The maintenance cost, however, is comparable for all three models. The salvage cost was found to be equally contributing to the LCC for all three models. Also, when in operation, the HEV emits the least GHG per year and also per unit kilometer traveled. However, it is observed that the production of HEV is responsible for the highest GHG emission among the three models studied (Fig. 7.2).

7.4.1 Sensitivity Analysis

We conducted a sensitivity analysis to verify the robustness of the findings. It is carried out by varying the input parameters and noticing the variations in the LCC and GHG emissions. How the same changes in inputs differentiate the types of vehicles (ICEV, HEV, EV) can be presented using a sensitivity analysis. The

Table 7.3 Results from economic and environmental analysis for different vehicles

Vehicle type		ICEV	HEV	EV
LCC	€	52,552	49,666	50,628
	% LCC	100%	100%	100%
	Comp to EV, €	1924	−962	0
AQC	€	31,340	39,990	34,105
	% LCC	59.64%	80.52%	67.36%
	Comp to EV, €	−2765	5885	0
OPC	€	12741.76602	2178.122656	9276.890997
	% LCC	24.25%	4.39%	18.32%
	Comp to EV, €	3464.875024	−7098.768341	0
MC	€	8480.677338	7508.151441	7256.636123
	% LCC	16.14%	15.12%	14.33%
	Comp to EV, €	1224.041215	251.5153182	0
SV	€	10.70968979	10.70968979	10.70968979
	% LCC	0.02%	0.02%	0.02%
	Comp to EV, €	0	0	0
FCR (scf/liter)		756 liter	504 liter	39384.5618 scf
GHG_{op} (kg) per year		1760.035827	1173.357218	2146.168354
GHG_{op} (kg/km) per year		0.146669652	0.097779768	0.178847363
GHG_{pr} (kg)		5265.52	9633.304	7095.472

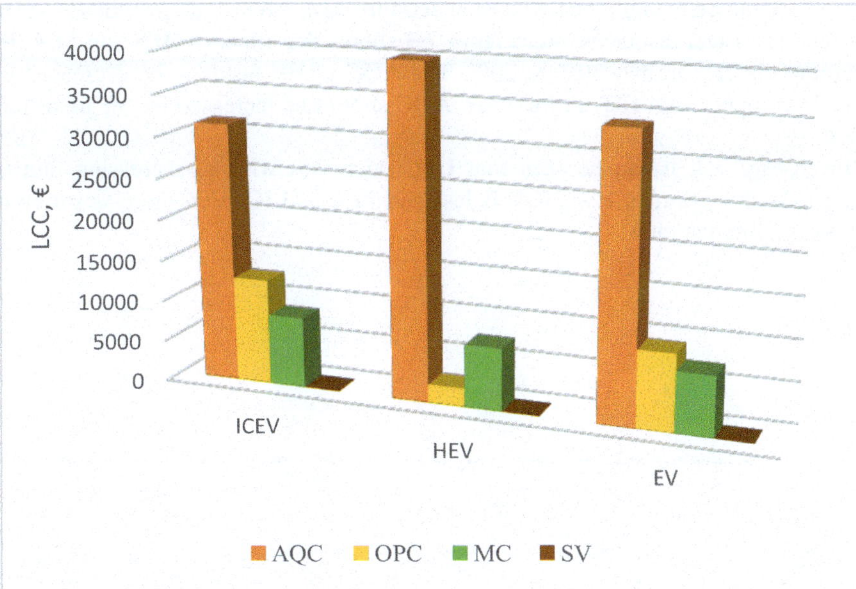

Fig. 7.1 LCC breakdown of different vehicles

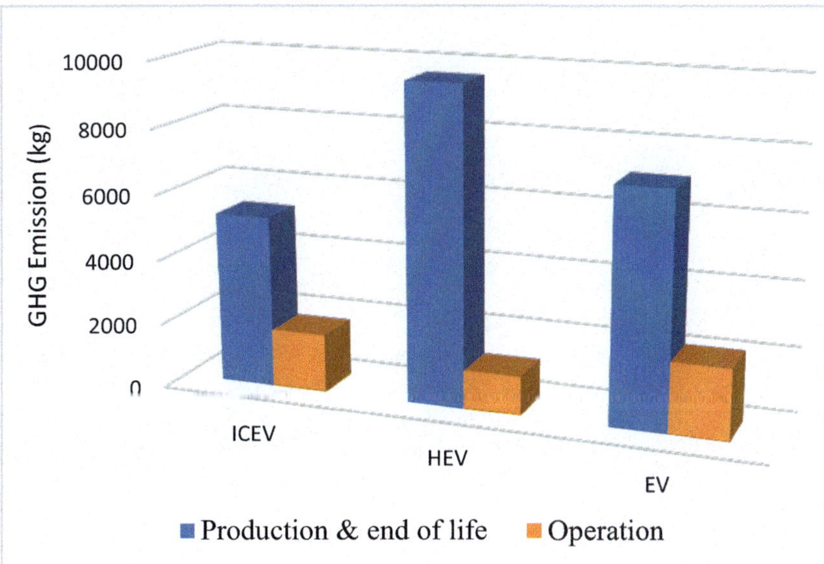

Fig. 7.2 GHG emission comparison of vehicles at different stages

sensitive parameters might be fuel cost, electricity cost, interest rate, various kinds of efficiencies, travel distances, etc. Figure 7.3 shows the effect of AQC on LCC by changing the cost from −30% to 20%. The AQC of EV needs to be decreased by 2.5% to compete with HEV. Moreover, AQC should be increased by 5% to sustain ICEV.

From Fig. 7.4, it can be seen that LCC decreases with an increasing rate of interest. With respect to EV, HEV is less sensitive and ICEV is more sensitive to changes in interest rates.

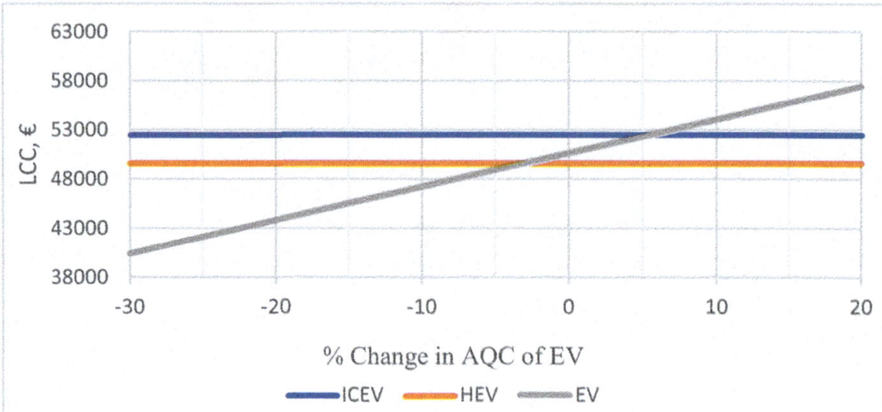

Fig. 7.3 Impact of varying AQC of EV on LCC

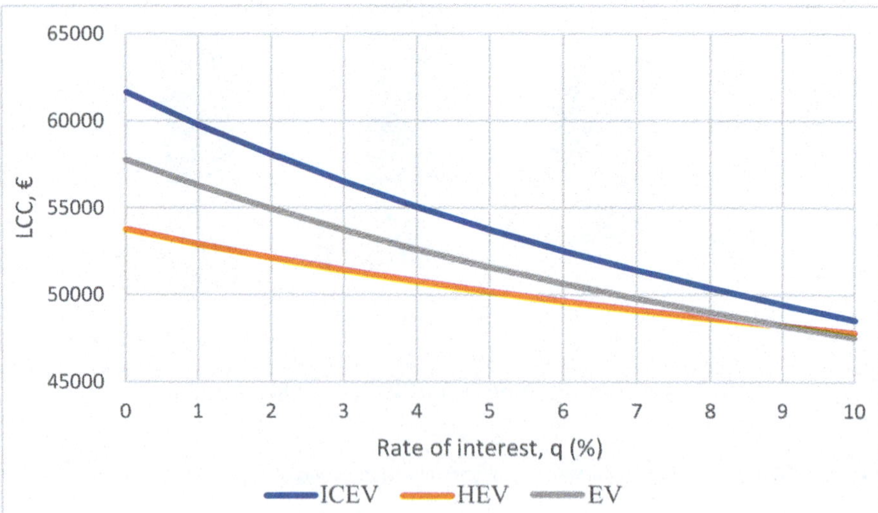

Fig. 7.4 Impact of interest rate on LCC

The distance traveled by a vehicle increases the fuel cost which in turn increases the LCC as shown in Fig. 7.5. HEV is least affected by TD compared to EV and ICEV. The fuel cost of ICEV is largely dependent on TD, and so in Fig. 7.5, it is more sensitive. Below 10,000 km annual TD, LCC is quite similar for all types of vehicles.

Reduction in battery prices decreases BRP which results in a reduction in LCC. This reduction rate is higher for EVs as shown in Fig. 7.6. This is because of the fall

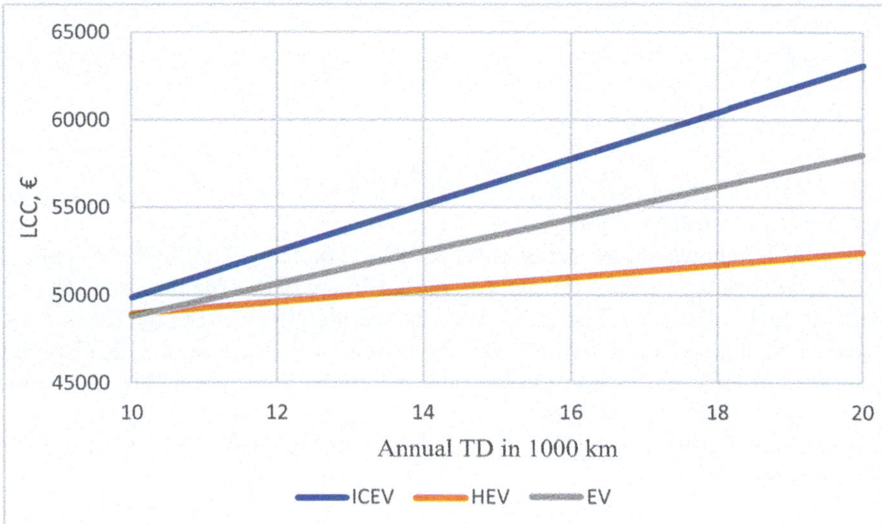

Fig. 7.5 Impact of TD on LCC

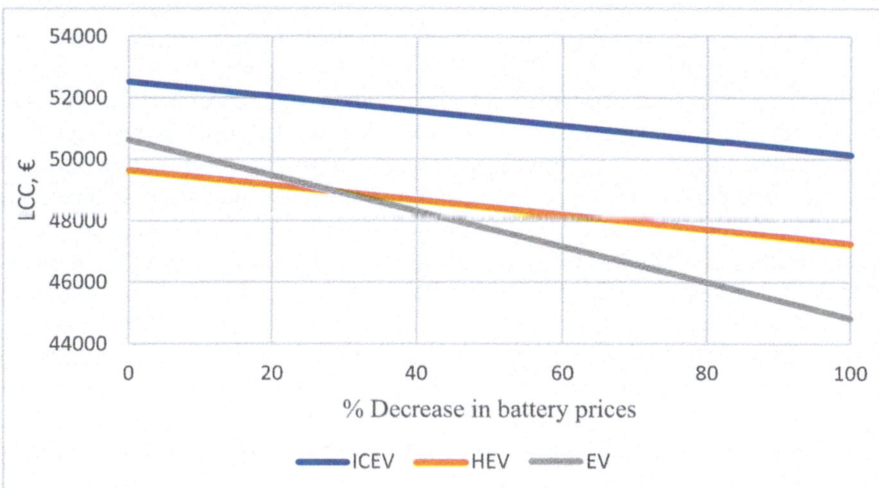

Fig. 7.6 Impact of battery price on LCC

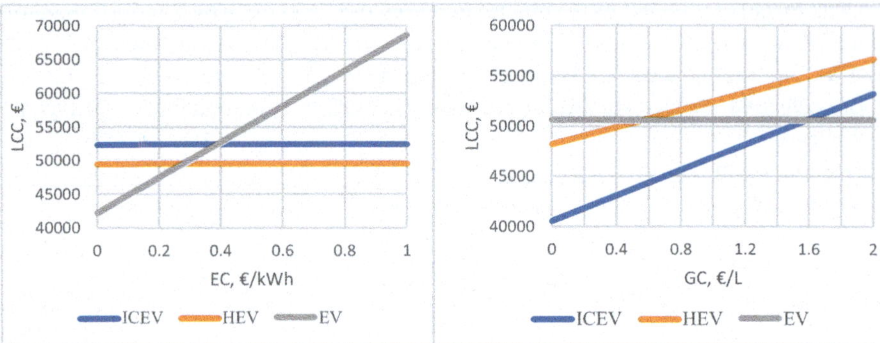

Fig. 7.7 Impact of EC or GC on LCC

in EC of EV. The fuel cost of ICEV and HEV does not include EC, and so they are less sensitive to battery prices.

Figure 7.7 shows the impact of varying GC/EC on the overall LCC of vehicles. Fuel cost depends on EC for EV and GC for ICEV and HEV. The rise in EC/GC can increase fuel costs as well as LCC. We can see the effect of EC by keeping GC constant in Fig. 7.7 (left figure). EC above 0.4€ will increase the LCC of EV compared to HEV and ICEV. On the other hand, the right part of Fig. 7.7 shows the changes in LCC due to GC by keeping EC fixed. GC above 1.5€ may increase the LCC of ICEV compared to EV, whereas the LCC of HEV will rise if the GC increases by 0.5€.

7.5 Conclusion

We have conducted an assessment of the three vehicle options—ICEV, HEV, and EV—for their potential as a sustainable and resilient transportation option in a supply chain. We have performed economic and environmental performance analyses of the three transport options and compared the findings. The findings have been validated through a sensitivity analysis by varying acquisition costs, interest rates, traveled distances, battery prices, and fuel costs. The findings of this research suggest that both EV and HEV show superior performance compared to that of traditional ICEV. The life cycle cost approach reveals the economic sustainability of electric vehicles with significantly lower LCC than that of ICEV. The results from GHG emissions indicate that in production and end-of-life use, GHG emissions are lower for EV than HEV. However, in operation, the GHG emissions from HEV are lower than EV. Therefore, there is a need for trade-offs in terms of environmental sustainability when choosing between EV and HEV options. The sensitivity analysis suggests that EV and HEV are less sensitive to input factors of decision-making. Finally, we conclude that both EV and HEV have tremendous potential to become sustainable and resilient transportation options in a supply chain.

References

Abas PE, Mahlia TMI (2019) Techno-economic and sensitivity analysis of rainwater harvesting system as alternative water source. Sustainability 11(8):2365. https://doi.org/10.3390/su11082365

Ambrose H, Kendall A (2016) Effects of battery chemistry and performance on the life cycle greenhouse gas intensity of electric mobility. Transp Res Part D: Transp Environ 47:182–194. https://doi.org/10.1016/j.trd.2016.05.009

Ascher S, Li W, You S (2020) Life cycle assessment and net present worth analysis of a community-based food waste treatment system. Bioresour Technol 305:123076

Ayodele BV, Mustapa SI (2020) Life cycle cost assessment of electric vehicles: a review and bibliometric analysis. Sustainability 12(6):2387

Compostella J, Fulton LM, De Kleine R, Kim HC, Wallington TJ (2020) Near-(2020) and long-term (2030–2035) costs of automated, electrified, and shared mobility in the United States. Transp Policy 85:54–66

Figliozzi MA, Boudart JA, Feng W (2011) Economic and environmental optimization of vehicle fleets: impact of policy, market, utilization, and technological factors. Transp Res Rec 2252(1): 1–6

Fiori C, Marzano V (2018) Modeling energy consumption of electric freight vehicles in urban pickup/delivery operations: analysis and estimation on a real-world dataset. Transp Res Part D: Transp Environ 65:658–673. https://doi.org/10.1016/j.trd.2018.09.020

Günther HO, Kannegiesser M, Autenrieb N (2015) The role of electric vehicles for supply chain sustainability in the automotive industry. J Clean Prod 90:220–233. https://doi.org/10.1016/j.jclepro.2014.11.058

Helmers E, Leitão J, Tietge U, Butler T (2019) CO_2-equivalent emissions from European passenger vehicles in the years 1995–2015 based on real-world use: Assessing the climate benefit of the European "diesel boom". Atmos Environ 198:122–132

Hu X, Murgovski N, Johannesson L, Egardt B (2013) Energy efficiency analysis of a series of plug-in hybrid electric buses with different energy management strategies and battery sizes. Appl Energy 111:1001–1009. https://doi.org/10.1016/j.apenergy.2013.06.056

Juan AA, Mendez CA, Faulin J, De Armas J, Grasman SE (2016) Electric vehicles in logistics and transportation: A survey on emerging environmental, strategic, and operational challenges. Energies 9(2):1–21. https://doi.org/10.3390/en9020086

Kennedy D, Philbin SP (2019) Techno-economic analysis of the adoption of electric vehicles. Front Eng Manag 6(4):538–550. https://doi.org/10.1007/s42524-019-0048-x

Liimatainen H, van Vliet O, Aplyn D (2019) The potential of electric trucks – an international commodity-level analysis. Appl Energy 236(December 2018):804–814. https://doi.org/10.1016/j.apenergy.2018.12.017

Maciel Fuentes DA, González EG (2021) Techno-economic analysis and environmental impact of electric vehicle introduction in taxis: A case study of Mexico City. World Electr Veh J 12(3):93

Moll C, Plötz P, Hadwich K, Wietschel M (2020) Are battery-electric trucks for 24-hour delivery the future of city logistics? – a German case study. World Electr Veh J 11(1). https://doi.org/10.3390/WEVJ11010016

Moon S, Lee D-J (2019) An optimal electric vehicle investment model for consumers using total cost of ownership: a real options approach. Appl Energy 253:113494

Moon H, Park SY, Jeong C, Lee J (2018) Forecasting electricity demand of electric vehicles by analyzing consumers' charging patterns. Transp Res Part D: Transp Environ 62:64–79

Nanaki EA, Koroneos CJ (2013) Comparative economic and environmental analysis of conventional, hybrid, and electric vehicles – the case study of Greece. J Clean Prod 53:261–266

Noori M, Gardner S, Tatari O (2015) Electric vehicle cost, emissions, and water footprint in the United States: development of a regional optimization model. Energy 89:610–625. https://doi.org/10.1016/j.energy.2015.05.152

Onat NC, Abousshaqrah NNM, Kucukvar M (2019) Supply chain linked sustainability assessment of electric vehicles: the case for Qatar. In 2019 IEEE 6th International Conference on Industrial eEngineering and Applications, ICIEA 2019, pp 780–785. https://doi.org/10.1109/IEA.2019.8715205

Osieczko K, Zimon D, Płaczek E, Prokopiuk I (2021) Factors that influence the expansion of electric delivery vehicles and trucks in EU countries. J Environ Manag 296(April). https://doi.org/10.1016/j.jenvman.2021.113177

Parker N, Breetz HL, Salon D, Conway MW, Williams J, Patterson M (2021) Who saves money buying electric vehicles? Heterogeneity in the total cost of ownership. Transp Res Part D: Transp Environ 96:102893. https://doi.org/10.1016/j.trd.2021.102893

Qiao Q, Zhao F, Liu Z, Hao H, He X, Przesmitzki SV, Amer AA (2020) Life cycle cost and GHG emission benefits of electric vehicles in China. Transp Res Part D: Transp Environ 86:102418

Sun Q, Ye Z-S, Zhu X (2020) Managing component degradation in series systems for balancing degradation through reallocation and maintenance. IISE Trans 52(7):797–810

Valladolid JD, Abril JS, Valdiviezo AD (2022) Analysis of energy efficient in traction and breaking for electric vehicles through laboratory experiments. In: Rocha Á, López-López PC, Salgado-Guerrero JP (eds) Communication, smart technologies and innovation for society. Springer, Singapore, pp 297–305

Yoon T, Cherry CR, Ryerson MS, Bell JE (2019) Carsharing demand estimation and fleet simulation with EV adoption. J Clean Prod 206:1051–1058

Yusof NK, Abas PE, Mahlia TMI, Hannan MA (2021) Techno-economic analysis and environmental impact of electric vehicles. World Electr Veh J 12(1):98565–98578. https://doi.org/10.3390/wevj12010031

Zhao Y, Noori M, Tatari O (2016) Vehicle to grid regulation services of electric delivery trucks: economic and environmental benefit analysis. Appl Energy 170:161–175. https://doi.org/10.1016/j.apenergy.2016.02.097

Zhao G, Nielsen ER, Troncoso E, Hyde K, Romeo JS, Diderich M (2019) Life cycle cost analysis: A case study of hydrogen energy application on the Orkney Islands. Int J Hydrog Energy 44(19):9517–9528

Chapter 8
Integrating Circular Economy and Reverse Logistics for Achieving Sustainable Dairy Operations

Mohammad Shamsuddoha, Tasnuba Nasir, and Niamat Ullah Ibne Hossain

Abstract In recent years, academic and industry circles have focused more on circular economy (CE) and reverse logistics (RL) to add more value to the industry, many studies focused on waste management, and the closed-loop process have helped to gain additional success. Accordingly, there have been numerous research opportunities in the dairy industry in developing nations (e.g., Bangladesh). This industry must introduce modern concepts (CE and RL) to extract more value from their processes and unused/leftover waste or products. These are the gray areas for the industry right now, and farmers are searching for an improved approach to gain further sustainable benefits. These research scopes include achieving sustainability, increasing productivity, waste management, genetic advancement, commercialization of dairy products, high-quality meat production, and a well-functioning supply chain. By creating a reverse logistics process and following environmental aspects, this research fills a significant research gap in dairy waste management to build sustainable dairy operations. Such a process will integrate CE and RL concepts to make the operation more profitable and sustainable. This chapter developed a supply chain model on a case dairy farm using the system dynamics method. The model was developed using a simulation application and the case method following a positivist and quantitative framework. This study creates a causal/qualitative model that is converted into quantitative algorithms in a simulation contest. Afterward, the simulation model was tested and expanded to use dairy wastes based on realistic and theoretical experience. This study reviewed sustainability, circular economy, and reverse logistics concepts to build a simulation model. Finally, the model outcomes supported the above theories to enrich the dairy industry by generating substantially diversified byproducts for renewable energy and household usage.

M. Shamsuddoha (✉) · T. Nasir
Western Illinois University, Macomb, IL, USA
e-mail: m-shamsuddoha@wiu.edu; tasnubanasir@gmail.com

N. U. Ibne Hossain
Engineering Management, College of Engineering and Computer Science, Arkansas State University, Jonesboro, AR, USA
e-mail: nibnehossain@astate.edu

© The Author(s), under exclusive license to Springer Nature Switzerland AG 2023
N. U. Ibne Hossain (ed.), *Data Analytics for Supply Chain Networks*, Greening of Industry Networks Studies 11, https://doi.org/10.1007/978-3-031-29823-3_8

Keywords Dairy wastes · Circular economy · Reverse logistics · System dynamics · Bangladesh

8.1 Introduction

In order to achieve sustainable economic, social, and environmental outcomes, numerous studies have illustrated reverse logistics in innumerable contexts (Nasir et al. 2013; Shamsuddoha 2015a, b). A plethora of research can be found in secondary sources to validate reverse supply chain and circular economy processes. A few studies are seen to be available from a systematic review of operational factors, process bottlenecks, and practices toward circular economy and reverse logistics perspectives (Govindan and Hasanagic 2018). Bernon et al. (2018) indicated that reverse logistics must incorporate CE concepts, which calls for a multifaceted strategy. Now, empirical work is necessary for the dairy supply chain field to analyze and explore the potential to gain through the circular economy and reverse logistics. This study developed a simulation model which includes cattle herds, mature milking cows, and culled cattle, milking and supplying milk to the value-added industry as well as utilizing dairy waste to convert into valuable byproducts. Thus, the objective of the research is threefold: first, the study develops a conceptual causal model for the dairy supply chain along with waste processing operation through reverse logistics; second, the causal model is converted into a quantitative simulation model; third, the study describes the model findings in light of CE and RL to support the research goals.

Most livestock policies encourage the dairy sector to play a multifunctional role in improving its socioeconomic and environmental sustainability (Shamsuddoha 2022; Shamsuddoha et al. 2021). Some European countries like Italy attract dairy farmers through incentives to utilize photovoltaic or biogas plants (Ghisellini et al. 2014). Such examples are common in developed countries like the USA, Europe, and rich Asian countries. Due to the rising demand for milk and dairy products over the past few decades, the worldwide dairy industry has significantly grown and intensified (Zhang et al. 2021). Such massive farming produces a considerable quantity of waste. It is time to convert these wastes into valuable byproducts using CE and RL theories. The interest in the sewage management strategy is snowballing due to environmental degradation and growing circular economy awareness (Zhang et al. 2021). A poultry industry study (Shamsuddoha 2013) describes achieving socioeconomic and ecological success by utilizing efficient forward and reverse supply chains.

Similarly, the dairy livestock subsector has a similar scope to contribute to the rural economy and an effective way to achieve additional income (Shamsuddoha et al. 2021; Shamsuddoha et al. 2009, 2021; Sheheli 2012). Dairy is a significant and growing sector in many developing nations (e.g., Bangladesh, Brazil, and Cambodia). The dairy industry is proving increasingly to be a viable income source for many households involved in milk production and trade (Shamsuddoha et al. 2013a, b). Although dairy farming is the earliest occupation in a rural setting, its progress has been unsatisfactory due to several economic, social, and environmental issues (Nasir et al. 2013). Although CE, RL, sustainability, governance, and supply chain management have gained extensive consideration from academicians and practitioners, the dairy farms in Bangladesh have yet to deploy such rigorous concepts in their operations. Implementing modern supply chain systems for animal waste renewables has piqued the industry's and academia's interest (Islam et al. 2021). Thus, there is a possible way to develop the dairy sector in light of sustainability, CE, and RL literature. Hence, this study attempts to integrate the theories mentioned above and deploy those concepts in dairy operations through a simulation model to quest for ideal conditions for dairy operations to gain more.

8.2 Methods

This study relies on empirical epistemology, positivist ontology, and quantitative methodology. The fundamental system dynamics (Sterman 2000; Wolstenholme 1982; Forrester 1961) methodology and case study (Yin 1994) methods are followed in this study. The design science method depends on "devising artifacts to achieve goals" (Meng 2009). Both primary and secondary data were used in this study. Primary data was collected through in-depth interviews with the farm managers and entrepreneurs. The primary data consist of year-long information from January to December 2020. The case farm is a renowned dairy farm with the highest number of cow heads (1600) based on individual farm capacity. They have several firm sites, and the researcher has chosen one of them to build a mental model or qualitative causal model. The respondents proposed flexible (open-ended) questions for understanding the appropriate procedure based on their comprehensive information and experiences. Every interview lasted around 2 h. Since it presented information on rearing, processing, and waste management, this study aided the interview. Secondary data came from various places, including reference journals, conference papers, books, company documents, statistical yearbooks, and accounts. Furthermore, secondary sources recognized the distinction between expected outcomes and case farm outputs. To build a dairy supply chain model and perform studies to inspect research goals, the simulation program Vensim was used and experimented to gain an adequate understanding.

8.3 Literature

8.3.1 Dairy Farming in Bangladesh

More or less than 50% of Bangladeshi people are involved in agriculture, though the number has been decreasing from 70 above in the last three decades. According to the Department of Livestock Services (DLS), the Bangladesh livestock contributed

13.44%, and GDP was 1.47 in 2020. Other sectors are booming, and the livestock sector holds less of the total GDP. It does not mean that the livestock sector is deteriorating in terms of GDP volume, as the growth is 3.04% (DLS 2020). Bangladesh is an agro-based country with a small piece of land of 57,321 square miles (FAOSTAT 2018). Crop, poultry, dairy, and fishery are the main professions of rural Bangladeshi people (Shahnaz et al. 2004). According to the DLS (2020), Bangladesh has 24.39 million cattle, 1.49 million buffaloes, 26.45 million goats, 3.60 million sheep, 296.60 million chickens, and 59.71 million ducks. The milk and meat productions are 10.68 and 7.67 million tons, respectively (DLS 2020). Small-scale dairy production makes a significant contribution to the society's well-being. The key emphasis is on determining the amount of milk generated by homestead dairy farms followed by renewable energy from dairy waste. Many constraints are faced by small-scale dairy farmers where farmers fail to produce enough milk for optimum profit (Shamsuddoha et al. 2009). Ninety-seven percent of the milk productions are sold to a casual local distributor; the rest of the productions go to the structured supply chain networks (Hemme and Khan 2004). The private owners are the majority in this industry and get a little help from the government and NGOs (Shamsuddoha et al. 2013a, b). They grow their own farm having indigenous resources.

New entrepreneurs have recently been motivated to be involved in small to medium dairy farming in rural and peri-urban areas (Saadullah 2011). Young entrepreneurs started investing in this sector as a start-up or diversified business unit. Most of the new businesses were successful in terms of profitability and reproduction. The traditional dairy industry has the main characteristics of small-scale and unorganized firms, low productivity, inadequate infrastructure, inappropriate feeding, and absence of qualified personnel (Raha and Talukder 2004). Around 15 structured industries run a dairy corporation, such as BRAC, Milk Vita, and PRAN, holding the topmost market shares (Shamsuddoha et al. 2013a, b). Under the name Milk Vita, the Bangladesh Milk Producers Cooperative Union Limited (BMPCUL) was first founded in 1973. Milk Vita had monopoly in the market before the new entrants, such as BRAC dairy and PRAN (PRAN-RFL 2007). Again, most milk sellers were not farming; instead, they built and maintained a supply chain network and collected and processed milk and milk-related products. Many studies were conducted based on production efficiency, productivity, genetics, disease control, and farm management. But only a few studies cover sustainability, RL and CE, and waste management.

8.4 Dairy Wastes

Dairy waste is produced in large quantities in Bangladeshi dairy farms. According to the in-depth interview, dairy wastage is mainly cow bone, blood, skins/hide, feed wastes, cow dung, feces, wastewater, and milk residue. Unfortunately, farmers do not collect all wastes; instead, they ignore blood, wastewater, urine, and milk residue. On the other hand, cow dung, skin, and bone are gathered and used in an unorganized manner. Most farms do not have structured waste management and recycling/remanufacturing facilities. The farm will get more convertible byproducts for additional values if dairy wastes are processed systematically. A mature cow, for example, produces 20–28 liters of waste per day, but farmers are lucky to get 3–4 kilograms due to unstructured waste collection procedures. The remainder of the waste drains out with the wastewater through sewage lines. Bangladesh is small in terms of land size and has an excessive number of people living per square mile. That creates a problem with unnecessary landfilling. Furthermore, the collected waste is not recycled correctly, resulting in a loss of value. This is where the researcher concentrated their efforts and discovered a way to reuse, recycle, and remanufacture different byproducts from the collected waste to make it more sustainable for dairy farmers. We need to get CE and RL implications in the dairy supply chain processes to do this.

8.5 Circular Economy

The circular economy is an emerging concept in different research fields (Kirchherr et al. 2017). Circular economics (EC) is more relevant than ever in the global sense to maintain the production rate of goods and services to satisfy the rising customer demand encumbering the atmosphere, community, and practitioners (Patwa et al. 2021). We use products until the end of life and discard them as waste (Ghisellini et al. 2018). In this approach, no one is thinking about the discarded things which may be useful for the economy. Everything that belongs to nature maintains cycles, such as the weather, water, day-night, and crop cycles. Enterprises are pursuing improved resource and process productivity at various output and consumption levels to support the circular economy (CE) ideals (Patwa et al. 2021). CE values encourage diminishing or removing waste and contamination, maximizing the use of resources and goods, and natural system regeneration (EMF 2020). Reduce, reuse, and recycling (3R) have increased in importance for companies and related stakeholders (Shamsuddoha 2013). Thus, a circular economy is an alternative to a conventional linear economy in which individuals carelessly purchase and dispose of the remaining.

According to a study, rural farming communities in North Ireland are ideally adapted to implementing the circular economy approach (Ghisellini et al. 2014). Figure 8.1 depicts biogas, electricity, and fish feed production using sophisticated technologies. Such diversified processes and technology can be helpful for farming and many other industries. The case farm also implemented some sub-projects to make their farm sustainable and eco-friendly. Thus, dairy farms, which include the circular economy model, can impact their stakeholders more than the conventional farming system.

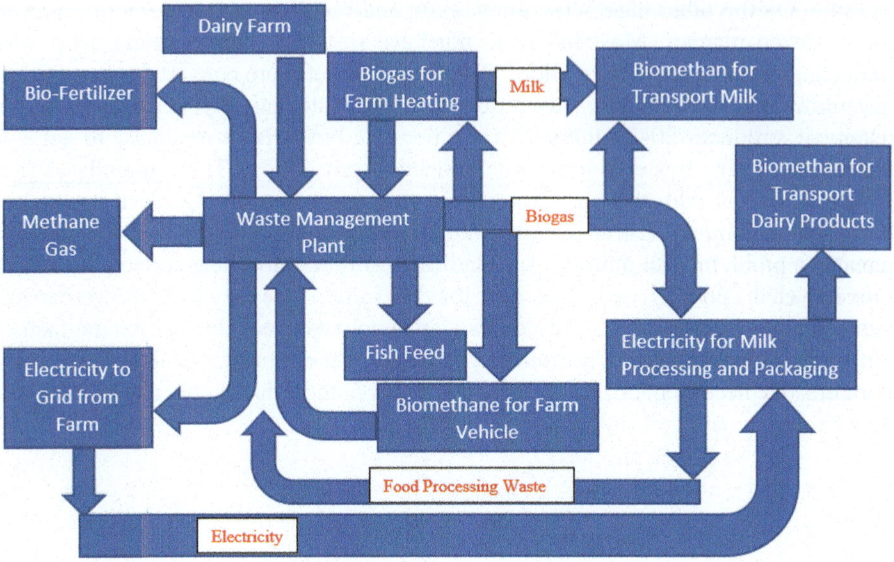

Fig. 8.1 CE and RL applied in a sustainable dairy farm and surrounding community. (Extracted from Ghisellini et al. 2014; Shamsuddoha 2022)

8.6 Reverse Logistics in the Dairy Industry

Environmental sustainability is an established concept for managing resources for present and future generations (Shamsuddoha 2015a, b; Shamsuddoha et al. 2013a, b). Supply chain management must broaden its traditional focus on the forward movement of resources and products to address recycling, remanufacturing, disposal, reconditioning, and end-life products (Kocabasoglu et al. 2007). The reverse supply chain is the sequence of activities required to collect a product from a consumer and either dispose of it or recover its value (Prahinski and Kocabasoglu 2006). Again, the RSC can potentially reduce the negative environmental impacts of extracting virgin raw materials and waste disposal (Kocabasoglu et al. 2007). This also denotes retrieval activities from disposal to regain values (Guide and Van 2002; Prahinski and Kocabasoglu 2006). Similar processes can accomplish five essential functions, that is, inspection, reverse logistics, remanufacturing or refurbishment disposition, and marketing (Blackburn et al. 2004).

For example, Walmart uses dedicated redistribution reverse logistics for return, inspection, repair, salvage, disposal, replacement, and rework such as upgrades (Krumwiedea and Sheub 2002). Also, the battery industry successfully implemented reverse logistics to extract sustainable benefits (Bansia et al. 2014). Besides, reverse logistics is evident in the poultry industry to recover wastes to generate valuable byproducts (Shamsuddoha 2022, 2015a, b; Shamsuddoha et al. 2011, 2013a, b). Notably, a Brazilian case study analyzed dairy property competitiveness through reverse logistics based on wastes produced and utilized accordingly (Gomes et al. 2014). Reverse logistics has had a substantial social, economic, and environmental influence on business and society during the previous decade (Cottrill 2000). Reverse logistics has features of environment-friendly manufacturing including disassembly (Edwards and Daniel 1992), remanufacturing, and recycling (Gungor and Gupta 1999).

A few studies addressed the handling and use of dairy wastes (Ahmad et al. 2019), biogas output from dairy wastes (El-Mashad and Zhang 2010), wastewater management in the dairy sector, and fertilizer derived from dairy cattle slurry (Oliveira et al. 2016). All of these areas have a tremendous deal of potential to meet social, economic, and environmental requirements that will increase the sustainability of the economy and society. Not a lot of studies can be found in the literature on the sustainable dairy operation and reverse logistics. There was no indication that reverse supply chain problems were being considered in dairy operations. As a result, the researcher has chosen the Bangladeshi dairy industry as a case study to see how to fit the reverse logistics perspective and achieve a sustainable operation.

8.7 Integrating Reverse Logistics and Circular Economy

According to Bernon et al. (2018), reverse logistics must incorporate CE concepts, which calls for a multifaceted strategy. Adopting the framework will affect industry people and support them in changing their reverse logistics practices to one that is more restorative and less impactful. The circular economy relies on production processes that take chances for reverse cycles not just of products but also of leftovers like packaging and other wastes which can be tied to key practices of environmental, economic, and social sustainability (Guarnieri et al. 2020). On the contrary, Kazancoglu et al. (2021) state that reverse logistics activities are becoming more and more significant in terms of scope and quantity considerations. These reverse logistics movements are dynamic and sophisticated in design (Shamsuddoha and Woodside 2022). Besides, the influence of RL actions on the environment has seldom been considered. Again, Makarova et al. (2021) revealed that RL fits naturally into a CE and focuses on returning goods for future use. The reverse logistics sector must adopt a fundamentally different mindset to transition to a CE. Working with logistics providers will be essential for businesses to adopt cost-effective circular strategies, as almost 98% of business leaders cite logistics as critical to moving to a circular economy. Sun (2017) believes that reverse logistics has caught entrepreneurs' attention for considering the environment, circular economy, and sustainable development. At the same time, RL is crucial for increasing low-carbon competitiveness. In addition, Schluter et al. (2021) mentioned that the remanufacturing sector makes up a minor portion of the entire European output. Here is the opportunity for the industry to focus more on remanufacturing to gain additional productivity and profitability.

For example, the auto industry makes substantial use of reverse logistics, and many other businesses are starting to do the same (Ravi and Shankar 2005). Apart from the auto industry, RL and CE are being considered by the processing industry (Kazancoglu et al. 2021), pharmaceutical industries (Javed et al. 2021), the Industry 4.0 Process (Rajput and Singh 2022), poultry industry (Shamsuddoha et al. 2021; De Giovanni 2022), dairy sector (Lavelli and Beccalli 2022), and many more. Understandably, there is not a plethora of research as these concepts are relatively new and developing. This study is another trial to involve RL and CE to get more positive gains for profitability and sustainability purposes.

The case industry clarifies its dairy supply chain networks by mentioning waste collection sources and procedures. They also indicate the practices of rearing, raising, feeding, and collecting waste for baby cattle, followed by heifers and milking and meat-producing cows. Several interviews with dairy entrepreneurs and upper-level managers helped the study to gain appropriate knowledge on dairy farming processes. They have indicated that dairy wastes can be collected from every generation stage like cattle, heifer, mature cow, milking cow, bull, culled cow, and the like. Interviewed participants provided information on various inputs for the dairy farms, milk processing units, and byproducts. An in-depth interview has given perceptions of various byproducts generated from dairy wastes. For example, a portion of raw cow dung directly goes to the crop/horticulture fields without further processing. At the same time, a part of dairy waste goes to the commercial plant of biogas and fertilizer. Produced biogas is mainly used in the farm area as a burner fuel source and for cooking for the employees. The above figure also represents fish/duck feed and artificial charcoal. A considerable quantity of biogas is blown away into the air as there is no additional farm-level usage. Farms can gain additional economic benefits if the excess biogas can be supplied to a pipeline network. Unfortunately, country rules and regulations are complex, so establishing a pipeline network that can provide the nearby households is challenging.

The above discussions reveal that dairy waste can be a research matter using RL, CE, and sustainability concepts. The concerned industry can fit the above theories into their operations and create an ideal situation of maximum benefits and environmental and social gains. To do this, the most considerable thing is to collect dairy wastes efficiently and utilize them through producing byproducts and encashing them to circulate their economy.

8.8 A System Dynamic-Based Dairy Model

The system dynamic model starts with causal loop diagramming. The dynamic behavior between and among the variables is represented by positive and negative feedback loops (Richardson, 1986). Figure 8.2 illustrates a simple causal loop diagram for the dairy farm. This diagram often depicts how interrelated variables interact and involves relevant nodes representing the connected variables (Aghalaya et al. 2012; Maani and Cavana 2007). Both positive and negative labels can be

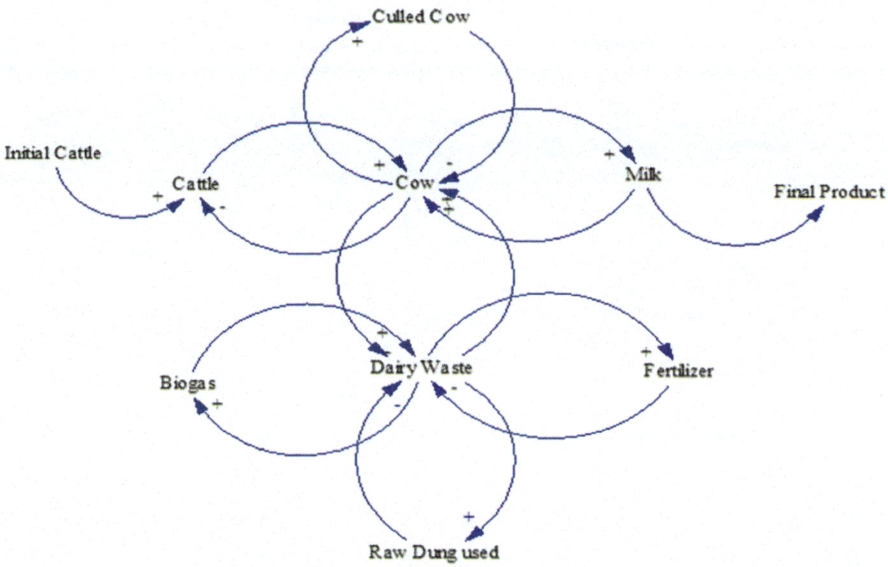

Fig. 8.2 A qualitative dairy model

applied to the arrows indicating the relationships between the variables. A positive (+) sign at the head of an arrow indicates a rise (or decrease) in a variable. If the impact is negative, a negative (−) sign is placed at the arrow's head (Aghalaya et al. 2012). Figure 8.2 depicts the relationship between/among critical dairy parameters. A wide range of variables controls dairy farming. However, this model only selected the most important variables to comprehend the dynamics over time. This qualitative/causal model has numerous loops, as seen in Fig. 8.2.

Some of the loops are as follows: Loop Number 1 of length 1, cattle grown, cattle to farm; Loop Number 2 of length 1, cattle grown, aging to cow; Loop Number 3 of length 3, cattle grown, aging to cow, mature cow, cattle to farm; and negative feedback loop, cattle and cow. After a certain time, the mature cow will improve if the supply of cattle increases. But when the number of mature cows increases, the number of cattle will temporarily decline until a new herd joins the production process. As the number of mature cow increases, so will the milk collection. Milk demand will increase and mature cows will be lined up for more milking. Similar things will happen in dairy waste collections. The higher the number of cattle and mature cows, the more the amount of waste will be that is collected at the farm level. More waste means more diversified byproducts. The result could be highly dynamic when the model adds numerous variables with diverse parameters. Based on linked variables' behavior, such dynamic behavior could be positive or negative. The modeler should consider composite behavior generated from different constant and auxiliary variables over time to track the complete model behavior. Such complex

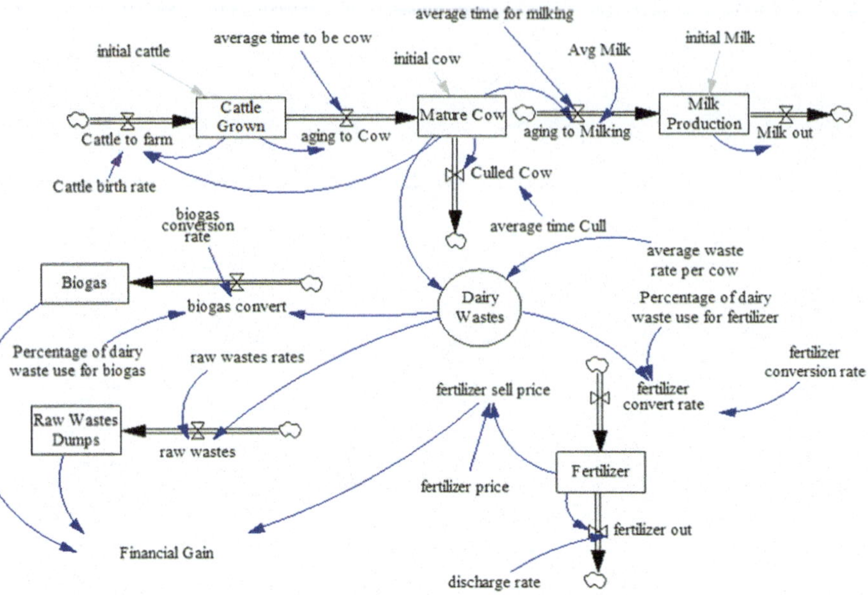

Fig. 8.3 Dairy supply chain simulation model with reverse logistics

behavior will give real-life operational behavior which can be tuned through changing values for the content variables.

Similarly, other loops exist in the model such as cow and culled cow, dairy wastes and biogas, as well as raw dung and fertilizer. The computer-generated simulation model was constructed based on seven separate loops in the above causal model, and the system runs dynamically depending on the input and output rate.

Figure 8.3 shows a basic dairy process modeled on Vensim simulation application. The model included the primary dairy process variables including livestock raised, mature livestock, milk processing, milk waste, biogas, fertilizer, and money gain. Reverse diary waste logistics help to produce useful biogas and fertilizer byproducts. Locals use milk waste for many items, such as human-made artificial charcoal, fish feed, protective layer for poor villagers' houses, and the like. This research shows how milk and slurry waste can produce other manufacturing byproducts (biogas and fertilizer) in the same industry as the supply (raw materials). The simulation model will evaluate the amount of waste that can be used as a byproduct input. Policymakers can simulate any input to monitor the output of each variable to compare and contrast different outputs. Dairy entrepreneurs and managers would select the correct test to give them the anticipated profitability and efficiency among the multiple outcomes. On the other hand, Fig. 8.2 is transformable by incorporating stock and flow diagrams from Fig. 8.1. A diagram like this made this model expression complete. The most crucial issue was the definition of all variables under different mathematical equations. All the equations built up by the interviewees were given data based on farmers' realistic experiences. The model was finally run for many tests, and possible results were achieved based on different scenario-based inputs.

8.9 Simulation Results

The study collected historical data on dairy cattle input. Milking cattle are raised until they are mature enough to be milked, which takes some time. These adult cows will be used to test future genetically altered milking cows for output. The majority of dairy breeds are imported from Australia, the United States, the Netherlands, and India. Parameters for various variables from the case farm via in-depth interviews have been collected. This study addresses a significant research gap in the dairy sector. Due to inappropriate governmental aid, the Bangladeshi dairy industry is unable to cope with new technology, equipment, and management. The private sector is doing its utmost to face the many challenges connected with different disasters, finances, and policies. After that, private entrepreneurs are curious enough to deal with these uncertainties in their farming.

Figure 8.4 shows the model run SyntheSim (Vensim application menu option) approach in which individual "time actions" can track and monitor their behavior over time. This model helps explain a variable's behavior pattern over time so that policymakers can detect solutions if necessary. The Vensim package synthesis mode

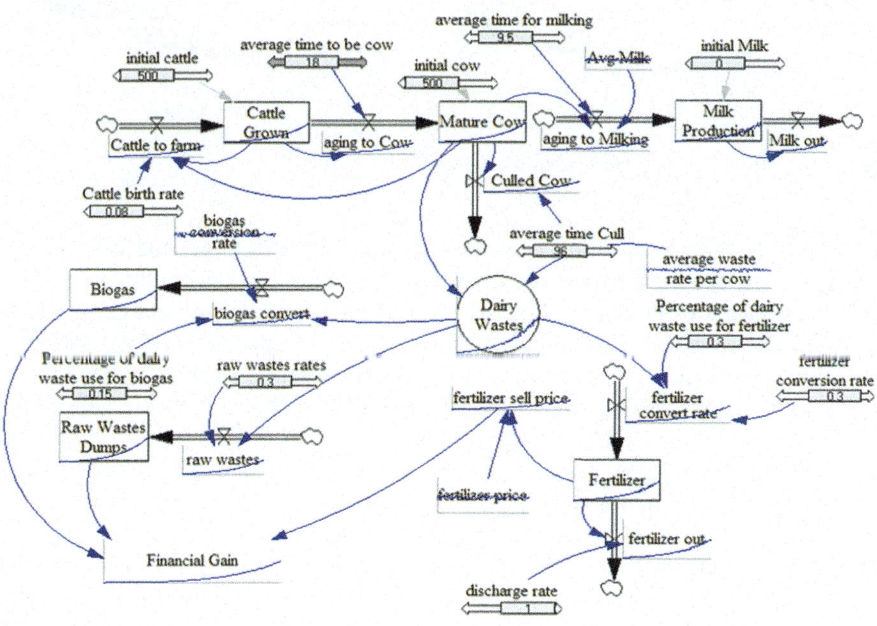

Fig. 8.4 Behavioral mode of the dairy supply chain model

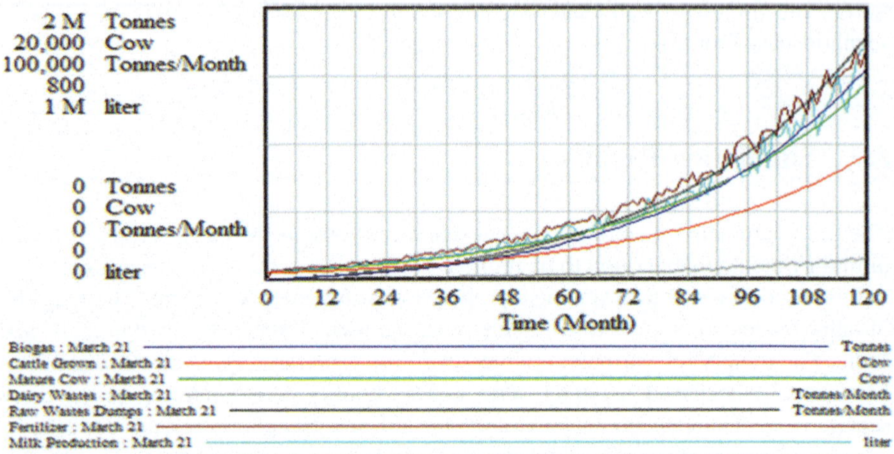

Fig. 8.5 Performance for selected dominating variables

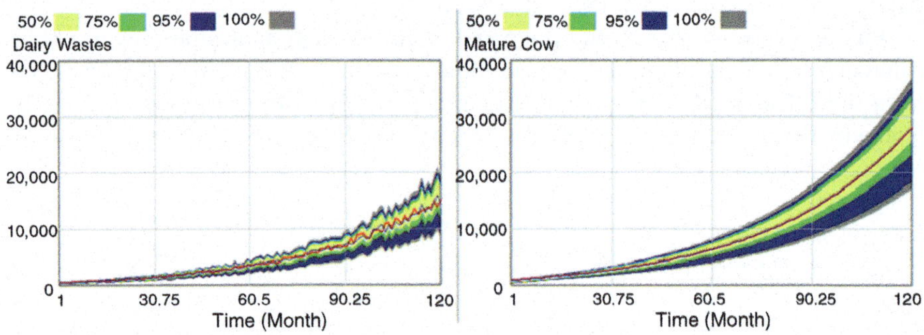

Fig 8.6 A sensitivity analysis on mature cows and dairy wastes

will experiment with adjusting constant variables in order to get the best results. Constant variables can be seen in a bar where a modeler can change the parameters and check instant change over time. It is a convenient way to find optimum results based on constant variables' inputs. Also, it helps the modeler to find absurd or unjustified results for any other variables through the graph's behavior. This figure shows an expected behavior from the model that is associated with a substantial number of constant and auxiliary variables.

Figure 8.5 shows the behavior of biogas, cattle, mature cows, raw wastes, fertilizer, and milk production. All the graphs follow normal behavior based on the given inputs. Inputs were taken from the primary data provided by the farmers. This model was simulated on March 21, 2022, and graph oscillations were verified by the farm management. This study runs a sensitivity test on two levels (stock) of variables to examine the model's reliability. A reliability test is important for model authenticity and consistency. Figure 8.6 illustrates the sensitivity analysis for the two primary variables of number of cows and milk production. The sensitivity behavior demonstrates the predicted behavioral changes over time. Several tests and analyses were carried out for this simulation model along with various inputs for finding the optimum results.

This study designed a simple model with a waste collection and process network supporting the reverse logistics theory. At the same time, the model includes byproduct processing, distribution, and revenues. Such financial gain from dairy waste can be used as further capital to run and extend the business. The model could even be developed in its byproduct supply chain network to understand how many small households or companies can get additional benefits. Such benefits understandably follow the sustainability perspectives of economic, social, and environmental. Thus, the results supported the reverse logistics and circular economy theory to gain more profitability and sustainable outcomes for dairy operations.

8.10 Concluding Remarks

This study provides a basic simulation model for the dairy supply chain, including cattle production followed by the number of mature cows, milk production, and waste collection through reverse logistics. Shamsuddoha and Woodside (2023) revealed six principles to achieve sustainability using system dynamics methodology. Similarly, this model can help decision-makers to estimate how much milk and waste a dairy operation generates for different operating conditions. The model is a basic adaptable, dynamic system showing how the key variables generate complex behavior. The model outputs demonstrated that CE and RL theories are handy for the dairy industry to modernize their operation along with various challenges. Different behaviors can help farmers understand the primary variables of futuristic behavior. Future output and its oscillating behavior are essential to predict the possible scenarios. This model will allow farmers to practice better than their previous practices. Future studies can incorporate additional micro- and macro-variables into the model to find the complete behavior of the industrial processes. Disease outbreaks, customer disregard, frequent disasters, value addition, uncertainty, and processing complexities are critical variables in real-life dairy operations.

References

Aghalaya SN, Elias AA, Pati RK (2012) Analysing reverse logistics in the Indian pharmaceuticals industry: a systems approach. Paper presented at the 26th Australian and New Zealand Academy of Management (ANZAM) Conference 2012 Perth

Ahmad T, Aadil RM, Ahmed H, Ur Rahman U, Soares BC, Souza SL et al (2019) Treatment and utilization of dairy industrial waste: a review. Trends Food Sci Technol 88:361–372

Bansia M, Varkey JK, Agrawal S (2014) Development of a reverse logistics performance measurement system for a battery manufacturer. Procedia Mater Sci 6:1419–1427

Bernon M, Tjahjono B, Ripanti EF (2018) Aligning retail reverse logistics practice with circular economy values: an exploratory framework. Prod Plan Control 29(6):483–497
Blackburn J, Guide DVR, Wassenhove LNV, Souza GC (2004) Reverse supply chains for commercial returns. Calif Manag Rev 46(2):1–20
Cottrill K (2000) Return to sender. TraNc World 262(7):17–28
De Giovanni P (2022) Leveraging the circular economy with a closed-loop supply chain and a reverse omnichannel using blockchain technology and incentives. Int J Oper Prod Manag. (ahead-of-print)
DLS (2020) Development and activities of livestock – an overview. Department of Livestock Services, Bangladesh, Dhaka
Edwards DR, Daniel TC (1992) Environmental impacts of on-farm poultry waste disposal-A review. Bioresour Technol 41(1):9–33
El-Mashad HM, Zhang R (2010) Biogas production from co-digestion of dairy manure and food waste. Bioresour Technol 101(11):4021–4028
EMF (2020) What is the circular economy? [Online]. Available at https://www.ellenmacarthurfoundation.org/circular-economy/what-is-the-circular-economy
FAOSTAT (2018) Sharing development knowledge about Asia and teh Pacific. ADB Institute. http://www.adbi.org/3rdpartycdrom/2005/06/01/1507.fao.agriculture.statistics/
Forrester JW (1961) Industrial dynamics. The MIT Press, Cambridge, MA
Ghisellini P, Protano G, Viglia S, Gaworski M, Setti M, Ulgiati S (2014) Integrated agricultural and dairy production within a circular economy framework. A comparison of Italian and Polish farming systems. J Environ Account Manag 2(4):367–384
Ghisellini P, Ripa M, Ulgiati S (2018) Exploring environmental and economic costs and benefits of a circular economy approach to the construction and demolition sector. A literature review. J Clean Prod 178:618–643
Gomes ACA, Lobo DS, Cardoso BF, Shikida PFA (2014) Analysis of competitiveness of a dairy property through reverse logistics: a case study. Int J Glob Energy Issues 37(5–6):253–266
Govindan K, Hasanagic M (2018) A systematic review on drivers, barriers, and practices towards circular economy: a supply chain perspective. Int J Prod Res 56(1–2):278–311
Guarnieri P, Cerqueira-Streit JA, Batista LC (2020) Reverse logistics and the sectoral agreement of packaging industry in Brazil towards a transition to circular economy. Resour Conserv Recycl 153:104541
Guide VDR, Wassenhove Van LN (2002) The reverse supply chain. Harv Bus Rev 80(2):25–26
Gungor A, Gupta SM (1999) Issues in environmentally conscious manufacturing and product recovery: a survey. Comput Ind Eng 36(4):811–853
Hemme TO, Khan AR (2004) A review of milk production in Bangladesh with particular emphasis on small scale producers. Retrieved from Rome
Islam KN, Sarker T, Taghizadeh-Hesary F, Atri AC, Alam MS (2021) Renewable energy generation from livestock waste for a sustainable circular economy in Bangladesh. Renew Sust Energ Rev 139:110695
Javed H et al (2021) Exploring disposition decision for sustainable reverse logistics in the era of a circular economy: applying the triple bottom line approach in the manufacturing industry. Int J Supply Oper Manag 8(1):53–68
Kazancoglu Y et al (2021) Performance evaluation of reverse logistics in food supply chains in a circular economy using system dynamics. Bus Strateg Environ 30(1):71–91
Kirchherr J, Reike D, Hekkert M (2017) Conceptualizing the circular economy: an analysis of 114 definitions. Resour Conserv Recycl 127:221–232
Kocabasoglu C, Prahinski C, Klassen RD (2007) Linking forward and reverse supply chain investments: the role of business uncertainty. J Oper Manag 25:1141–1160
Krumwiedea DW, Sheub C (2002) A model for reverse logistics entry by third-party providers. Omega 30:325–333

Lavelli V, Beccalli MP (2022) Cheese whey recycling in the perspective of the circular economy: modeling processes and the supply chain to design the involvement of the small and medium enterprises. Trends Food Sci Technol 126:86–98
Maani KE, Cavana RY (2007) Systems thinking, system dynamics: managing change and complexity, vol 2nd. Pearson Education (NZ)/Prentice Hall, Auckland
Makarova I, Shubenkova K, Buyvol P, Shepelev V, Gritsenko A (2021) The role of reverse logistics in the transition to a circular economy: case study of automotive spare parts logistics. FME Trans 49(1):173–185
Meng JCS (2009) Donald Schön, Herbert Simon and the sciences of the artificial. Des Stud 30(1): 60–68
Nasir T, Shamsuddoha M, Quaddus M (2013) Dairy supply chain: a simulation study. Int J Econ Soc Sci 2:89
Oliveira V, Ottosen LM, Labrincha J, Dias-Ferreira C (2016) Valorisation of phosphorus extracted from dairy cattle slurry and municipal solid wastes digestates as a fertilizer. Waste Biomass Valorization 7(4):861–869
Patwa N, Sivarajah U, Seetharaman A, Sarkar S, Maiti K, Hingorani K (2021) Towards a circular economy: an emerging economies context. J Bus Res 122:725–735
Prahinski C, Kocabasoglu C (2006) Empirical research opportunities in reverse supply chains. OMEGA Int J Manag Sci 34(6):519–532
PRAN-RFL (2007) The development of dairy industry in Bangladesh. Retrieved from Dhaka
Raha SK, Talukder RK (2004) Vertical integration in the dairy sector in Bangladesh the case of Bangladesh Milk Producers Co-operative Union Ltd. Bangladesh J Polit Econ 20(1):41–52
Rajput S, Singh SP (2022) Industry 4.0 model for integrated circular economy-reverse logistics network. Int J Log Res Appl 25(4–5):837–877
Ravi V, Shankar R (2005) Analysis of interactions among the barriers of reverse logistics. Technol Forecast Soc Chang 72(8):1011–1029
Richardson GP (1986) Problems with causal-loop diagrams. Syst Dyn Rev 2(2):158–170
Saadullah M (2011) Smallholder dairy production and marketing in Bangladesh. Retrieved from http://www.ilri.cgiar.org/InfoServ/Webpub/fulldocs/South_South/ch06.htm
Schlüter M, Lickert H, Schweitzer K, Bilge P, Briese C, Dietrich F, Krüger J (2021) AI-enhanced identification, inspection and sorting for reverse logistics in remanufacturing. Procedia CIRP 98: 300–305
Shahnaz P, Shimazaki K-I, Kato I (2004) Milk and milk products in Bangladesh. J Rakuno Gakuen Univ 29(1):9–16
Shamsuddoha M (2013) Socio-economic and environmental success through effective poultry forward and reverse supply chain process. Annals of Philosophy, Social and Human Disciplines, University Ştefan cel Mare, Suceava, Series of Philosophy and Social Human
Shamsuddoha M (2015a) Integrated supply chain model for sustainable manufacturing: a system dynamics approach, vol 22, Part A. Emerald Group Publishing Limited
Shamsuddoha M (2015b) Reverse supply chain in the poultry industry. Amity Bus J 4(2):21–26
Shamsuddoha M (2022) Sustainable outcomes through the structured forward supply chain: a system dynamic approach. In: Qudrat Ullah H (ed) Understanding the dynamics of new normal for supply chains: post COVID opportunities and challenges. Springer, Cham, pp 55–80
Shamsuddoha M, Woodside AG (2022) Achieving radical process innovations by applying technology-mindset transformations via second-order system-dynamics engineering. J Bus Res 147:37–48
Shamsuddoha M, Woodside AG (2023) Sustainability process innovations resulting in new value-added byproducts: principal lessons from second-order systemdynamics engineering (SOSDE). J Bus Ind Mark ahead-of-print (ahead-of-print). https://doi.org/10.1108/JBIM-05-2022-0193
Shamsuddoha M, Nedelea A, Veronica G (2009) Dairy farming – an alternative income generating activity. Bull Univ Agric Sci Vet Med Cluj-Napoca Hortic 66(2):352–355

Shamsuddoha M, Quaddus M, Klass D (2011) Incorporating reverse supply chain in the poultry process of Bangladesh. Australian New Zealand Marketing Academy Conference 2011, Perth, Western Australia, 28–30 November 2011

Shamsuddoha M, Klass D, Quaddus M (2013a) Poultry wastes reuse through reverse supply chain process to achieve environmental sustainability. Paper presented at the Australian and New Zealand Academy of Management (ANZAM), Auckland, New Zealand

Shamsuddoha M, Nasir T, Quaddus M (2013b) Dairy supply chain: a vensim based conceptual model. Paper presented at the 3rd international forum & conference on logistics and supply chain management (LSCM), Bali, Indonesia, June 26–29

Shamsuddoha M, Quaddus MA, Woodside AG (2021) Environmental sustainability through designing reverse logistical loops: case research of poultry supply chains using system dynamics. J Bus Ind Mark 37(4):823–840. https://doi.org/10.1108/JBIM-02-2021-0119

Sheheli S (2012) Improving livelihood of rural women through income generating activities in Bangladesh

Sterman JD (2000) Business dynamics: systems thinking and modeling for a complex world. Irwin McGraw-Hill, Boston

Sun Q (2017) Research on the influencing factors of reverse logistics carbon footprint under sustainable development. Environ Sci Pollut Res 24(29):22790–22798

Wolstenholme EF (1982) System dynamics in perspective. J Oper Res Soc 33:547–556

Yin RK (1994) Case study research: design and methods, 2nd edn. Sage, Newbury Park

Zhang J, Wang M, Yin C, Dogot T (2021) The potential of dairy manure and sewage management pathways towards a circular economy: a meta-analysis from the life cycle perspective. Sci Total Environ 779:146396

Chapter 9
The Impact of Big Data Analytics Capabilities on the Sustainability of Maritime Firms

Ziaul Haque Munim, Mariia Dushenko, Hyeran Raaness, and Tine Viveka Westerberg

Abstract Big data analytics (BDA) allow enterprises to be proactive and forward-looking. Previous research has overlooked the impact of big data analytics capabilities (BDAC) on the sustainability performance of maritime firms. This chapter proposes a multi-criteria decision-making (MCDM) framework using 3 criteria, 11 sub-criteria, and 3 performance indicators. The performance indicators are rooted in the three dimensions of sustainability: economic, environmental, and social performance. The proposed MCDM framework was validated using the best-worst method (BWM) by analysing the data collected from experts in a Norwegian maritime cluster. Results show that respondents perceive that firms' BDAC has the most influence on economic performance and the least on social performance. The most implemented sub-criteria for BDAC are connectivity, decision-making, and modularity. Implications of the findings for the maritime industry are discussed.

Keywords Big data analytics · Maritime industry · Sustainability performance · Multi-criteria decision-making · Sustainable shipping

9.1 Introduction

Big data evolution has transformed conventional business models (McAfee and Brynjolfsson 2012). Reportedly, managers have been using big data to monitor and control the performance of companies operating in various industries (Akter et al. 2016). The maritime industry is not an exception. Nowadays, a massive amount of data is being generated and collected in the maritime sector, for example, ship performance and navigation-related data from a wide variety of sensors in a ship and the automatic identification system (AIS) data. Maritime organizations are working

Z. H. Munim (✉) · M. Dushenko · H. Raaness · T. V. Westerberg
Faculty of Technology, Natural and Maritime Sciences, University of South-Eastern Norway, Horten, Norway
e-mail: Ziaul.h.munim@usn.no

© The Author(s), under exclusive license to Springer Nature Switzerland AG 2023

N. U. Ibne Hossain (ed.), *Data Analytics for Supply Chain Networks*, Greening of Industry Networks Studies 11, https://doi.org/10.1007/978-3-031-29823-3_9

towards utilizing such data to improve the efficiency of daily operations and decision-making (Zhang and Lam 2019).

Despite its unexplored potential, big data analytics (BDA) can be a game-changer by assisting organizations to improve business efficiency (Wamba et al. 2017). BDA can handle large volumes of data from business systems and operations including non-traditional data types such as video, audio, and texts (Hagel 2015). Various applications of BDA are evident in the maritime industry (Munim et al. 2020).

Several ports, for example, the Port of Rotterdam, Singapore, Hamburg, and Port of Cartagena, have successfully introduced BDA to sustain a competitive advantage over other ports by predicting vessel arrival time, preservation of port machinery, and illegal bunkering (Zhang and Lam 2019; Noll and Hogeweg 2015; Yeo et al. 2019; Trelleborg Marine Systems 2018). Jafarzadeh and Schjølberg (2018) demonstrated the advantages of using big data from AIS in emission reduction of container, Ro-Ro, as well as offshore and passenger ships. Lee et al. (2018) showed improving vessel fuel efficiency by utilizing weather archive-related data. Hence, exploring the impact of big data analytics capabilities (BDAC) on the performance of maritime firms through the three sustainability pillars or the triple bottom line (TBL)—the financial, social, and environmental performance, is of great value.

9.2 Big Data Analytics Capability (BDAC) and Sustainability Performance

There has been a growing interest in developing BDAC with the aim of enhancing firms' performance in the recent years (Kallinikos and Constantiou 2015; Akter et al. 2016). The interest for attaining competitive advantage has resulted in constant growth in BDAC investment, which exceeded US$ 3.8 trillion in 2014 (Akter et al. 2016; Lunden 2013). Rialti et al. (2019a, b) asserted that it is necessary to make significant investments in BDAC in order to leverage the benefits of BDA. However, there is not only scarce research on the relation between BDAC and performance but also a lack of support for the claim that these investments lead to measurable business values (Mikalef et al. 2018). Some of the literature only outlines the benefits and challenges in utilizing BDA (Mirović et al. 2018; Wamba et al. 2015; Zhang and Lam 2019).

Only a few studies have explored the potential consequences of BDAC on a firm's social, environmental, and financial performance (Wamba et al. 2017; Akter et al. 2016; Gupta and George 2016). Song et al. (2018) further addressed that a firm can accomplish more sustainable production and advance its sustainable competitive advantage by improving its use of natural resources, energy, and environmental efficiency with the help of BDA. Similarly, Dubey et al. (2019) suggested that BDAC is one of the capabilities that would allow organizations to enhance their social and environmental performance. Dubey et al. (2019) highlighted that data-driven decision-making skills, management skills, capability of organizational

Table 9.1 Performance alternatives

Performance alternatives	Description
Economic performance	The economic performance in maritime industry can be measured by: Reduction in fuel consumption Cost-benefit from proper and timely maintenance Reduction in operating cost
Social performance	The social performance in maritime can be measured by: Improvement of the safety and well-being of crew members Investment in training and courses for employees' development Fulfilment of social mission Relations and contribution to the surrounding community
Environmental performance	The environmental performance in maritime can be measured by: Reduction of carbon footprint by using alternative energy sources Reduction of oil spill quantity Reduction of greenhouse gas (GHG) emissions from shipping traffic

Adapted from Kenton (2020), Slaper and Hall (2011) and Trelleborg Marine Systems (2018)

learning, and technical skills are essential precedents of BDAC, which have an impact on social and environmental performance.

Sustainability has become significant to both academia and industry due to a rapid depletion of natural resources and growing concerns over corporate social responsibility (Govindan et al. 2013). However, measuring to which degree an organization achieves sustainable growth can be challenging (Slaper and Hall 2011). The concept of the triple bottom line (TBL) is an approach to measure sustainable development. The concept was developed by Elkington (1994) and is also known as "people, planet, and profits". Elkington (1998) proposed that businesses should measure their outcome not only by the traditional financial bottom line (i.e. ROI and profit) but also by their influence on society, the environment, and the broader economy. Table 9.1 presents the description of TBL with measurements applicable to the maritime industry.

9.2.1 *Dimensions of Big Data Analytics Capabilities*

The literature largely agrees that the fundamental components of IT capability, i.e. physical, human, and organizational, and physical capability consists of IT infrastructure and human capability including skill or knowledge (Barney 1991; Grant 1991; Bhatt and Grover 2005; Chen and Wu 2011). It is therefore fundamental to include infrastructure and human assets as key factors of capability. Accordingly, this study, similar to Kim et al. (2012), perceives that the key determinants of a firm's BDAC are flexibility of infrastructure, capability of management, and expertise of the personnel.

9.2.2 BDA Infrastructure Flexibility

BDA infrastructure flexibility is defined by its aptitude to assist firm personnel to improve, deploy, and back a firm's resources quickly. The BDA infrastructure flexibility allows staff to increase their agility in developing and establishing sufficient information and data-sharing channels across various functions that boost functional integration, which in return leads to enhanced business processes (Bharadwaj 2000). Predictive analytics, automatic identification system (AIS), vessel traffic services (VTS), maritime surveillance, voyage data recorder (VDR), and machine learning are examples of enhancing BDA infrastructure flexibility in the maritime sector.

The flexibility of BDA infrastructure consists of three main components: connectivity, compatibility, and modularity (Table 9.2). Connectivity is the ability of BDA infrastructure to analyse and utilize various types of big data across different business units (Akter et al. 2016), for example, optimized planning and scheduling process with big data (Jimenez et al. 2022). *Compatibility* allows multiple analytics platforms to share a variety of transparent information and data within the firm (Kim et al. 2012). *Modularity* enables staff to add, eliminate, and amend software components by expanding the concepts of shareability and reusability to both applications and data (Duncan 1995).

9.2.3 BDA Management Capabilities

BDA management capability is a crucial aspect of BDAC (Akter et al. 2016). Kim et al. (2012) defined BDA management capabilities as the ability to manage BDA resources in a structured manner according to functional requirements and priorities. BDA management capability ensures that decision-making is done in accordance with the proper management framework. Various practices are important for BDA management capability; however, Kim et al. (2011) stated that the essential foundations for BDA management capability are planning, investment, coordination, and control. These four core elements, according to Kim et al. (2012), are necessary to compose a framework for BDA management capability, and they are presented in Table 9.2 as sub-criteria of BDA management capabilities.

9.2.4 BDA Personal Expertise

BDA personal expertise refers to the ability to undertake assigned tasks in a big data environment, which is critical to get an insight and manage variety of data (Akter et al. 2016; Barton and Court 2012; Kim et al. 2012). Akter et al. (2016) highlighted that BDA staff should be proficient in four knowledge classifications, (1) technical, (2) technology management, (3) business, and (4) relational (see Table 9.2), to align

Table 9.2 Criteria and sub-criteria

Criteria	Sub-criteria	Description
BDA infrastructure flexibility	Connectivity	Extent to which internal and external BDAC components are connected
	Compatibility	Extent to which various data types and software application across multiple analytics platforms are shared
	Modularity	Extent to which system and its elements can be added, removed, and modified
BDA management capabilities	Planning	Extent to which the planning of business analytics and utilization are organized in accordance with procedures both formal and informal
	Decision-making	Extent to which investment decision-making about BDA is organized in accordance to procedures both formal and informal
	Coordination	Extent to which coordination efforts between business and data analysts and business clients are structured according to formal and informal procedures
	Control	Extent to which analytics control activities (e.g. development, monitoring performance, and establishing clear performance criteria) are structured according to procedures: formal and informal
BDA personnel expertise	Technical knowledge	Analytics team's programming skills as well as knowledge in managing project life cycles, data management, distributed computing, and decision support systems
	Technological management knowledge	Analytics team's deep understanding of technological trends and the ability to learn new technologies as well as knowledge of key factors of the organization's success
	Business knowledge	Analytics team's insight into of various business functions and business environment with its problems in order to develop appropriate solutions
	Relational knowledge	Analytics team's communication and cooperation with people from other business functions and assistance to the end users and clients

Adapted from Kim et al. (2012) and Wamba et al. (2017)

BDA strategy with business strategy. Besides, they take connectivity, compatibility, and modularity into account when developing analytics system as well as supporting business demands.

9.3 The MCDM Framework

A multi-criteria decision-making (MCDM) framework has been developed to investigate the influence of BDAC on the TBL utilizing 3 criteria, 11 sub-criteria, and 3 TBL performance alternatives. The BDAC dimensions are established based on

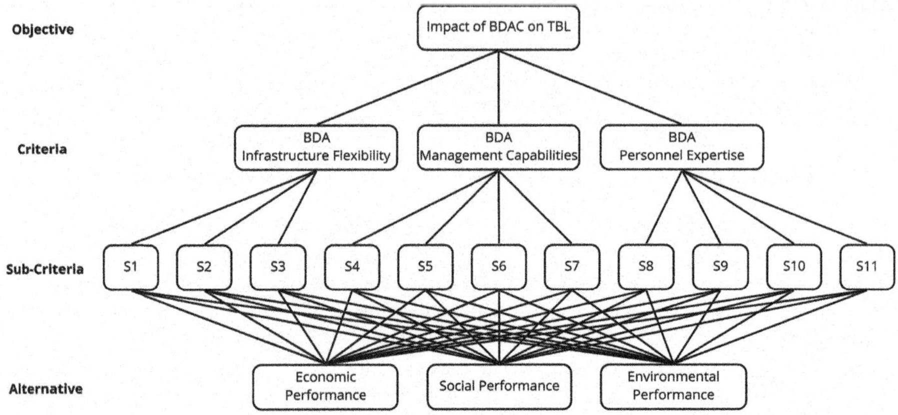

Fig. 9.1 Proposed multi-criteria decision-making (MCDM) framework. *S1* connectivity, *S2* compatibility, *S3* modularity, *S4* planning, *S5* decision-making, *S6* coordination, *S7* control, *S8* technical knowledge, *S9* technological management knowledge, *S10* business knowledge, *S11* relational knowledge

three capabilities: (1) infrastructure flexibility, (2) management capability, and (3) personnel expertise. Each capability consists of three, four, and four elements, respectively (see Table 9.2 and Fig. 9.1). The best-worst method (BWM) (Rezaei 2015) was used for validating the framework using collected data from industry experts in Norwegian maritime firms. Data were collected from individuals working at Norwegian ports, shipping companies, and offshore oil and gas industries located in the Oslo fjord maritime cluster during the period of April to May 2020.

9.4 DAC Implementation in Norwegian Maritime Firms

9.4.1 The Most Implemented Criteria and Its Preference Over All Criteria

The most implemented criteria and sub-criteria to other criteria vectors (best to others) are presented in Table 9.3. The respondents are asked the following questions: "Which of the following criteria is the 'most implemented' for big data analytics capabilities?" "How much more implemented is the 'most implemented criterion' compared to others on a scale of 1–9?". Similarly, the questions for sub-criteria are asked one by one in the survey.

Four out of nine respondents answered that *infrastructure flexibility (IF)* is the most implemented capability among the three big data capabilities. On the sub-criteria level, the most implemented management capability and personal expertise are *decision-making (DM)* and *technical knowledge (TK)*, respectively. For IF, connectivity (CN), compatibility (CP), and modularity (MD) are reported most implemented by three respondents each.

Table 9.3 The most implemented-to-others vector

Respondent	Big data analytics capabilities				Infrastructure flexibility				Management capability					Personnel expertise capability				
	Best	IF	MC	PE	Best	CN	CP	MD	Best	PL	DM	CD	CR	Best	TK	TM	BK	RK
1	IF	1	4	3	CP	2	1	2	CR	2	2	2	1	TK	1	4	2	2
2	MC	6	1	5	CP	8	1	3	DM	4	1	5	6	TK	1	3	5	6
3	PE	5	5	1	MD	3	3	1	CR	1	3	3	1	RK	2	2	3	1
4	IF	1	3	6	CN	1	5	5	DM	4	1	2	5	TK	1	3	4	6
5	MC	9	1	3	CN	1	7	7	DM	5	1	5	5	TK	1	6	6	6
6	IF	1	4	4	CN	1	4	4	PL	1	4	4	4	TK	1	4	4	4
7	IF	1	7	4	MD	8	5	1	CR	2	3	1	1	TK	1	2	5	3
8	PE	5	5	1	CP	1	1	5	PL	1	3	7	5	BK	5	4	1	4
9	MC	3	1	2	MD	2	2	1	DM	2	1	3	1	RK	6	6	4	1

IF infrastructure flexibility, *MC* management capability, *PE* personnel expertise, *CN* connectivity, *CP* compatibility, *MD* modularity, *PL* planning, *DM* decision-making, *CD* coordination, *CR* control, *TK* technical knowledge, *TM* technological management knowledge, *BK* business knowledge, *RK* relational knowledge

9.4.2 *The Least Implemented Criteria and Preference of All Criteria Over It*

Others-to-the-least implemented vectors (others to worst) are presented in Table 9.4. Respondents are asked: "Which of the following criteria is the 'least implemented' for big data analytics capabilities? "How much more implemented are other criteria compared to the 'least implemented criteria' on a scale of 1–9?" Consequently, the questions for sub-criteria are asked one by one in the survey.

Personal expertise (PE) and *management capability (MC)* are reported as the least implemented big data capabilities by four respondents each. On the sub-criteria level, the least implemented infrastructure flexibility is modularity (MD). Control (CR) and coordination (CD) are reported as least implemented management capabilities by three respondents each. Moreover, relational knowledge (RK) is reported as least implemented personal expertise by four respondents.

Surprisingly, it was observed that some respondents rated the others-to-worst vectors in reverse order (respondents 2, 4, and 5). Therefore, their values were transformed using $(10 - a_{jW})$, where a_{jW} represents the rating by the respondent (see Table 9.4).

9.4.3 *The Optimal Weights of Criteria*

A single weight vector was obtained for each respondent using BWM Excel solver optimization on the criteria and sub-criteria level. Table 9.5 presents the optimal weights for each implemented sub-criteria. The higher value indicates the greater degree of implementation, while the lower shows the weaker degree of implementation. On the aggregate level, connectivity (mean 0.164) is the most implemented criteria followed by decision-making (mean 0.128) and modularity (mean 0.124) among all the sub-criteria of BDAC.

9.4.4 *Aggregate Impact Scores*

The aggregated impact of BADC on the TBL performance indicators was calculated the BWM algorithm (Rezaei 2015). Respondents rated the impacts of each of the 11 sub-criteria on the three TBL performance indicators. The impacts of BDAC on the TBL performance indicators for all the respondents is shown in Table 9.6, revealing that BDAC has the most impact on economic performance (mean 0.956), followed by the environmental performance (mean 0.824) and social performance (mean 0.721).

Additionally, priorities of the TBL alternatives by oil and gas industry, shipping industry, and port are compared as shown in Fig. 9.2. The respondents from the oil

Table 9.4 Others-to-the-least implemented vector

Respondent	Big data analytics capabilities				Infrastructure flexibility				Management capability					Personnel expertise capability				
	Worst	IF	MC	PE	Worst	CN	CP	MD	Worst	PL	DM	CD	CR	Worst	TK	TM	BK	RK
1	MC	4	1	2	MD	2	2	1	PL	1	1	1	2	TM	4	1	2	2
2	PE	*4*	*5*	1	MD	2	*5*	1	CR	*5*	*8*	*4*	1	RK	*7*	*4*	*5*	1
3	IF	1	3	6	CN	1	2	3	CD	3	2	1	3	BK	5	5	1	4
4	MC	*8*	1	*6*	CP	5	1	5	CD	*4*	2	1	5	RK	*6*	*3*	*4*	1
5	PE	*1*	*3*	1	MD	9	8	1	CR	5	5	5	1	RK	4	4	4	1
6	PE	4	4	1	MD	4	4	1	CR	4	4	4	1	RK	4	4	4	1
7	MC	7	1	4	CN	1	5	8	DM	2	1	3	3	BK	5	4	1	3
8	MC	3	1	6	MD	5	3	1	CD	7	5	1	3	TK	1	3	5	3
9	PE	2	3	1	CN	1	2	2	PL	1	3	2	2	TK	1	7	6	6

IF infrastructure flexibility, *MC* management capability, *PE* personnel expertise, *CN* connectivity, *CP* compatibility, *MD* modularity, *PL* planning, *DM* decision-making, *CD* coordination, *CR* control, *TK* technical knowledge, *TM* technological management knowledge, *BK* business knowledge, *RK* relational knowledge. Italic numbers were transformed using $(10 - a_{jW})$

Table 9.5 Optimal weights

Respondent[a]	CN	CP	MD	PL	DM	CD	CR	TK	TM	BK	RK
1 (0.039)	0.189	0.314	0.126	0.029	0.029	0.029	0.057	0.102	0.025	0.051	0.051
2 (0.211)	0.018	0.109	0.025	0.139	0.434	0.111	0.052	0.061	0.027	0.016	0.007
3 (0.116)	0.019	0.026	0.067	0.065	0.025	0.018	0.065	0.190	0.190	0.058	0.277
4 (0.234)	0.548	0.070	0.149	0.014	0.035	0.007	0.011	0.091	0.037	0.028	0.012
5 (0.274)	0.078	0.016	0.006	0.098	0.359	0.098	0.046	0.186	0.043	0.043	0.029
6 (0.202)	0.444	0.148	0.074	0.121	0.040	0.040	0.020	0.061	0.020	0.020	0.010
7 (0.098)	0.051	0.130	0.528	0.016	0.009	0.029	0.029	0.098	0.056	0.017	0.037
8 (0.104)	0.083	0.070	0.019	0.057	0.023	0.007	0.014	0.067	0.123	0.415	0.123
9 (0.154)	0.048	0.072	0.120	0.072	0.201	0.086	0.201	0.011	0.028	0.042	0.118
Mean	**0.164**	0.106	0.124	0.068	0.128	0.047	0.055	0.096	0.061	0.077	0.074

CN connectivity, *CP* compatibility, *MD* modularity, *PL* planning, *DM* decision-making, *CD* coordination, *CR* control, *TK* technical knowledge, *TM* technological management knowledge, *BK* business knowledge, *RK* relational knowledge

[a]Average consistency ratio (CR) in parenthesis

Table 9.6 Priority for performance

Respondent	Economic performance	Social performance	Environmental performance
1 (Shipping)	0.933	0.983	0.977
2 (Shipping)	0.939	0.721	0.724
3 (Port)	0.968	0.809	0.800
4 (Oil and gas)	1.000	0.342	0.481
5 (Shipping)	1.000	0.610	0.610
6 (Oil and gas)	0.963	0.745	0.996
7 (Oil and gas)	0.975	0.323	0.996
8 (Shipping)	0.861	0.958	0.870
9 (Port)	0.966	1.000	0.966
Mean (sample)	**0.956**	**0.721**	**0.824**

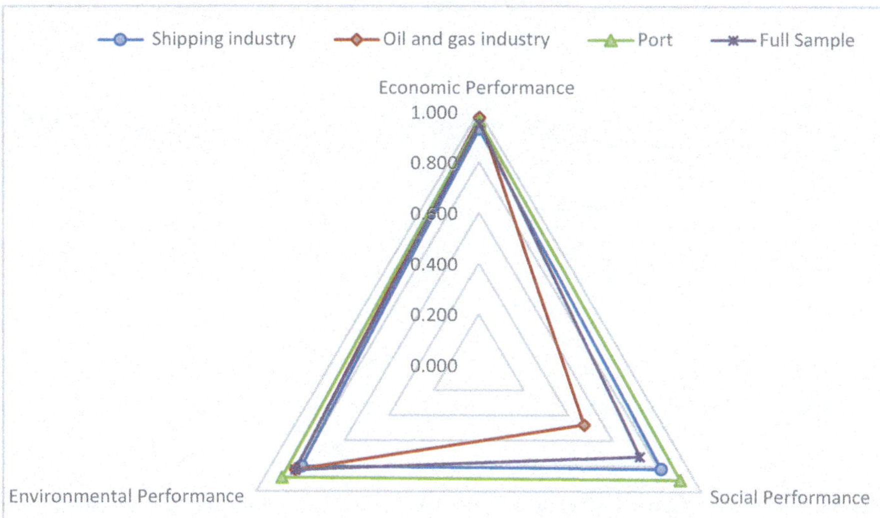

Fig. 9.2 Aggregated impact scores on the TBL indicators

and gas sector (n=3) perceive that BDAC has the most impact on economic performance (mean 0.979), followed by environmental (mean 0.824) and social (mean 0.470) performance. Similarly, respondents from the shipping company and port industry (n=6) perceive that BDAC has the most impact on economic performance (mean 0.945). However, while the respondents from the shipping and port sector perceive that BDAC has the second most influence on social performance (mean 0.847), respondents of oil and gas perceive social performance as the least impacted by BDAC (mean 0.470). In the full sample aggregated score, BDAC also has the least impact on social performance (mean 0.721).

Figure 9.3 presents the aggregated optimal weights of the whole sample and for each industry. The respondents believe that the connectivity (mean 0.164) is the

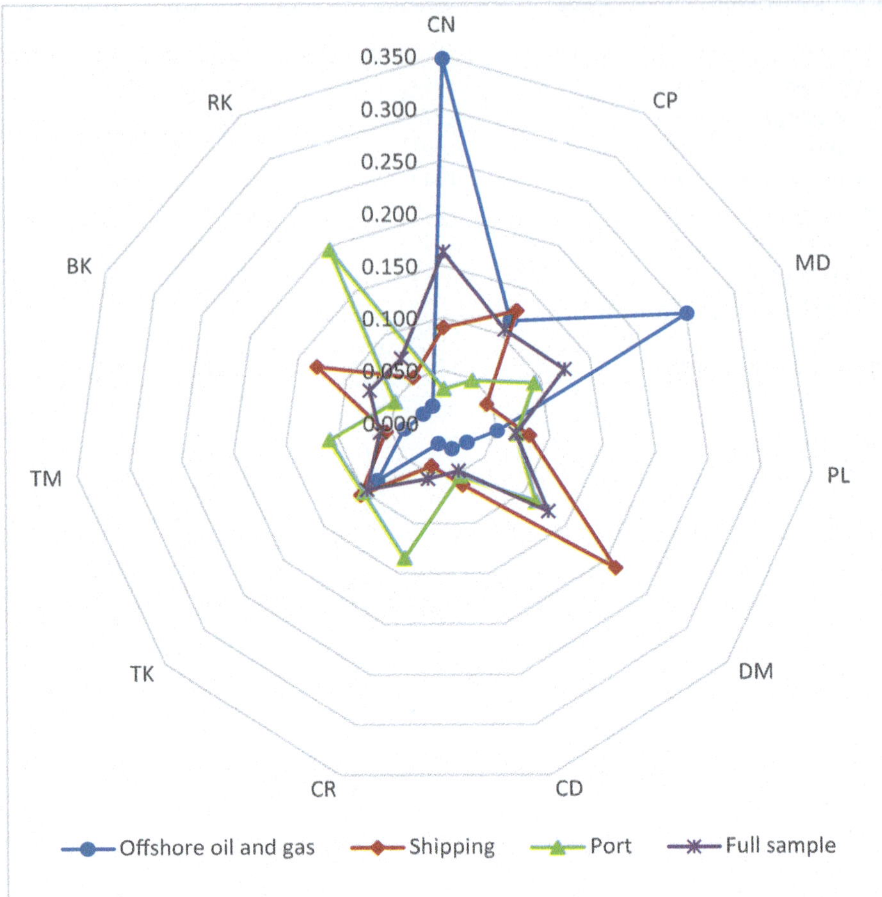

Fig. 9.3 Aggregated optimal weights. *CN* connectivity, *CP* compatibility, *MD* modularity, *PL* planning, *DM* decision-making, *CD* coordination, *CR* control, *TK* technical knowledge, *TM* technological management knowledge, *BK* business knowledge, *RK* relational knowledge

most implemented criteria, followed by decision-making (mean 0.128) and modularity (0.124) among all the sub-criteria of BDAC. Similar to the mean of all industries, the respondents from the oil and gas industry perceive that connectivity (mean 0.348) is the most implemented criteria, followed by modularity and compatibility (mean 0.250 and 0.116 correspondingly). Interestingly, the respondents from the shipping industry do not view connectivity as the most implemented criteria for BDAC, and instead, they believe that decision-making (mean 0.211), followed by business knowledge (mean 0.131), is the most implemented element for BDAC as shown in Fig. 9.3. Finally, respondents representing the port industry believe that rational knowledge (mean 0.198) is the most implemented criteria for BDAC, while connectivity (mean 0.033) is the least implemented one.

9.5 Discussions and Implications

This chapter explores the influence of BDAC on maritime firms' sustainability performance through the lens of TBL. The result shows that connectivity, decision-making, and modularity are the most implemented sub-criteria for BDAC and the firm's BDAC has the most influence on economic performance (mean 0.956) among the three performance indicators. According to a report from Trelleborg Marine Systems (2018), there have been significant increases in partnerships to develop technology capabilities; the number of maritime companies, investing in infrastructure that enables big data implementation as well as promotes application of big data in maritime sector, has been growing. This trend shows that the maritime industry has been investing in BDA connectivity, compatibility, and modularity, which are three core elements of infrastructure flexibility.

Furthermore, a comparison among sectors in the maritime industry (Fig. 9.2) reveals that BDAC affects more economic performance in the oil and gas industry (mean 0.979) than the port (mean 0.967) and the shipping industry (mean 0.933). This outcome does not represent each sector (due to relatively low sample size), nor guarantees the same output, but the proposed MCDM framework can be used by a firm to analyse how their BDAC impacts the TBL indictors. Also, the result can be used to support the rising investment in the implementation of BDAC given trending development of data-collecting sensors in the areas of exploration, drilling, and extraction of the offshore oil and gas industry. With the advancement of BADC, there is a potential for reservoir modelling and simulation, drilling time reduction and increase in drilling safety, performance optimization of pump production, and improvement in asset management of petrochemicals. As a result, it leads to cost reduction (Mohammadpoor and Torabi 2018).

The positive effect of BDAC on social and environmental performance is emphasized in extant literature (Dubey et al. 2019; Song et al. 2018). Besides, corporate social responsibility (CSR) in the shipping industry is evolving and transforming market conditions as there are new requirements from regulators, investors, NGOs, and customers emerging (Froholdt 2018). One reason for the noticeable difference of social performance preference between the shipping and the oil and gas industry may indicate that respondents from the shipping company are more aware of the benefit from social sustainability than the oil and gas industry. Although the difference of priority by industry should not be generalized to each industry, it indicates the difference in the perspective on the impact of BDAC on TBL.

9.5.1 *Implications for Practice*

The result shows the importance of developing BDAC to improve maritime firms' sustainability performance because it does not only provide managers with the direction when implementing BDAC but also helps a firm to prioritize and invest in relatively more important capabilities, which are connectivity, modularity, and decision-making. As an example, connectivity, which is one of the crucial components for infrastructure flexibility (Kim et al. 2012), can be utilized for improving customer relationship management by enabling smart personalization and customization to customers (Anshari et al. 2019).

Additionally, firms could strengthen decision-making capability by considering the impact of teamwork on productivity and efficiency, and the training cost of staffs that is required, when making decisions on BDA investment (Kim et al. 2012; Wamba et al. 2017). Moreover, this chapter helps maritime managers to consider the impact of BDAC on social and environmental performance as well as economic performance, since BDAC not only plays a significant role in improving firms' sustainable competitive advantages but also is one of the organizational capabilities which allow organizations to enhance social and environmental performance (IGEL 2014; Song et al. 2018; Dubey et al. 2019). Moreover, the findings also allow maritime firms to set their strategy in a specific perspective according to their objectives in order to strengthen their competitive advantage in the industry.

9.5.2 Implications for Research

The impact of BDAC on a firm's performance has been addressed in existing literature (Kim et al. 2012; Wamba et al. 2015; Gupta and George 2016; Rialti et al. 2019a, b; Akter et al. 2016). Some studies (Song et al. 2018; Dubey et al. 2019) emphasize the positive influence on environmental, social, and environmental performance by utilizing BDAC. However, studies on the impact of BDAC on the sustainability of maritime firms through the lens of TBL are limited. Hence, this chapter provides not only a guideline that facilitates developing BDAC of maritime firms but also an overview of aspects lacking examination in previous BDAC research.

9.6 Conclusions

The impact of big data analytics capabilities on the sustainability of maritime firms is evaluated in this chapter through utilizing the best-worst MCDM method. Data are collected from a sample of nine industry experts from a Norwegian maritime cluster. The findings reveal that the maritime firm's BDAC has the most impact on economic performance among the three dimensions of sustainability.

With the growing interest in BDAC, this chapter extends BDAC's role on sustainability in the sense that it provides direction of when should firms implement BDAC by providing them a framework to prioritize and invest in relatively more important capabilities. Although this chapter covered several aspects of BDAC and its impact on sustainability, it can be extended to explore the impact of BDAC on firms' performance among different industries or economies. Furthermore, a repeated measure study design, where the ranking of implemented BDAC changes over time is assessed, can trace the dynamic impacts on sustainability performance.

References

Akter S, Wamba SF, Gunasekaran A, Dubey R, Childe SJ (2016) How to improve firm performance using big data analytics capability and business strategy alignment? Int J Prod Econ 182:113–131

Anshari M, Almunawar MN, Lim SA, Al-Mudimigh A (2019) Customer relationship management and big data enabled: personalization & customization of services. Appl Comput Inform 15(2): 94–101

Barney J (1991) Firm resources and sustained competitive advantage. J Manag 17(1):99–120

Barton D, Court D (2012) Making advanced analytics work for you. Harv Bus Rev 90(10):78–83

Bharadwaj AS (2000) A resource-based perspective on information technology capability and firm performance: an empirical investigation. MIS Q 24:169–196

Bhatt GD, Grover V (2005) Types of information technology capabilities and their role in competitive advantage: an empirical study. J Manag Inf Syst 22(2):253–277

Chen Y-C, Wu J-H (2011) IT management capability and its impact on the performance of a CIO. Inf Manag 48(4–5):145–156

Dubey R, Gunasekaran A, Childe SJ, Papadopoulos T, Luo Z, Wamba SF, Roubaud D (2019) Can big data and predictive analytics improve social and environmental sustainability? Technol Forecast Soc Chang 144:534–545

Duncan NB (1995) Capturing flexibility of information technology infrastructure: a study of resource characteristics and their measure. J Manag Inf Syst 12(2):37–57

Elkington J (1994) Towards the sustainable corporation: win-win-win business strategies for sustainable development. Calif Manag Rev 36(2):90–100

Elkington J (1998) Partnerships from cannibals with forks: the triple bottom line of 21st-century business. Environ Qual Manag 8(1):37–51

Froholdt LL (2018) Corporate social responsibility in the maritime industry, vol 5. Springer

Govindan K, Khodaverdi R, Jafarian A (2013) A fuzzy multi criteria approach for measuring sustainability performance of a supplier based on triple bottom line approach. J Clean Prod 47: 345–354

Grant RM (1991) The resource-based theory of competitive advantage: implications for strategy formulation. Calif Manag Rev 33(3):114–135

Gupta M, George JF (2016) Toward the development of a big data analytics capability. Inf Manag 53(8):1049–1064

Hagel J (2015) Bringing analytics to life. J Account 219(2):24

IGEL (2014) Sustainability in the age of big data. Retrieved from http://d1c25a6gwz7q5e.cloudfront.net/reports/2014-09-12-Sustainability-in-the-Age-of-Big-Data.pdf

Jafarzadeh S, Schjølberg I (2018) Operational profiles of ships in Norwegian waters: an activity-based approach to assess the benefits of hybrid and electric propulsion. Transp Res Part D: Transp Environ 65:500–523

Jimenez VJ, Kim H, Munim ZH (2022) A review of ship energy efficiency research and directions towards emission reduction in the maritime industry. J Clean Prod 366:132888

Kallinikos J, Constantiou ID (2015) Big data revisited: a rejoinder. J Inf Technol 30(1):70–74

Kenton W (2020) Triple Bottom Line (TBL). Retrieved from https://www.investopedia.com/terms/t/triple-bottom-line.asp

Kim G, Shin B, Kim KK, Lee HG (2011) IT capabilities, process-oriented dynamic capabilities, and firm financial performance. J Assoc Inf Syst 12(7):1

Kim G, Shin B, Kwon O (2012) Investigating the value of sociomaterialism in conceptualizing IT capability of a firm. J Manag Inf Syst 29(3):327–362

Lee H, Aydin N, Choi Y, Lekhavat S, Irani Z (2018) A decision support system for vessel speed decision in maritime logistics using weather archive big data. Comput Oper Res 98:330–342

Lunden I (2013) Forrester: $2.1 trillion will go into IT spend in 2013. Apps and the U.S. Lead the Charge

McAfee A, Brynjolfsson E (2012) Big data: the management revolution. Harv Bus Rev 90(10): 60–68

Mikalef P, Pappas IO, Krogstie J, Giannakos M (2018) Big data analytics capabilities: a systematic literature review and research agenda. IseB 16(3):547–578

Mirović M, Miličević M, Obradović I (2018) Big data in the maritime industry. NAŠE MORE: znanstveno-stručni časopis za more i pomorstvo 65(1):56–62

Mohammadpoor M, Torabi F (2018) Big data analytics in oil and gas industry: an emerging trend. Petroleum 6(4):321–328

Munim ZH, Dushenko M, Jimenez VJ, Shakil MH, Imset M (2020) Big data and artificial intelligence in the maritime industry: a bibliometric review and future research directions. Marit Policy Manag 47(5):577–597

Noll G, Hogeweg M (2015) Big data management at Port of Rotterdam using a GIS platform to streamline IT at growing maritime hub. Seal Technol 56(5):31-+

Rezaei J (2015) Best-worst multi-criteria decision-making method. Omega 53:49–57

Rialti R, Marzi G, Ciappei C, Busso D (2019a) Big data and dynamic capabilities: a bibliometric analysis and systematic literature review. Manag Decis 57(8):2052–2068

Rialti R, Zollo L, Ferraris A, Alon I (2019b) Big data analytics capabilities and performance: evidence from a moderated multi-mediation model. Technol Forecast Soc Chang 149:119781

Slaper TF, Hall TJ (2011) The triple bottom line: what is it and how does it work. Indiana Bus Rev 86(1):4–8

Song P, Zheng C, Zhang C, Yu X (2018) Data analytics and firm performance: an empirical study in an online B2C platform. Inf Manag 55(5):633–642

Trelleborg Marine Systems (2018) Use of big data in the maritime industry report. Retrieved from https://www.patersonsimons.com/wp-content/uploads/2018/06/TMS_SmartPort_InsightBee_Report-to-GUIDE_01.02.18.pdf

Wamba SF, Akter S, Edwards A, Chopin G, Gnanzou D (2015) How 'big data' can make big impact: findings from a systematic review and a longitudinal case study. Int J Prod Econ 165: 234–246

Wamba SF, Gunasekaran A, Akter S, Ren SJ-F, Dubey R, Childe SJ (2017) Big data analytics and firm performance: effects of dynamic capabilities. J Bus Res 70:356–365

Yeo G, Lim SH, Wynter L, Hassan H (2019) MPA-IBM project SAFER: sense-making analytics for maritime event recognition. INFORMS J Appl Analyt 49(4):269–280

Zhang X, Lam JSL (2019) A fuzzy Delphi-AHP-TOPSIS framework to identify barriers in big data analytics adoption: case of maritime organizations. Marit Policy Manag 46(7):781–801

Chapter 10
Smart Transportation Logistics: Achieving Supply Chain Efficiency with Green Initiatives

Mohammad Shamsuddoha, Mohammad Abul Kashem, and Tasnuba Nasir

Abstract Global economies have long relied on the efficient flow of people and products, despite inefficiencies such as time waste, delays, increased costs, and pollution. Smart transportation systems can be built using various technologies and techniques to tackle modern challenges. Modern autos and sophisticated transportation networks represent important technology developments that increase driving comfort and safety. There are a number of new issues that need to be faced that require significant modifications in the models of transportation systems to abundantly realize their potential and stimulate the expansion of smart movement applications and services. This chapter aims to identify the latest trend in transportation logistics based on a systematic literature review along with a multidimensional overview of smart logistical apparatus. The chapter accentuated the sustainable issues for improving road safety, traffic management, reduced carbon emission, and ways for real-time tracking and decision-making within modern technological upgradation through this advanced transportation model. The ultimate improvement of this literature-based model was to smarten the real transportation system with the advent of IoT and artificial intelligence. This review ultimately powered the policymakers and academicians for new thinking and a new dimension for the vehicle to everything (V2X) for a sustainable future. Accordingly, the need for empirical proof of the sustainability profits of transportation logistics efforts, data-driven decision-making, and optimization techniques should be systematically considered for practices and outlining universally relevant metrics.

Keywords Smart transportation · Intelligent transportation · Logistics · Sustainability · Supply chain · Green initiative

M. Shamsuddoha (✉) · T. Nasir
Western Illinois University, Macomb, IL, USA
e-mail: m-shamsuddoha@wiu.edu; tasnubanasir@gmail.com

M. A. Kashem (✉)
Feni University, Feni, Bangladesh
e-mail: mak.mktg@yahoo.com

© The Author(s), under exclusive license to Springer Nature Switzerland AG 2023

N. U. Ibne Hossain (ed.), *Data Analytics for Supply Chain Networks*, Greening of Industry Networks Studies 11, https://doi.org/10.1007/978-3-031-29823-3_10

10.1 Introduction

Intelligent transportation systems (ITS) are a collection of state-of-the-art information and communication technologies incorporated into traffic and transportation management systems (Singh et al. 2022b). This system is ardently engaged to improve transportation networks' security, dependability, and environmental management and reduce congestion and pollution that drivers experience (Sharifi et al. 2022). In addition, the transportation industry relies heavily on data analytics to use the resources at hand more effectively, such as reducing traffic and improving the travel experience (Jain et al. 2022). One more time, in China, transportation is at the forefront of innovative applications for smart cities and highway safety (Liu and Ke 2022). However, it takes into account all forms of connectivity and intelligent mobility, such as ride-sharing, autonomous and partially autonomous vehicles, bike-sharing initiatives, traffic signals, habitations, parking spaces, and humans (Komninos et al. 2022). Evidently, conventional transportation systems seek to improve mobility, particularly for vehicles, but may fail to consider wider consequences (McLeod and Curtis 2022) and even less consideration for passenger choice which ultimately be influential in order to preserve the sustainable growth of the traveling infrastructure and services (Kashem 2020). Therefore, during the past few decades, smart transportation structures have been idealized to increase vehicle and travel safety and security, improve transportation efficiency, and modernize facilities for drivers and passengers.

In today's world, fostering links between environmental protection and economic viability, social advancement is necessary when applying sustainable development to transportation networks (Palit et al. 2022). Yet, environmentally friendly transportation promotes efficiency, safety, and environmental advancements for long-term development (Khalili et al. 2019). Sustainable innovation encompasses technological progress that contributes to energy conservation, pollution reduction, waste recovery, the production of eco-friendly goods, and corporate environmental management (Aftab et al. 2022). Hence, an efficient knowledge representation scheme on sustainability is required for an effective transportation system that assists businesses in creating a balance between the economic viability of their decisions and the ecological and social consequences.

On the other hand, logistics and sustainability issues are especially concerned about energy usage and CO_2 emissions owing to the urgency of transformation into transportation metrics with the advent of the latest technology (Vilathgamuwa et al. 2022). Meanwhile, natural habitat effects are causing additional pressure on supply systems by these transport effects (Lalendle et al. 2021). While logistics performance and environmental degradation were highly correlated with CO_2 emissions, supply chain empowerment has gradually raised the demand for international freight transit (Kong et al. 2022). In addition, green logistics management significantly improved business performance and environmental quality (Aftab et al. 2022), while new regulations encouraged green logistics management and increased ecological profitability (Vienažindienė et al. 2021). The integration of transportation and logistics systems was deemed advantageous as it resulted in improved tracking capabilities, greater cost savings, enhanced warehouse productivity, and reduced inventory levels (Khalili et al. 2019).

In management science, the green supply chain is a growingly prevalent issue, and green logistics is its subfield (Nimsai et al. 2020). For instance, European countries' biggest challenges in the next millennium have long been considered environmental issues in logistics management and green supply chain procedures (Shekarian et al. 2022). Additionally, logistics and freight transportation, which are concerned with the movement and storage of resources and goods in the supply chain, are significant components in this regard (Sindi and Roe 2017). So, as a component of sustainable movement, green logistics attempts to reduce the environmental externalities of logistics operations and establish a sustainable balance between economy, environment, and societal advantages (Agyabeng-Mensah et al. 2020). Therefore, in addition to lowering resource consumption, energy use, and waste, integrating green technologies, optimizing the order fulfillment process, and energy-saving solutions can improve internal supply chain operations.

To this end, various analyses of various aspects have been steered, and a range of environmentally friendly measures are being considered in this recent paradigm shifting into intelligent transportation (Palit et al. 2022). These measures may be technological, logistical, or market-based and have significant consequences for the supply chain's economics and logistics (Khan et al. 2021). The number of research that has addressed the actual application of this transportation logistics notion may be limited, even though many studies have discussed the only ideas of transportation logistics and sustainable challenges associated with supply chain operations (Gayialis et al. 2022). Furthermore, fewer case studies have addressed sustainability issues in addition to the advantages of intelligent transportation systems (Gohar and Nencioni 2021). Thus, the gap exists specifically in the context of supply chain sustainable operation where artificial intelligence, IoT, and modern technologies contribute toward smart transportation for the entire supply chain entities.

10.2 Transportation Practice and Carbon Emission

Transportation sustainability historically includes the efficiency and efficacy of the public transport system as well as the transportation system's consequences for the environment and climate (Zhou et al. 2019b). Naturally, the main environmental problems stem from carbon emissions and energy use. With significant evidence, urban air quality significantly deteriorates and climate change exacerbates by carbon emissions from transport (Sultana et al. 2019). In terms of efficiency, transportation has the quickest growth rate in greenhouse gas emissions compared to other energy-consuming industries (Kazancoglu et al. 2021; Khalili et al. 2019). It generated roughly 31% of worldwide emissions and 24% of emissions in the EU in 2019. The idea is uncomplicated but still impactful for the environment owing to 20–25% of both world's carbon dioxide emissions and energy usage (Zhou et al. 2019a).

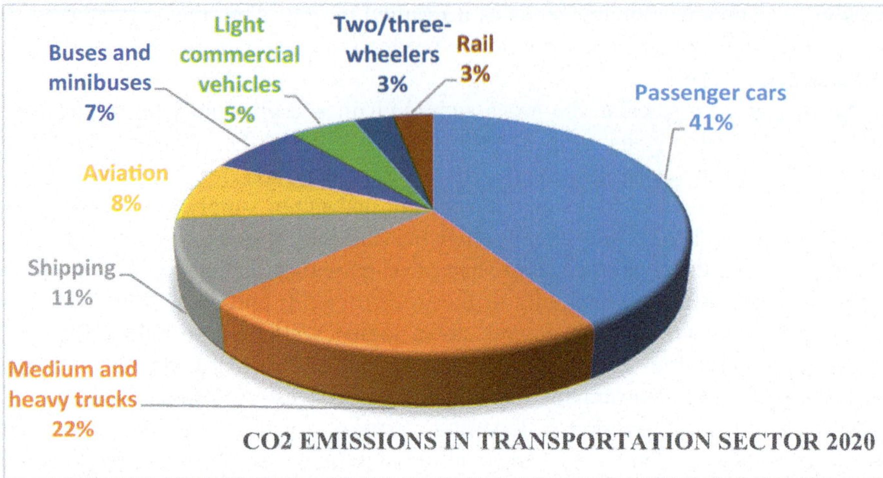

Fig. 10.1 CO_2 emission in the transportation sub-sectors 2022. (Tiseo 2021; Statista)

However, in 2019, fossil fuels accounted for almost 95% of energy, whereas direct burning of fossil fuels was to blame for nearly 97% of the emissions (Babu et al. 2022).

Figure 10.1 illustrates how the global transportation industry is a significant polluter and is generated roughly (Tiseo 2021). Zahoor et al. (2023) stated that passenger cars accounted for 41% of global transportation and CO_2 emissions of 3.2 billion tons in 2019. In 2020, car emissions were reduced drastically due to the COVID-19 pandemic. Heavy and medium trucks account for the second largest contaminators with 22% of emissions. Smart transportation has the potential to make significant contributions towards creating an environmentallysustainable supply chain, particularly in the areas of carbon emission reduction.

10.3 Transportation Logistics and Sustainability

Supply chain provides prompt and precise product distribution to the most suitable places, which is essential for any successful company (Bvuchete et al. 2021). Efficient transportation can enhance these by decreasing material and time loss (Bozorgi and Fahimnia 2021). By adopting a comprehensive strategy, a sustainable supply chain minimizes waste reductions and environmental effects and improves safety, working conditions, health, and labor abuses (Wilhelm and Villena 2021). As a result, a company must address environmental, social, economic, and legal challenges at every step in its supply chain in order to be sustainable (Allen et al. 2021). However, economic and social growth depends heavily on transportation (Strulak-Wójcikiewicz and Wagner 2021) due to the connection of people with the transportation infrastructure. Hence, an efficient transportation system reduces delays, gives visibility into freight shipping, and helps the business save money, resulting in a sustainable balance.

10.4 Sustainability and Transportation

From the theories of sustainable development, transportation and sustainable development are related (Kazancoglu et al. 2021) and essential to a city's economic and social growth. For implication, shared, automated, electric vehicles have become the mainstay of urban transportation systems. Therefore, leading to more resilient and sustainable urban futures, integrating disruptive transportation technologies and services helps address several issues in the current transportation industry (Strulak-Wójcikiewicz and Wagner 2021). So, policies that address innovation, infrastructure development, and energy efficiency, use of alternative fuels, pollution prevention, and intelligent transportation systems are illustrative of environmentally friendly transportation technologies (Sultana et al. 2019). Thus, sustainability will be the prime issue for future transportation development.

10.5 Green Supply Chain, Logistics, and Sustainability

For sustainable development, reduce, reuse, recycle, reclaim, and degradable options are required in supply chains' manufacturing, operations, and end-of-life management over traditional processes (Palacios-Mateo et al. 2021). However, green reverse logistics is also one of the familiar dimensions of a green supply chain (Shekarian et al. 2022) which involves employing more environmentally friendly and sustainable techniques (Strulak-Wójcikiewicz and Wagner 2021). Transportation plays a crucial role in logistics, which involves moving resources from supply sites to manufacturing facilities (Sindi and Roe 2017), inventory repositioning across multiple operations and distribution centers, and final delivery to customers (Firouzi et al. 2020). Again, transportation strategies ensure that deliveries to and from the plant go off without a hitch and arrive on time (Badidi 2022). So, it is crucial to incorporate transportation into the supply chain management strategy for a company's success, boosting supply chain efficiency and lowering inefficiencies (Singh et al. 2022a). But due to CO_2 emissions, pollution, and other factors, transportation has many adverse environmental effects causing global warming (Kazancoglu et al. 2021). As opposed to that, transportation improvements can affect the commodity and labor markets by increasing access to supply, components, customers, and labor (Firouzi et al. 2020). As a result, existing businesses become more effective and marketable, which boosts their output and creates more jobs. It shows a sustainable tendency toward local economic growth.

Society and the economy are getting the benefit when transportation networks are effective and efficient. For instance, such improvements create more jobs, accessible market access, and investment (Bharathidasan et al. 2022). Less air pollution, for example, the absence of greenhouse gases from biking and walking, reduced noise and traffic, reduced demand for new parking lots and roads, and the protection of irreplaceable green space from development are just a few of the benefits related to sustainable transportation (Dugan et al. 2022). Therefore, this chapter accommodated the ideology linking smart transportation with logistic movement, which contributed toward the operation of the green supply chain as a possible approach to reduce hazards like pollution and noise.

10.6 Methodology

This literature review concentrated on smart transportation systems with the advent of technological upgradation and amalgamation. More specifically, the chapter entails the roles of recent technological advancements in transportation systems to address ways to mitigate the sustainability issues on existing transportation systems resulting in smooth supply chain operation. A foretelling literature-based model was also developed and justified here under the advancement of the related technology on intelligent transportation systems along with the systematic literature survey. So, the research questions for this chapter are as follows: "What is the recent trend of research on transportation logistics for sustainability, and how to smarten transportation systems by incorporating modern technological advancement?" This research question addresses the trends of recent research on transportation systems from year to year based on the citation. Later, a sophisticated model of the smart transportation system draws through entailing insights from highly cited research papers on transportation and sustainability.

We used the scholarly articles from Google Scholar dataset predominately due to its ease of use for the citation and broad exposure for all types of research. Non-English articles, strictly technical papers, and papers focusing on legitimate factual contributions rather than business are prohibited. For the bibliometric data, author information, publication titles, abstracts, keywords, and other information were collected.

According to the systematic review, the development of smart transportation and sustainability issues between 2014 and 2022 is far ahead of schedule. As a result, the initial search was limited to those years and the Google Scholar database. Using the keywords "smart transportation" OR "Intelligent transport" OR "intelligent transportation systems" OR "smart transport," a total of 16,700 records were retrieved (Table 10.1). A total of 204 articles were considered after focusing on keywords, abstracts, and manuscripts (Fig. 10.2). However, based on "smart transportation sustainability," the topic is still fanciful because of the citation metrics displayed in Table 10.2 (2014–2022 citations). It shows 5 citations per document, 2.86 citations per paper, an h-index of 4, a g-index of 6, and an hA-index of 3.

Lastly, a plausible model was proposed based on the available research on smart transportation networks. In reality, this model was developed to answer the chapter

Table 10.1 Systematic literature search and bases of inclusion and exclusion

Keywords/search string	Search engine	No. of papers	Inclusion and exclusion parameters
"Smart transportation" OR "intelligent transport" OR "intelligent transportation systems" OR "smart transport"	Google Scholar	16,700	In the title Year: 2014–2022
"Smart transportation system"	Google Scholar	204	In the title of the article
"Smart transportation and sustainability"	Google Scholar	14	In the title of the article

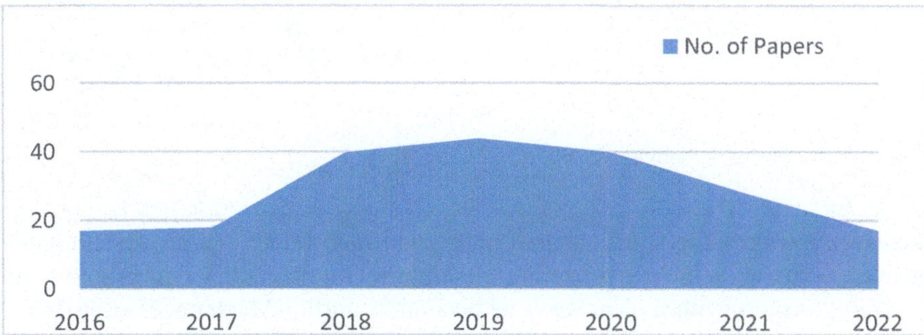

Fig. 10.2 Number of papers cited from 2016 to 2022

Table 10.2 Citation metrics

Publication year	Papers	Citations	Cites/per year	Cites/paper	Author/paper	h-index	g-index	hA-index
2014–2022	14	40	5	2.86	2.07	4	6	3

issues and may be applied to changes in transportation systems based on cutting-edge technological advancements in the future. Therefore, the draft model is a forward-looking transportation transformation with new facilities with the advent of artificial intelligence, IoT, and other modern instruments. More particular, this model will widen the sustainability movement toward improved transportation logistics, enhancing the supply chain's capabilities.

10.7 Future Smart Transportation Model and Mechanism

Smart transportation developments are defined as vehicular ad hoc network (VANET) applications involving mobile vehicles, network communications, and transportation systems to produce significant services (Giang et al. 2022). Again, future technology is known as intelligent transport systems (ITS), intended to

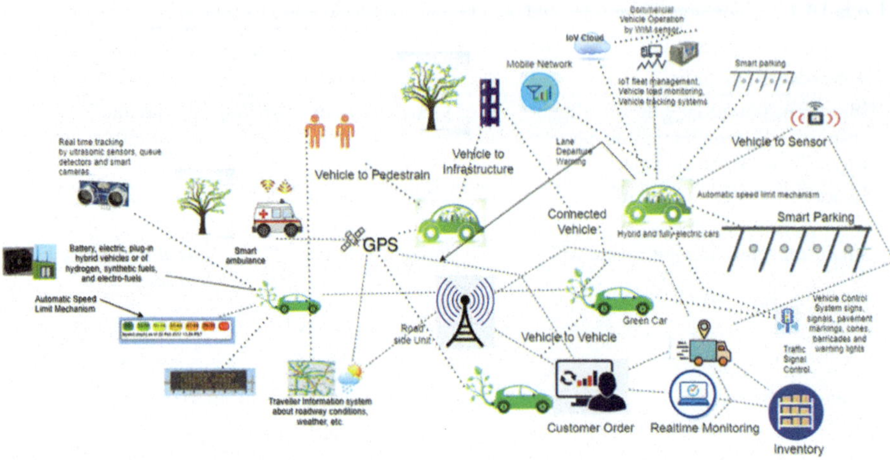

Fig. 10.3 Smart transportation logistics

increase driver comfort, traffic management, and road safety. Such a system would enable pervasive vehicle interaction, creating an Internet of Vehicles (IoV) and providing drivers with a clear view of oncoming traffic (Maglaras et al. 2016). A sustainable urban paradigm also involves collecting data from connectivity roadside units (RSUs) and other factors justified on roads, buildings, and people as shown in Fig. 10.3 (Tak et al. 2020). According to the rigorous system application, the gathered info is directed to the vehicle cloud that helps traffic control at the city level and makes sure that emergency alerts are sent out on time, giving the wireless transceiver of the vehicle crucial information for controlling congestion and choosing security algorithms, as well as taking actions to quantify the data quality and reliability (Aljohani et al. 2021).

Only a few ITS-based applications used in smart cities include cooperative awareness, safe lane discipline, intersection crossings and warning alerts, optimal traffic control, intelligent parking, and retrieving digital content from the World Wide Web (Sarkar and Jain 2017). The model created here and altered versions with support for logistics and supply chain dynamics are largely based on the findings of a chapter by Javed et al. (2016) (Singh et al. 2022a). As shown in Fig. 10.3, vehicle-to-infrastructure (V2I) and vehicle-to-vehicle (V2V) communication are the main ITS communication routes. Also, wireless data transmission improves neighborhood vision and security (Arena and Pau 2019). So, these models encompass two primary ideologies, initially on how to smarten the existing transportation system, whereas the supply chain logistics interlink with real-time monitoring and inventory management for better sustainable operation.

A thorough understanding of economic and environmental goals may influence a sustainable transportation system. In order to make decisions in a safe and efficient environment, smart traffic control ultimately aims to develop a system that allows all automobiles, warning signals, and command bases to communicate data with one another (Daniel et al. 2016). Also, this movement diversified under a smart community through driverless vehicles and automated traffic control management (Bechtsis et al. 2018). Very specifically, hardware modules are installed in smart cars and nearby roadside units (RSUs) to gather, process, analyze, and disseminate real-time data and statistics (sensors, processing units, etc.) (Ferreira et al. 2018). As a result, a mobile ad hoc network is formed and originated.

Again, smart cities are a whole other sector where the Internet of Things could have a considerable impact. Its attributes hold the potential for a variety of applications, such as pedestrian and traffic management (such as tracking traffic jams, parking lots, smart infrastructures (roads and highways), providing instant data, traffic hitches, weather monitoring, or road accidents), threat and risk forecasting (such as vibrations, materials quality, and strength in buildings, bridges, and historic constructions), noise, and lighting (such as adaptive lighting) (Korczak and Kijewska 2019). So, the transportation system ought to reflect on embedding the latest technological advancement for solving traffic, congestion, and relevant issues.

Along the main roadways, the rise in accidents is a significant issue. In developing cities, the road network needs to be carefully planned. Transportation in the USA is undergoing a critical revolution, made possible by cellular vehicle-to-everything (V2X) connections. Through V2X communications, remote sensors, and autonomous vehicles, the automotive industry is entering a new era of linked and automated vehicles (Raj and Appadurai 2022). The fundamental building blocks for ubiquitous smart mobility involve metro optical networks and high-speed wireless access technologies (Rodriguez et al. 2017). The introduction of technological advancements such as the Internet of Things (IoT), artificial intelligence (AI) for smart traffic management, real-time data transmission, and safe pedestrian movement without accidents can transform the entire transportation system.

In the second phase of this model, viz., logistics, operational participation with intelligent transportation, inventory, warehouse management, predictive maintenance, real-time information flow, tracking, and monitoring consider to improve the situation (Singh et al. 2022a). Initially, the starting point for the deployment of IoT in smart transport focuses on fleet management. For fleet management, it is possible to gather client information about vehicle conditions and fuel and driver behavior through sensors and in-vehicle devices (Maglaras et al. 2016). In advance, vehicle condition monitoring can provide valuable advance information to minimize breakdowns and logistical delays (Gutschi 2022). Therefore, real-time access to this critical information ensures that required maintenance can be easily detected and carried out, even for real-time tracking and managing goods (Sarkar and Jain 2017). In addition, IoT solutions on driver behaviors also allow companies to analyze speed and braking patterns while monitoring time between journeys (Maglaras et al. 2016). It also provides real-time statistics about the exact location while identifying any anomalies that may signal theft, delays, or driver related faults (Singh et al. 2022b), assisting with space management (Firouzi et al. 2020; Lee 2019). Through RFID sensors tagged on products, warehouse staff are automatically notified of dispatch and can continue monitoring its ongoing journey (Arena and Pau 2019; Jiang et al. 2020) and stock's current position (Singh et al. 2022b). Since ensuring all stock is in the right truck and on the way to the correct destination is the bread and butter of logistics (Boysen et al. 2021), the entire process can be optimized from stock order to final shipment to reduce costs and boost overall profits. Overall, the benefits of upgrading to smart logistics can be felt at every level of the supply chain.

10.8 Tentative Prediction About the Future Transportation

Predictions state that automated vehicles will be merged with a range of sensing technologies, guidance, control, and movement planning algorithms during the next 10 years to support drone transportation (Raj and Appadurai 2022). For instance, to actively foster an environment that would support the growth of smart transportation, the Chinese government has aggressively proposed new policies. Indifferently, sustainable mobility relates to transportation modes and planning systems consistent with general sustainability issues (Kazancoglu et al. 2021; Shamsuddoha et al. 2021). In essence, it will promote equity within and between succeeding generations while enabling basic access and development needs of individuals, businesses, and society to be safe and sustainable (Dugan et al. 2022). Thus, the following table encompasses the summarized view of transportation practice and innovation toward sustainability (Table 10.3).

According to the literature, sustainability encompasses more than just operational effectiveness and emissions, for instance, sustainable driving habits (Ercan et al. 2022; Shamsuddoha and Woodside 2022; Shamsuddoha 2015) and energy efficiency (Gutschi 2022; Sharifi et al. 2022). On the other hand, green vehicles are designed for lower ecological impact than regular vehicles. However, this may not always be the case when a vehicle's environmental impact is evaluated throughout its life cycle (Sasaki et al. 2022). For this green movement, the electric vehicle technique significantly reduces transportation-related CO_2 emissions, depending on how much energy is incorporated in the vehicle and where the electricity is generated (Bharathidasan et al. 2022). In countries where coal makes up a sizable portion of electricity production, adopting electric vehicles has little to no positive effects on the environment. For instance, a Nissan Leaf emitted one-third fewer greenhouse gases than the average internal combustion vehicle in the UK in 2019 (Sotnyk et al. 2020). So, the benefits to the environment vary widely and depend on several factors, including vehicle size, electricity emissions, driving style, and even the weather (Dong et al. 2022).

Green vehicles are marginally more fuel-efficient than conventional ones, yet they still contribute to traffic congestion and auto accidents. In reality, many people who have experienced the highest levels of vehicle noise, pollution, and safety risk are those who don't own cars, are unable to drive cars, and are severely burdened financially by the cost of car ownership (Luz and Portugal 2022). Traditionally, diesel buses used in well-utilized public transportation networks consume less fuel per passenger and take up less space on the road than private vehicles do (Yeh et al. 2022). As a result, the use of sustainable transport has a beneficial effect on social benefits as long as there are economic benefits, leading to smart logistics in particular and efficient supply chain operations in general.

Table 10.3 Transportation practice and innovation toward sustainability

Author	Findings/subject matter/focus/ suggested measures	Sustainability issues
Kou et al. (2022)	Solar energy	Reduce carbon emissions, sustainable production, and consumption
Sarkar et al. (2022) and Liu et al. (2022)	Biofuel	
Ishaq et al. (2022)	Hydrogen	
Odilov (2022)	Synthetic fuel	
Agarwal and Valera (2022)	Electro-fuel	
Jiang et al. (2022) and Sun et al. (2022)	Alternative energy usage	
Wang et al. (2022) and Cheng and Hu (2022)	Efficient use of renewable energy	
McLeod and Curtis (2022)	Counterintuitive road safety dynamics	Health safety (accident issues, increased road safety)
Abou El Hassan et al. (2022)	5G automotive association	
Abduljabbar et al. (2021)	Micro-mobility transportation	
Raj and Appadurai (2022)	Combine ICT and the internet of things (IoT)	
Sharma and Maherchandani (2022)	Traffic management system (TMS) by artificial intelligence and data analysis	Traffic management for decent work and economic growth
Paul and Mitra (2022)	Traffic signal management	
Nguyen and Mogaji (2022)	Big data-based intelligent transportation systems that interact with vehicles and people	
Zhang et al. (2022)	A method based on deep reinforcement learning	Real-time tracking and decision-making for resilient infrastructure and sustainable cities
Xu et al. (2022)	Electric vehicles with power transportation networks	
Nie and Farzaneh (2022)	Artificial neural network-based energy consumption model	
Badidi (2022)	Blockchain-based trustworthy communications and trading between vehicles and smart energy trading	

10.9 Conclusion

Smart transportation mobility offers substantial social and economic benefits that could speed sustainable development. For instance, communities that successfully improve the sustainability features of their transportation systems do so as part of a more significant effort to create livelier and more vibrant cities. But a real option to save time and money from home and government budgets is also advisable for investing in sustainable transport for a "win-win" situation. Owing to that fact, this chapter identifies the sophisticated development of transportation networks by reviewing the literature to include applications and standards that address most environmental pollution problems. Moreover, the revised model incorporates the most revolutionary scenario for the advancement of transportation systems with multifaceted images supposed to support sustainable development. Ultimately, this review has inspired academics and policymakers to adopt a new perspective on vehicle-to-everything (V2X) for a sustainable future. Such theoretical development encourages the concerned transportation agencies (manufacturer and user) to think about smart and intelligent technologies and system applications for a better and sustainable future. Also, such green initiatives will help the business processes to gain logistical and supply chain efficiency. Future research should consider the complexity of supply chains covering all aspects of supply chain management and operational excellence. As a result, the ultimate solution will emphasize tracking and enhancing sustainability throughout the transportation system's advanced movement.

References

Abduljabbar RL, Liyanage S, Dia H (2021) The role of micro-mobility in shaping sustainable cities: a systematic literature review. Transp Res D Transp Environ 92:102734. ISSN 1361-9209, Elsevier

Aftab J, Abid N, Sarwar H, Veneziani M (2022) Environmental ethics, green innovation, and sustainable performance: exploring the role of environmental leadership and environmental strategy. J Clean Prod 378:134639. ISSN 0959-6526, Elsevier

Agarwal AK, Valera H (2022) Introduction of greener and scalable E-fuels for decarbonization of transport. In: Greener and scalable E-fuels for decarbonization of transport. Springer, Singapore, pp 3–8

Agyabeng-Mensah Y, Afum E, Ahenkorah E (2020) Exploring financial performance and green logistics management practices: examining the mediating influences of market, environmental and social performances. J Clean Prod 258:120613. ISSN 0959-6526, Elsevier

Aljohani M, Olariu S, Alali A, Jain S (2021) A survey of parking solutions for smart cities. IEEE trans Intell Transp Syst 23(8):10012–10029

Allen SD, Zhu Q, Sarkis J (2021) Expanding conceptual boundaries of the sustainable supply chain management and circular economy nexus. Clean Logist Supply Chain 2:100011. ISSN 2772-3909, Elsevier

Arena F, Pau G (2019) An overview of vehicular communications. Future Internet 11(2):27. ISSN 1999-5903, MDPI

Babu JM, Chandra MS, Ganesh PRC, Jayaprakash P, Kumar KS, Nagappan M (2022) Experimental evaluation of direct injection diesel engine performance and emissions with acacia biodiesel. Int J Ambient Energy 43:7038–7038. ISSN 0143-0750, Taylor & Francis

Badidi E (2022) Edge AI and Blockchain for Smart Sustainable Cities: Promise and Potential. Sustainability 14(13):7609. ISSN 2071-1050, MDPI

Bechtsis D, Tsolakis N, Vlachos D, Srai JS (2018) Intelligent autonomous vehicles in digital supply chains: a framework for integrating innovations towards sustainable value networks. J Clean Prod 181:60–71. ISSN 0959-6526, Elsevier

Bharathidasan M, Indragandhi V, Suresh V, Jasiński M, Leonowicz Z (2022) A review on electric vehicle: technologies, energy trading, and cyber security. Energy Rep 8:9662–9685. ISSN 2352-4847, Elsevier

Boysen N, Fedtke S, Schwerdfeger S (2021) Last-mile delivery concepts: a survey from an operational research perspective. Or Spectr 43(1):1–58. ISSN 1436-6304, Springer

Bozorgi A, Fahimnia B (2021) Transforming the vaccine supply chain in Australia: opportunities and challenges. Vaccine 39(41):6157–6165. ISSN 0264-410X, Elsevier

Bvuchete M, Grobbelaar SS, Van Eeden J (2021) A network maturity mapping tool for demand-driven supply chain management: a case for the public healthcare sector. Sustainability 13(21): 11988. ISSN 2071-1050, MDPI

Cheng Z, Hu X (2022) The effects of urbanization and urban sprawl on CO2 emissions in China. Environ Dev Sustain, 1–17. ISSN 1573-2975, Springer

Daniel A, Paul A, Ahmad A, Rho S (2016) Cooperative intelligence of vehicles for intelligent transportation systems (ITS). Wirel Pers Commun 87(2):461–484. ISSN 1572-834X, Springer

Dong P, Zhao J, Liu X, Wu J, Xu X, Liu Y, Wang S, Guo W (2022) Practical application of energy management strategy for hybrid electric vehicles based on intelligent and connected technologies: development stages, challenges, and future trends. Renew Sustain Energy Rev 170: 112947. ISSN 1364-0321, Elsevier

Dugan A, Mayer J, Thaller A, Bachner G, Steininger KW (2022) Developing policy packages for low-carbon passenger transport: a mixed methods analysis of trade-offs and synergies. Ecol Econ 193:107304. ISSN 0921-8009, Elsevier

El Hassan AA, Kerrakchou I, El Mehdi A, Saber M (2022) Road safety enhancement of intelligent transportation systems: from cellular LTE-V2X toward 5G-V2X. In: International conference on digital technologies and applications, pp 745–754, Springer

Ercan T, Onat NC, Keya N, Tatari O, Eluru N, Kucukvar M (2022) Autonomous electric vehicles can reduce carbon emissions and air pollution in cities. Transp Res D Transp Environ 112: 103472. ISSN 1361-9209, Elsevier

Ferreira J, Alam M, Fernandes B, Silva L, Almeida JÃ£o, Moura L, Costa R, Iovino G, Cordiviola E (2018) Cooperative sensing for improved traffic efficiency: the highway field trial. Comput Netw 143:82–97. ISSN 1389-1286, Elsevier

Firouzi F, Farahani B, Weinberger M, DePace G, Aliee FS (2020) Iot fundamentals: definitions, architectures, challenges, and promises. In: Intelligent internet of things. Springer, pp 3–50

Gayialis SP, Kechagias EP, Konstantakopoulos, GD (2022) A city logistics system for freight transportation: Integrating information technology and operational research. Oper Res 1–30. ISSN 1866-1505, Springer

Giang AT, Tran HT, Le HT, Doan NQ, Nguyen MH (2022) Jamming attack in vehicular networks: adaptively probabilistic channel surfing scheme. Wirel Commun Mob Comput, 2022. ISSN 1530-8669, Hindawi

Gohar A, Nencioni G (2021) The role of 5G technologies in a smart city: the case for intelligent transportation system. Sustainability 13(9):5188. ISSN 2071-1050, Multidisciplinary Digital Publishing Institute

Gutschi A (2022) End-of-life vehicle recycling in the European Union: Analysing changing material flows of end-of-life steel, aluminum, copper and plastics due to the transition toward zero-emission vehicles, Wien

Ishaq H, Dincer I, Crawford C (2022) A review on hydrogen production and utilization: challenges and opportunities. Int J Hydrogen Energy 47(62):26238–26264. ISSN 0360-3199, Elsevier

Jain DK, Neelakandan S, Veeramani T, Bhatia S, Memon FH (2022) Design of fuzzy logic based energy management and traffic predictive model for cyber physical systems. Comput Electr Eng 102:108135. ISSN 0045-7906, Elsevier

Javed MA, Hamida EB, Znaidi W (2016) Security in intelligent transport systems for smart cities: From theory to practice. Sensors 16(6):879. ISSN 1424-8220, MDPI

Jiang J, Wang H, Mu X, Guan S (2020) Logistics industry monitoring system based on wireless sensor network platform. Comput Commun 155:58–65. ISSN 0140-3664, Elsevier

Jiang T, Yu Y, Yang B (2022) Understanding the carbon emissions status and emissions reduction effect of China's transportation industry: dual perspectives of the early and late stages of the economic "new normal". Environ Sci Pollut Res 29(19):28661–28674. ISSN 1614-7499, Springer

Kashem MA (2020) Passenger Service Dominance through Brand Loyalty: An Idiosyncratic View of Starline Bus Service (Pvt) Limited. Kelaniya J Manag 9(1)

Kazancoglu Y, Ozbiltekin-Pala M, Ozkan-Ozen YD (2021) Prediction and evaluation of greenhouse gas emissions for sustainable road transport within Europe. Sustain Cities Soc 70:102924. ISSN 2210-6707, Elsevier

Khalili S, Rantanen E, Bogdanov D, Breyer C (2019) Global transportation demand development with impacts on the energy demand and greenhouse gas emissions in a climate-constrained world. Energies 12(20):3870. ISSN 1996-1073, MDPI

Khan SAR, Razzaq A, Yu Z, Miller S (2021) Industry 4.0 and circular economy practices: a new era business strategies for environmental sustainability. Bus Strateg Environ 30(8):4001–4014. ISSN 0964-4733, Wiley Online Library

Komninos N, Tsampoulatidis I, Kakderi C, Nikolopoulos S, Kompatsiaris I (2022) Projects for smart cities: ecosystems, connected intelligence and innovation for the radical transformation of cities. In: Smart cities and smart communities. Springer, pp 33–68

Kong Y, Liu J, Chen J (2022) Exploring the carbon abatement measures in maritime supply chain: a scenario-based system dynamics approach. Int J Prod Res. 1–22. ISSN 0020-7543, Taylor & Francis

Korczak J, Kijewska K (2019) Smart logistics in the development of smart cities. Transp Res Procedia 39:201–211

Kou G, YÃ¼ksel S, DinÃ§er H (2022) Inventive problem-solving map of innovative carbon emission strategies for solar energy-based transportation investment projects. Appl Energy 311:118680. ISSN 0306-2619, Elsevier

Lalendle C, Goedhals-Gerber L, van Eeden J (2021) A monitoring and evaluation sustainability framework for road freight transporters in South Africa. Sustainability 13(14):7558. ISSN 2071-1050, MDPI

Lee I (2019) The Internet of Things for enterprises: An ecosystem, architecture, and IoT service business model. Internet of Things 7:100078. ISSN 2542-6605, Elsevier

Liu C, Ke L (2022) Cloud assisted internet of things intelligent transportation system and the traffic control system in the smart city. J Cont Decis, 1–14. ISSN 2330-7706, Taylor & Francis

Liu Z, Wang J, Nielsen J (2022) Yeast synthetic biology advances biofuel production. Curr Opin Microbiol 65:33–39. ISSN 1369-5274, Elsevier

Luz G, Portugal L (2022) Understanding transport-related social exclusion through the lens of capabilities approach. Transp Rev 42(4):503–525. ISSN 0144-1647, Routledge

Maglaras LA, Al-Bayatti AH, He Y, Wagner I, Janicke H (2016) Social internet of vehicles for smart cities. J Sens Actuator Netw 5(1):3. ISSN 2224-2708, MDPI

McLeod S, Curtis C (2022) Integrating urban road safety and sustainable transportation policy through the hierarchy of hazard controls. Int J Sustain Transp 16(2):166–180. ISSN 1556–8318, Taylor & Francis

Nguyen NP, Mogaji E (2022) Information technology for enhancing transportation in developing countries. In: Management and information technology in the Digital Era, vol 29, pp 81–94. ISSN 1803822961, Emerald Publishing Limited

Nie Z, Farzaneh H (2022) Real-time dynamic predictive cruise control for enhancing eco-driving of electric vehicles, considering traffic constraints and signal phase and timing (SPaT) information, using artificial-neural-network-based energy consumption model. Energy 241:122888. ISSN 0360-5442, Elsevier

Nimsai S, Yoopetch C, Lai P (2020) Mapping the knowledge base of sustainable supply chain management: a bibliometric literature review. Sustainability 12(18):7348. ISSN 2071-1050, MDPI

Odilov O (2022) Synthetic fuel. Central Asian J STEM 3(1):65–69. ISSN 2181-1628

Palacios-Mateo C, van der Meer Y, Seide G (2021) Analysis of the polyester clothing value chain to identify key intervention points for sustainability. Environ Sci Eur 33(1):1–25

Palit T, Bari ABMM, Karmaker CL (2022) An integrated principal component analysis and interpretive structural modeling approach for electric vehicle adoption decisions in sustainable transportation systems. Decis Anal J 4:100119. ISSN 2772-6622, Elsevier

Paul A, Mitra S (2022) Application of DRL in transformer network for traffic signal management using fog computing in ITS. In: 2022 IEEE 11th international conference on communication systems and network technologies (CSNT), pp 464–469. ISSN 1665480386, IEEE

Raj EFI, Appadurai M (2022) Internet of things-based smart transportation system for smart cities. In: Intelligent systems for social good, pp 39–50. Springer

Rodriguez J, Radwan A, Barbosa C, Fitzek FHP, Abd-Alhameed RA, Noras JM, Jones SMR, Politis I, Galiotos P, Schulte G (2017) SECRET – secure network coding for reduced energy next generation mobile small cells: a European training network in wireless communications and networking for 5G. In: 2017 Internet technologies and applications (ITA), pp 329–333. ISSN 1509048154, IEEE

Sarkar PK, Jain AK (2017) Intelligent transport systems. ISSN 8193593847, PHI Learning Pvt. Ltd.

Sarkar B, Mridha B, Pareek S (2022) A sustainable smart multi-type biofuel manufacturing with the optimum energy utilization under flexible production. J Clean Prod 332:129869. ISSN 0959-6526, Elsevier

Sasaki Y, Orikasa T, Nakamura N, Hayashi K, Yasaka Y, Makino N, Shobatake K, Koide S, Shiina T (2022) Optimal packaging for strawberry transportation: evaluation and modeling of the relationship between food loss reduction and environmental impact. J Food Eng 314:110767. ISSN 0260-8774, Elsevier

Shamsuddoha M (2015) Integrated supply chain model for sustainable manufacturing: a system dynamics approach. In: Sustaining competitive advantage via business intelligence, knowledge management, and system dynamics. Emerald Group Publishing Limited

Shamsuddoha M, Woodside AG (2022) Achieving radical process innovations by applying technology-mindset transformations via second-order system-dynamics engineering. J Bus Res 147:37–48

Shamsuddoha M, Quaddus MA, Woodside AG (2021) Environmental sustainability through designing reverse logistical loops: case research of poultry supply chains using system dynamics. J Bus Ind Mark 37(4):823–840

Sharifi F, Meitiv A, Shelton J, Xu X, Burris M, Vallamsundar S, Xu YA (2022) Regional traffic operation and vehicle emission impact assessment of lane management policies. Res Transp Econ 91:101067. ISSN 0739-8859, Elsevier

Sharma B, Maherchandani JK (2022) Review of recent developments in sustainable traffic management system. In: Intelligent manufacturing and energy sustainability. Springer, pp 401–409

Shekarian E, Ijadi B, Zare A, Majava J (2022) Sustainable supply chain management: a comprehensive systematic review of industrial practices. Sustainability 14(13):7892. ISSN 2071-1050, MDPI

Sindi S, Roe M (2017) The evolution of supply chains and logistics. In: Strategic supply chain management. Springer, pp 7–25

Singh N, Charak A, Tyagi M, Walia RS (2022a) Effect of IoT in supply chain Management – a review. In: Recent advances in operations management applications. Springer, pp 283–297

Singh P, Elmi Z, Lau Y-y, Borowska-Stefańska M, Wiśniewski S, Dulebenets MA (2022b) Blockchain and AI technology convergence: applications in transportation systems. Veh Commun 38:100521. ISSN 2214-2096, Elsevier

Sotnyk I, Hulak D, Yakushev O, Yakusheva O, Prokopenko OV, Yevdokymov A (2020) Development of the US electric car market: macroeconomic determinants and forecasts. Polityka Energ 23(3):147–164

Strulak-Wójcikiewicz R, Wagner N (2021) Exploring opportunities of using the sharing economy in sustainable urban freight transport. Sustain Cities Soc 68:102778

Sultana S, Salon D, Kuby M (2019) Transportation sustainability in the urban context: a comprehensive review. Urban Geogr 40(3):279–308

Sun Y-F, Zhang Y-J, Su B (2022) How does global transport sector improve the emissions reduction performance? A demand-side analysis. Appl Energy 311:118648. ISSN 0306-2619, Elsevier

Tak S, Yoon J, Woo S, Yeo H (2020) Sectional information-based collision warning system using roadside unit aggregated connected-vehicle information for a cooperative intelligent transport system. J Adv Transp 2020:1–12. ISSN 0197-6729, Hindawi

Tiseo I (2021) Breakdown of CO2 emissions in the transportation sector worldwide 2020, by subsector. https://www.statista.com/statistics/1185535/transport-carbon-dioxide-emissions-breakdown/#:~:text=Breakdown%20of%20CO2%20emissions%20in%20the%20transportation%20sector%20worldwide%202020%2C%20by%20subsector&text=The%20global%20transportation%20sector%20is,carbon%20dioxide%20(CO2)%20emissions. Accessed 01 Sept 2022

Vienažindienė M, Tamulienė V, Zaleckienė J (2021) Green logistics practices seeking development of sustainability: evidence from Lithuanian transportation and logistics companies. Energies 14(22):7500

Vilathgamuwa M, Mishra Y, Yigitcanlar T, Bhaskar A, Wilson C (2022) Mobile-Energy-as-a-Service (MEaaS): sustainable Electromobility via integrated Energy – Transport – urban infrastructure. Sustainability 14(5):2796. ISSN 2071-1050, MDPI

Wang M, Li TF, Yuan C, Tian H, Tian S (2022) Research on vehicle renewable energy use in cities with different carbon emission characteristics. Energy Rep 8:343–352. ISSN 2352-4847, Elsevier

Wilhelm M, Villena VH (2021) Cascading sustainability in multi-tier supply chains: when do Chinese suppliers adopt sustainable procurement? Prod Oper Manag 30(11):4198–4218

Xu Y, Li JP, Chu CC, Dinca G (2022) Impact of COVID-19 on transportation and logistics: a case of China. Econ Res-Ekon Istraz 35(1):2386–2404

Yeh S, Gil J, Kyle P, Kishimoto P, Cazzola P, Craglia M, ... Teter J (2022) Improving future travel demand projections: a pathway with an open science interdisciplinary approach. Progress in Energy 4(4):043002

Zahoor A, Mehr F, Mao G, Yu Y, Sápi A (2023) The carbon neutrality feasibility of worldwide and in China's transportation sector by E-car and renewable energy sources before 2060. J Energy Storage 61:106696

Zhang X, Zhao Y, Min G, Miao W, Huang H, Ma Z (2022) Intelligent video ingestion for real-time traffic monitoring. In: ACM Transactions on Sensor Networks (TOSN). ISSN 1550-4859. ACM, New York

Zhou N, Price L, Yande D, Creyts J, Khanna N, Fridley D, Lu H, Feng W, Liu X, Hasanbeigi A (2019a) A roadmap for China to peak carbon dioxide emissions and achieve a 20% share of non-fossil fuels in primary energy by 2030. Appl Energy 239:793–819. ISSN 0306-2619, Elsevier

Zhou Y, Wang J, Yang H (2019b) Resilience of transportation systems: concepts and comprehensive review. IEEE Trans Intell Transp Syst 20(12):4262–4276. ISSN 1524-9050, IEEE

Index

A

Analytic hierarchy process (AHP), 19–21, 25, 30, 43, 109, 172–175, 179
Artificial intelligence (AI), 2, 6–8, 245, 249, 253
Automobile industry, 130–141, 198, 199

B

Bangladesh, 16, 17, 31, 109, 212–215
Best and worst criteria, 179, 182
Best-to-others vectors, 180, 182, 183
Best worst method (BWM), 141, 170–176, 179, 182–184, 187, 188, 232, 234
Big data, 2, 6–8, 11, 227, 228, 230, 232, 234, 239
Big data analytics (BDA), 2, 6, 7, 228, 230–235, 239, 240

C

Capability, 6, 19, 40, 41, 52, 54–57, 59–61, 63–65, 69–71, 73, 74, 77, 81, 93, 94, 96, 132, 134, 139, 141, 170, 174, 228–235, 239, 240, 249
Carbon emission, 17, 56, 168, 198, 245–246, 253
Circular economy (CE), 7, 8, 11, 52, 139, 212–223
Citation metrics, 248
Closed-loop supply chain, 8, 11
Close-loop, 7, 8, 11
Compatibility, 230, 231, 238, 239
Connectivity, 230, 231, 234, 237–240, 244, 250
Consistency index (CI), 21, 181
COVID-19, 3, 11, 246

D

Dairy farming, 212–214, 218, 219
Dairy wastes, 212, 214–215, 217–220, 222, 223
Data analytics, 2–11, 244
Data fusion, 21, 33, 42
Decision-making, 6, 8, 11, 18, 20, 38, 42, 56, 80, 95, 96, 108, 109, 134, 170, 179, 188, 208, 228, 230, 231, 234, 238–240, 253
Decision-making trial and evaluation laboratory (DEMATEL), 109, 110, 113–116, 172–175, 179
Dempster-Shafer theory (DST), 17, 19–21, 23, 25, 26, 30–41, 43

E

Electric vehicle (EV), 198–202, 208, 247, 252, 253
Emissions, 8, 28, 52, 54, 83, 91, 94, 111, 120, 187, 198, 199, 202–205, 208, 228, 229, 244–247, 252

F

Farming, 212, 214, 215, 221
Fashion industry, 16
Flexibility, 19, 53, 55–57, 63, 65, 67–69, 73, 75, 83, 86, 87, 91, 178, 185, 229–235, 239, 240

© The Author(s), under exclusive license to Springer Nature Switzerland AG 2023
N. U. Ibne Hossain (ed.), *Data Analytics for Supply Chain Networks*, Greening of Industry Networks Studies 11, https://doi.org/10.1007/978-3-031-29823-3

Foretelling literature-based model, 248
Fuel, 198, 200–203, 206–208, 218, 228, 229, 246, 247, 251–253
Fuzzy decision matrix, 136
Fuzzy TOPSIS, 131, 134–136, 138–142

G
Greenhouse gas (GHG), 111, 120, 198, 199, 202–205, 208, 229, 245
Green initiative, 254
Green supply chain management (GSCM), 2, 16–20, 23–43, 108, 109, 171, 187

H
Hierarchical evidential reasoning (HER), 25, 26, 29, 33
Hierarchical sustainability framework, 187
Hybrid electric vehicle (HEV), 198–204, 206–208

I
Independent risks (IR), 115, 118, 119
Indicators, 50–59, 61, 63–66, 70–74, 77, 80, 82–84, 86–88, 90–96, 131, 176, 186, 234, 237, 239
Indirect risks (InR), 115, 118–120
Industry 4.0, 6–8, 11, 218
Intelligent transportation, 245, 247–249, 251, 253
Intuitionistic fuzzy sets (IFS), 111–113, 115, 116, 121

K
Key risks, 109, 120

L
Life cycle cost (LCC), 199, 200, 202, 204–208
Linguistic scale, 135, 136
Livestock, 212–214, 220
Logistics, 2, 6, 27, 28, 33, 87, 90, 108, 169, 172–174, 176, 177, 181, 198, 199, 217, 220, 244–252

M
Maritime industry, 227–229, 239
Minor key risks (MKR), 115, 118, 119
Mobility, 198, 244, 251–253
Modularity, 230, 231, 234, 238–240
Multi-criteria decision-making (MCDM), 20, 109, 120, 131, 170, 171, 179, 231–232, 239, 240

N
New adopters, 51, 52, 66, 96
Normalized fuzzy decision, 136

O
Others-to-worst vectors, 180

P
Policy-making, 96
Preferred Reporting Items for Systematic Reviews and Meta-Analyses (PRISMA), 170
Prominence-relation map (PRM), 114–116, 118, 119
Prominence vector, 116, 119

R
Relation vector, 117
Relative closeness, 137, 138
Resilience, 3, 6–8, 23, 50, 51, 55–61, 63–66, 70, 71, 74, 77–82, 84, 86, 88, 90, 92–96, 108, 109, 179
Resilient, 7, 8, 57, 64, 111, 171, 185, 208, 247, 253
Resilient and sustainable supply chains, 3
Reverse logistics, 7, 28, 29, 109, 178, 185, 212, 216–218, 220, 223, 247
Risk, 2, 6, 8, 57, 64, 70, 77, 82, 84–86, 88, 90–92, 95, 108–116, 118–120, 168, 169, 171–175, 177–179, 182, 184, 185, 198, 251, 252

S
Scalability, 58, 66, 96
Simulation, 3, 8, 212, 213, 220–223, 239
Small and medium enterprises (SMEs), 7, 18, 51, 52, 66, 95, 96
Smart transportation, 244, 245, 248–253
Step-wise weight assessment ratio analysis (SWARA), 141, 172, 175, 176
Supply chain (SC), 2, 16, 50, 107, 130, 168, 198, 212, 244

Supply chain management (SCM), 2, 11, 108–110, 130–134, 139, 141, 168–178, 181, 185–187, 213, 216, 247, 254
Supply chain network, 71, 130, 132, 135, 138, 141, 174, 214, 218, 223
Supply chain resilience, 2, 12, 50–96, 132, 171
Sustainability, 2, 27, 50, 109, 130, 168, 198, 212, 228, 244
Sustainability evaluations, 178, 187, 188
Sustainability performance, 19, 51, 54, 61, 66, 199, 239
Sustainable shipping, 247
Sustainable supply chain, 109, 168, 172, 175–177, 185, 246
Sustainable supply chain management (SSCM), 130–135, 138–141, 170, 181
System dynamics, 199, 213, 218

T
Techno-economic, 199
Textile, 16, 17, 24, 25, 31, 33, 34, 39, 173–175
Transportation, 2, 11, 28, 31, 56, 86, 130, 131, 173, 181, 187, 198, 199, 208, 244–254
Transportation practice innovation, 252, 253
Transportation sub-sectors, 246
Triangular fuzzy numbers (TFN), 113, 116, 136
Triple bottom line (TBL), 51, 56, 59, 60, 81, 85, 228, 229, 234, 237, 239, 240

V
Viability index, 23, 25–30, 33–40

W
Weighted normalized matrix, 136

Y
Yager's rule, 26, 30, 36–41, 43

Z
Zero defect zero effect (ZED), 140

The manufacturer's authorised representative in the EU is Springer Nature Customer Service Centre GmbH, Europaplatz 3, 69115 Heidelberg, Germany. If you have any concerns regarding our products, please contact ProductSafety@springernature.com

Printed and bound by CPI Group (UK) Ltd, Croydon, CR0 4YY
28/09/2026
02224077-0005